Strange Chemistry

The Stories Your Chemistry Teacher Wouldn't Tell You

Steven Farmer

Sonoma State University, Rohnert Park, California, USA

This edition first published 2017
© 2017 John Wiley & Sons, Inc.

The right of Steven Farmer to be identified as the author of this work has been asserted in accordance with law.

Registered Office
John Wiley & Sons, Inc., 111 River Street, Hoboken, NJ 07030, USA

Editorial Office
111 River Street, Hoboken, NJ 07030, USA

For details of our global editorial offices, customer services, and more information about Wiley products visit us at www.wiley.com.

Wiley also publishes its books in a variety of electronic formats and by print-on-demand. Some content that appears in standard print versions of this book may not be available in other formats.

Library of Congress Cataloging-in-Publication Data

Names: Farmer, Steven C., author.
Title: Strange chemistry : the stories your chemistry teacher wouldn't tell you / by Steven Farmer.
Description: Hoboken, NJ : John Wiley & Sons, 2017. | Includes
    bibliographical references and index. |
Identifiers: LCCN 2017016092 (print) | LCCN 2017026188 (ebook) | ISBN 9781119265290 (pdf) |
    ISBN 9781119265283 (epub) | ISBN 9781119265269 (pbk.)
Subjects: LCSH: Chemistry–Popular works.
Classification: LCC QD37 (ebook) | LCC QD37 .F37 2017 (print) | DDC 540–dc23 LC record
    available at https://lccn.loc.gov/2017016092

Paperback ISBN: 9781119265269

Cover image: Courtesy of the author; (Background) © P Wei/iStockphoto
Cover design by Wiley

Set in 10/12pt WarnockPro by SPi Global, Chennai, India

10 9 8 7 6 5 4 3 2 1

Dedication

I would like to dedicate this book to my parents James and Margaret.
Throughout my whole life whenever I looked, you were there; ready to give me love and support, guidance and security, and praise and encouragement. You filled me with your dreams and showed me what it takes to succeed in life. Without you both none of the things I have accomplished would have been possible. I am truly blessed to have such incredible parents, and I love you both.

Contents

Preface

Growing up in Northern California was much more curious than one might think. Napa, being part of Northern California, was affected by the LSD (lysergic acid diethylamine) counterculture centered in Berkeley and San Francisco. LSD was everywhere and I recall multiple instances in high school where a classmate would admit to attending class under the influence of LSD and try to describe the effects. This seems very rebellious, but in one of the most tragic events of my life, a high school friend jumped in front of a car on the highway after ingesting LSD. He was killed instantly. This event had such a profound effect on me that it eventually drove me toward a career in chemistry – I needed to understand what had happened to my friend. How could the ingestion of a molecule cause such profound effects? Is awareness really just a fragile chemical process that can be so easily tricked?

After the mass closures of the 1980s, Napa State Hospital was one of the few remaining state run mental hospitals in California. If you have seen the movie *One Flew Over the Cuckoo's Nest*, it was filmed at Napa State Hospital. As a child, I would often wonder about the causes of mental illness. I was told that mental illness was the result of a "chemical imbalance" in the brain, but what did that really mean? Could a slight change in a chemical really change our perception of the world?

Similar to many scientists before me, my career in chemistry was driven by a quest to better understand some of the questions that haunted my childhood. Surely, obtaining a degree in chemistry would allow me to understand how hallucinogens work, or what causes mental illnesses. Unfortunately, I was wrong. Chemistry courses seemed to steer clear of any topic of an edgy, dangerous, or unusual nature. In fact, initially learning about these fascinating topics required a course outside the chemistry department. Eventually, a graduate elective course from a psychology department, called "Psychopharmacology," explained the chemical basis for the effect of hallucinogens and the causes of mental illness (I share what I learned in this book).

Later, when I became a chemistry instructor, I made it a point to share these and other stories. It was delightful to find that almost everyone found these topics just as interesting as I did. As I collected new stories, I realized how much of this material was never discussed as part of the numerous chemistry courses required for my Ph.D. Roughly 90% of these stories contained in this book were learned after I graduated. This is where the subtitle of this book, "The stories your chemistry teacher wouldn't tell you" comes from. It seems that there is an overwhelming push to teach the fundamentals of chemistry while neglecting to show the utility of learning the material by connecting it to the real world. Particularly for organic chemistry, there seems to be an aversion of some of these topics, which I feel is because chemists do not want their science associated with anything that poisons you, blows you up, or gets you high. However, these are the topics that many people find exciting (as can be seen by looking at the plot of almost any action movie). Ask a nonchemist where chemicals appear in everyday life and inevitably the answer involves pharmaceuticals, toxins, or illicit drugs.

To share these stories with my students, I usually would take about 5–10 minutes each week to present one of the stories described in this book. For those of you who are teachers or who plan to be, I can say that these stories have been the largest source of positive feedback I have received from my students. Although there is an enormous amount of material that needs to be covered in a typical chemistry course, I say make the time for these extras. It is that important! On multiple occasions, students admitted to me that they only came to class that day so that they could hear the story. Many times, students would speak to me after the lecture to share how that day's story had touched them in some way. One student had been to the emergency room for an acetaminophen overdose, another had a stepfather who was addicted to opioids, and yet another was prescribed amphetamine to treat their attention deficit hyperactivity disorder (ADHD).

You will note that most of the presented stories are short and involve a question or a defined idea. This is done for two reasons: First, I love presenting these questions to my students and trying to evoke an answer from them. Putting students on the spot drives home how little they actually know about the world and how learning chemistry helps them understand their lives. I admit, few things have made me feel more educated than seeing a single simple question stump a classroom with over 400 students. Try it. You will find that very few people know the answers to the questions posed in this book. In addition, some of the cheeky answers I receive have become the highlights of my teaching. Second, I present the stories in a simple format because they will be easy to remember. Jokingly, I tell students to share these stories with their friends and family members so that they can prove that they are receiving an education at Sonoma State University. I am pleased to say that they do just that. An informal poll of my students showed that 90% of them had shared a story at least once, and 75% said that they shared these stories on a regular basis.

Students, like all human beings, want to understand the world around them – they may just not realize it. Telling stories that help students understand and connect to the world they see inspires them in a primal way, making them want to learn and keep coming back for more. This book contains the best stories I have collected over the last 10 years. If you are a teacher, try some of them out and see the profound effect they have on students. Even if you are not a teacher, read on, better understand the world around you, and see how truly strange chemistry can be.

Acknowledgments

To my loving wife, Joy: You are still the most beautiful woman I have ever seen. You are my muse, my life, and the air that I breathe. You are the personification of everything that makes me happy in this world. It was only your love that allowed me to face the adversity I have seen. You have been with me since the start of this journey and I cannot wait to see where life takes us.

To my brother, Richard: Thanks for being the oldest friend I have and for being the funniest person I know.

To my first college chemistry professor, Dr Steven Fawl: Thanks for all of those long talks in your office. Thanks for taking time out for someone who had absolutely no idea what he was going to do with his life. Of all my science professors, you seemed the most worldly and grounded. Your knowledge of chemistry seemed to let you understand the world and how it works. It was because of you that I decided to become a chemist.

To the students of Sonoma State University: Thanks for listening to all of my crazy stories and for continually reminding me why I love teaching so much.

To my colleagues in the chemistry department: Thanks for your help in vetting these stories.

To my agent, Priya Doraswamy of Lotus Lane Literary (lotuslit.com): Thanks for being one of the nicest people I have ever worked with and for helping me realize my dream.

To my editor, Christine Miller (http://tellmewhatyouwanttosay.com): Thanks for all of your encouragement and for helping me find my voice.

To Michelle Sanner: Thanks for your help with the acetaminophen story and helping to start me down the chemical education path.

1

If You Do Not Know Any Chemistry, This Chapter Is For You

As a professor, I regularly teach college-level chemistry courses. These courses present various materials, which are important for students who wish to continue their careers in chemistry. Although most people reading this book will not need all the information covered in these courses, understanding a few key concepts will allow them to understand various ways in which chemistry shows up in everyday life. In fact, one of the driving forces of compiling these stories is to show that even a basic understanding of chemistry can help us comprehend how the world and society work. In particular, I would like to bring readers up to speed on a few key chemical concepts that are referred to in this book (Figure 1.1).

Representing Atoms and Molecules in Chemistry

The first concept concerns the representation of atoms and molecules. Often, the structure of molecules can provide insight into its properties or the ways in which it will affect a human being, if ingested. Certain structural features will imbue molecules with particular properties. In addition, molecules with similar structures will often have similar properties. A detailed understanding of chemistry is not required to make this connection, but only the ability to see similarities.

Chemists represent an individual element with a capital letter, such as "C" for carbon, "H" for hydrogen, and "Fe" for iron, as listed in a periodic table. This letter represents all of the protons and neutrons in the atom's nucleus plus any electrons not involved in bonding. During most chemical reactions, the nucleus of atoms remains unchanged, so this simple representation of elements is helpful to chemists. If an oxygen atom is involved in a chemical reaction, it will remain an oxygen atom. Its structure and bonding may change, but the nucleus will be the same. An important exception is radioactive decay, where a nucleus can be changed and one element can change into another. This will be discussed later.

Strange Chemistry: The Stories Your Chemistry Teacher Wouldn't Tell You, First Edition. Steven Farmer.
© 2017 John Wiley & Sons, Inc. Published 2017 by John Wiley & Sons, Inc.

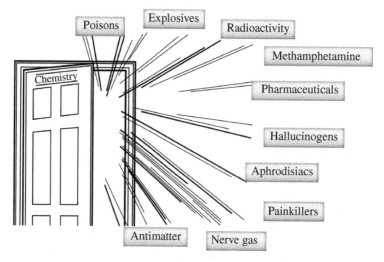

Figure 1.1 Look at what hides behind the door of understanding chemistry.

In the case of some metals and gases, the atom is not bonded (connected) to any other atoms; hence, the bulk material can be represented with the elemental symbol. A block of iron is made up entirely of iron atoms that can be represented by symbol Fe. Similarly, a balloon filled with helium can represented with the symbol He.

Although individual elements are important, chemistry truly becomes interesting when atoms start bonding together to form more complex structures. Two major types of bonds are *ionic* and *covalent*. In an ionic bond, one atom gives up one or more electrons, giving it a positive charge, while another atom gains one or more electrons, giving it a negative charge. Electrostatic forces bring the positive and negative ions together. However, when ionic compounds are placed in an appropriate solvent, such as water, the compounds break apart into their ionic species. The classic ionic compound is common table salt sodium chloride (NaCl). In the crystals of table salt, the sodium and chlorine atoms are being held together by the attraction of a positive and negative charge. When placed in water, table salt tends to break apart into its ionic species, in this case Na^+ and Cl^- (Scheme 1.1).

Ionic compounds are generally made with ionic bonds. Ionic bonds are easily identified because they are made by combining a metal (elements on the left-hand side of the periodic table) with a nonmetal (elements found in the upper right-hand corner of the periodic table). Ionic bonds are typically not formally drawn; rather, the ions are drawn together in a molecular formula where the overall compound is neutral. For example, $FeCl_3$ means a Fe^{3+} ion bonds to three Cl^- ions using ionic bonds. This simple discussion will allow for a better understanding of many ionic compounds with which you may be familiar (Table 1.1).

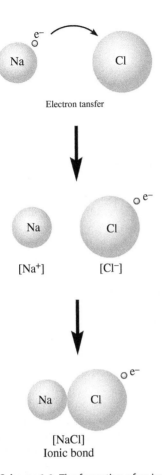

Electron tansfer

[Na⁺]   [Cl⁻]

[NaCl]
Ionic bond

Scheme 1.1 The formation of an ionic bond in NaCl.

Table 1.1 Some common ionic compounds.

Compound	Name	Ions involved	Common use
KI	Potassium iodide	K^+ & I^-	Treatment of hyperthyroidism
PbO_2	Lead (IV) oxide	Pb^{+4} & O^{-2}	Found in car batteries
$CaCl_2$	Calcium chloride	Ca^{+2} & Cl^-	Road deicing

Covalent bonds differ from ionic bonds in that electrons are shared rather than stolen to form a bond between two atoms. This means that covalent bonds are not easily broken into ionic species and do not break apart when dissolved in water. The sharing of two electrons between two atoms to form a covalent bond is represented with a single line. The water molecule is made up of two H—O single covalent bonds. Similarly, if four electrons are shared between two

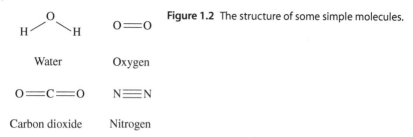

Figure 1.2 The structure of some simple molecules.

atoms, the covalent bond is shown with a double line and is called a double bond. Six shared electrons are depicted by three lines and called a triple bond. Molecular oxygen is made up of a double bond between the two oxygen atoms, and molecular nitrogen is made up of a triple bond between the two nitrogen atoms. Single, double, and triple bonds all have different properties and reactivity that are dependent on the types of atoms involved in the covalent bond. Even now, this basic description of covalent bonds can help you understand the structure of multiple simple molecules (Figure 1.2).

What makes covalent bonds so interesting is their ability to combine to form large molecular structures. Inorganic compounds do not have this ability. Literally, thousands of atoms can be linked together by covalent bonds to create such complex molecules as polymers, proteins, and even deoxyribonucleic acid (DNA).

This book focuses mostly on *organic molecules*, which are typically constructed with covalent bonds. Organic molecules were originally called "organic" because it was believed that these types of compounds could only come from living, organic sources, such as plants or animals. Once it was shown that organic molecules could be made from inorganic materials, the definition was expanded. The current definition states that organic molecules contain the element carbon. *Organic chemistry* is the study of carbon-containing molecules. For the purposes of this book, we will be focusing on the conversion of one organic molecule into another using reactions. Using these reactions, organic chemists create many pharmaceuticals, many plastics, and a multitude of other molecules.

The versatility of covalent bonds creates virtually limitless possible combinations of organic molecules, which is why organic chemistry is such a broad field of study. In college, an entire year of study is devoted to organic chemistry to obtain a typical chemistry degree. At this point, millions of organic compounds are known, with new ones being generated every day. One of the more interesting aspects of organic chemistry is the ability to combine atoms in new ways to make new organic molecules, many of which have never been seen in nature.[1]

---

1 I am formally trained as an organic chemist. During my career, I estimate that I have created roughly 50 novel organic molecules. These include anticancer drugs, novel polymers, linkers for nanoparticles, and supramolecular agents, which allow for the controlled ordering of molecules.

Because of the large numbers of variations, organic molecules are commonly represented by structures as well as their formal names. In addition, due to a large and complex nature of organic molecules, they are often drawn using a condensed form. Because organic molecules typically have a large number of hydrogens in their structures, it is particularly common to represent hydrogens in an abbreviated form. In a condensed structure, the bonds attached to the hydrogens are omitted and the number of H's is represented with a subscript. Examples of these abbreviations are represented below using some simple organic molecules (Figure 1.3).

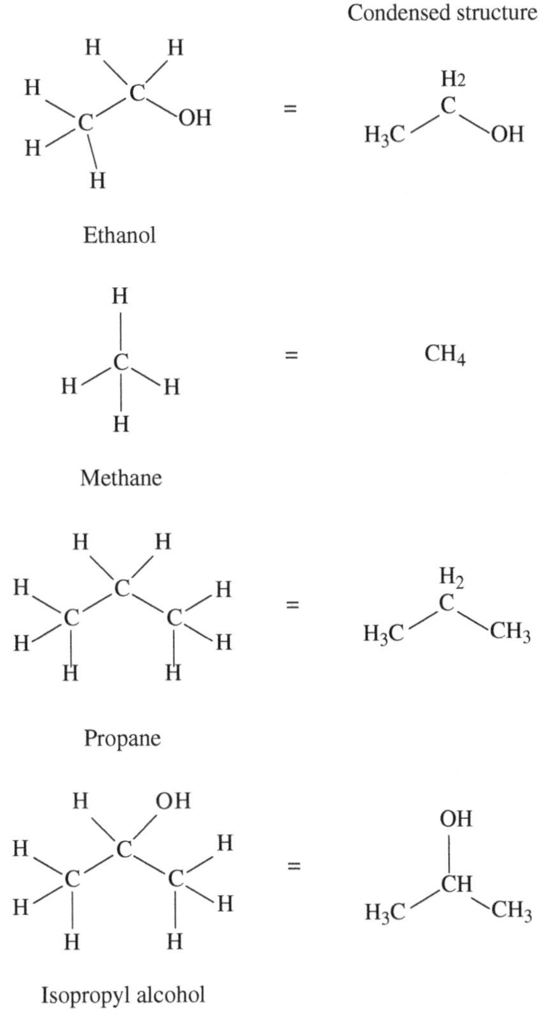

Figure 1.3 The condensed structure of some simple organic molecules.

Condensed structure

Ethanol

Methane

Propane

Isopropyl alcohol

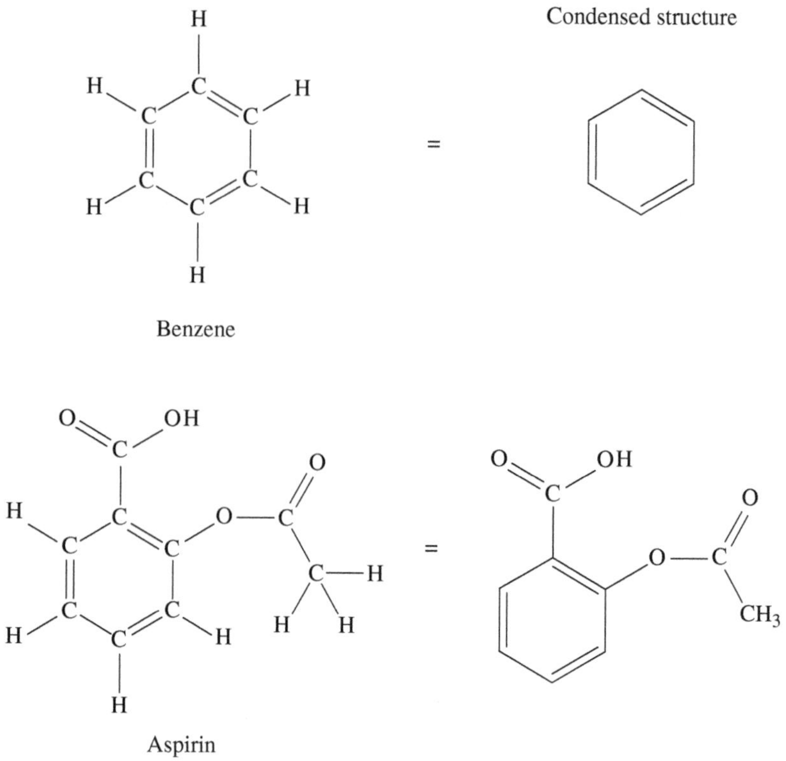

Condensed structure

Benzene

Aspirin

Figure 1.4 The condensed structure of the benzene ring.

Another important way in which hydrogens are abbreviated involves the *benzene* ring. This ring is immensely important in organic chemistry, and its presence can be seen in many important organic molecules. To simplify the structure, the hydrogens at the points of the benzene ring are commonly omitted. Moreover, the carbon atoms in the benzene ring are represented simply by lines denoting the covalent bonds (Figure 1.4).

Lastly, the structures of polymers are usually represented using a type of abbreviation. Small molecules called *monomers* are connected in large numbers during a polymerization reaction to create large molecules called *polymers*. This process is represented in the name "polymer," which means many monomers. Because polymers are made up of a repeating monomer subunit, they are represented by the subunit surrounded by brackets. The monomer subunit is repeated a variable number of times, which is represented by the

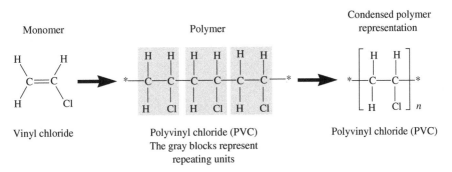

Figure 1.5 How polymers are represented.

subscript "*n*." The actual number of monomers subunits in a polymer is usually unknown, which is why it is represented by a variable (Figure 1.5).

Neurotransmitters

In this book, neurotransmitters are the most important molecules used to describe the function of organic molecules in the body. Virtually everything we do involves neurons communicating with one another. Everything from movement, breathing, and even awareness are brought about by electrical impulses moving across our nervous system. Anyone who has seen a Taser in action knows that neurons are affected by electricity; however, certain chemicals also play an important role in how neurons operate. Many neurons are separated by a small gap called the *synaptic cleft*. During a typical nerve impulse, specific molecules called *neurotransmitters* bridge this gap. When an electrical impulse reaches the end of a presynaptic neuron, neurotransmitters are released and subsequently diffused across the synaptic cleft, binding to the receptors on the receiving postsynaptic neuron. *Receptors* are typically proteins on the surface of the neurons, which recognize and bind to specific neurotransmitters. This binding usually brings about a chemical change that creates an electrical impulse in the receiving postsynaptic neuron. In short, neurotransmitters allow for electrical impulses to be transmitted between adjacent neurons despite the presence of a synaptic gap. By repeating this process, electrical nerve impulses can be sent across the body or across the brain. Neurotransmitters that cause a neuron to fire are considered "*excitatory*" and are responsible for motion, mental

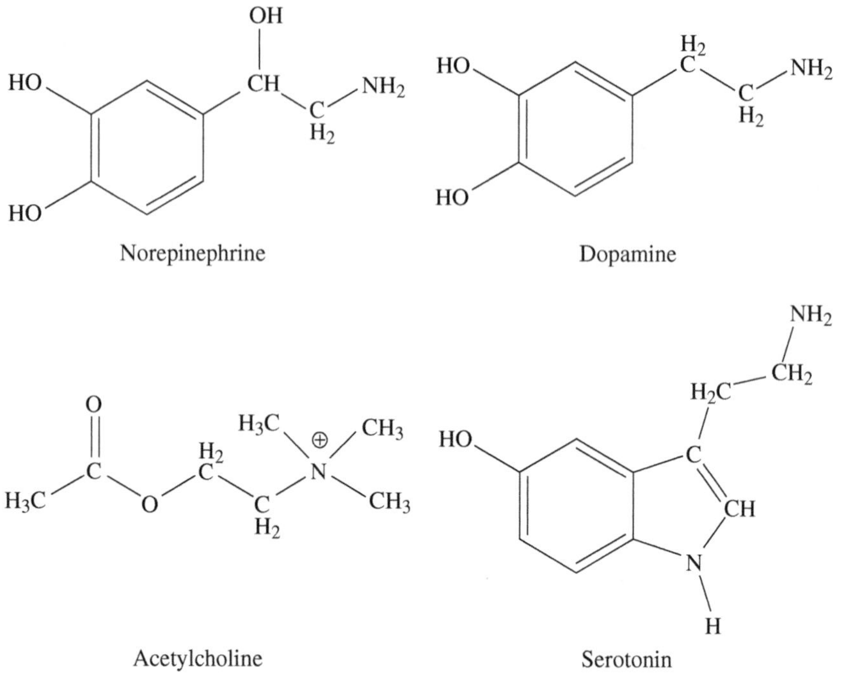

Figure 1.6 Various neurotransmitters.

cognition, and other activities that require the brain and body to be active (Figure 1.6).

In addition, certain neurotransmitters can also be "*inhibitory*" and actually impede the transmission of impulses in neurons. The effect of inhibitory neurotransmitters in these neurons causes a chemical change within the neuron that opposes the effects of excitatory neurotransmitters. In general, inhibitory neurotransmitters are responsible for inducing sleep and filtering out unnecessary excitatory signals.

In short, neurotransmitters send chemical messages between neurons and act as the on and off switches of the nervous system. By understanding that chemicals can affect how neurons work, many interesting concepts can be discussed. Many mental illnesses are believed to be caused by a "chemical imbalance" of neurotransmitters in certain areas of the brain. Many medications used to treat mental illnesses, as well as many psychoactive drugs and neurotoxins, obtain their effects by changing the ways in which neurotransmitters are released and absorbed or by simply mimicking the structure

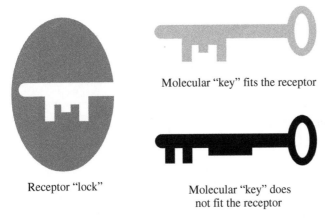

Molecular "key" fits the receptor

Receptor "lock"

Molecular "key" does
not fit the receptor

Figure 1.7 A representation of the lock-and-key model of receptors.

of a neurotransmitter. The key receptors in neurons designed to recognize neurotransmitters look for specific structural features. This is called the lock-and-key model. Receptors proteins are typically wadded into a ball-like structure that has small pockets. Certain structural features of molecules allow them to fit into these pockets, activating the receptors. Because the receptors are looking for specific structural features, molecules that have similar structural features can fool these receptors (Figure 1.7).

An excellent example is seen with the molecules dopamine and methamphetamine. Dopamine is one of the most important neurotransmitters in the parts of the brain involving motion and alertness. Key receptors in neurons recognize the benzene ring connected to two carbons and a nitrogen found in dopamine. Methamphetamine also has a benzene ring connected to two carbons and nitrogen, so it can also fit into these receptors, which tricks the neurons into thinking that it is dopamine. The presence of methamphetamine causes the areas of the brain, which utilize dopamine to become excited, causing the hyperactivity and insomnia associated with methamphetamine use. Now that we understand the structural features that can allow molecules to mimic dopamine, we can look for them in other molecules. Ritalin® has these structural features, and it is used to treat attention deficit hyperactivity disorder (ADHD) by stimulating the parts of the brain associated with attention. In addition, the common decongestant pseudoephedrine has these structural features and has the side effects of causing restlessness and insomnia, which has to be stated on the packaging (Figures 1.8 and 1.9).

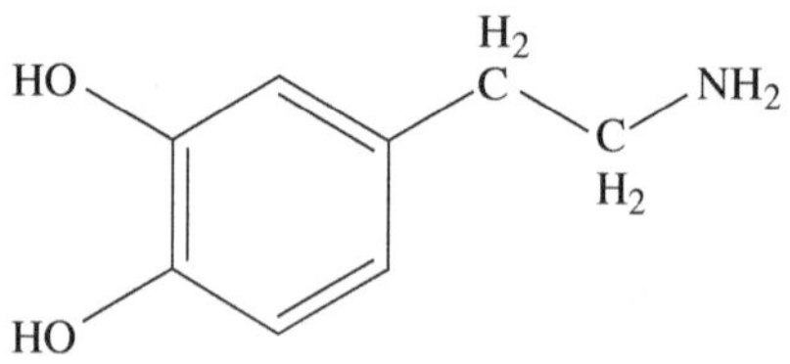

Dopamine

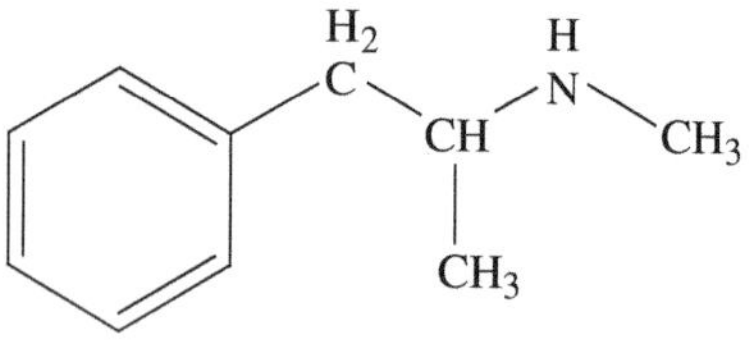

Methamphetamine

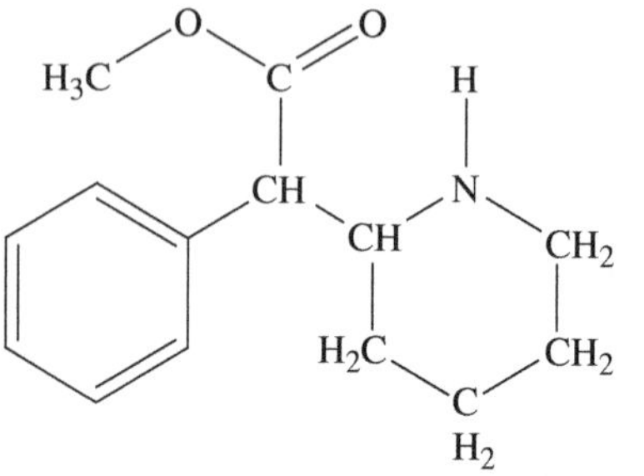

Methylphenidate
(Ritalin®)

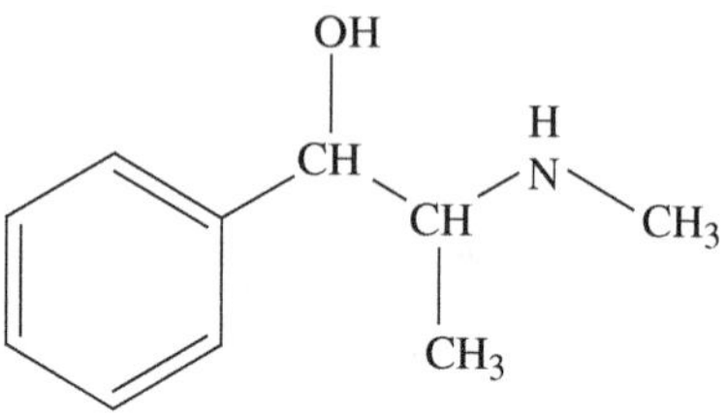

Pseudoephedrine

Figure 1.8 Molecules with structures similar to dopamine.

Dopamine receptor

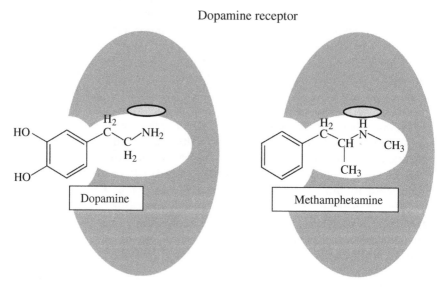

Figure 1.9 A representation of how dopamine and methamphetamine both fit in the dopamine receptor.

Intermolecular Forces

Have you ever wondered why some molecules like oxygen are a gas at room temperature and ambient pressure while other molecules like water are liquid and still others like table salt (NaCl) are a solid? This has to do with a concept called *intermolecular forces* (IMF) or the forces between individual molecules. In general, molecules with relatively strong IMFs tend to hold together better and form solids and those with weak IMFs tend not to hold together and form gases. Having a basic understanding of IMFs allows one to have a better understanding of the world. Why a solid is a solid, why does one substance stick to another, and why do some liquids mix while others separate into layer?

To have a basic understanding of IMFs, you just have to remember that positive and negative charges are attracted to each other. The ways in which these positive and negative charges are generated in molecules determine the strength of the IMFs between them.

This book discusses four major IMFs. The strongest is called the *ionic* IMF. As discussed previously, when a compound contains an ionic bond, one or more electrons are shared to form positive and negatively charged species. These charged species are then attracted to each other by the ionic IMF. These compounds, which contain ionic bonds (metals bonded to nonmetals), are typically solids under normal conditions. Salts, such as common table salt (NaCl) or potassium chloride (KCl), as well as most minerals, such as chalk ($CaCO_3$) and iron pyrite (FeS), all contain ionic bonds and ionic IMFs.

The next strongest IMF is called a *dipole* IMF. Dipole IMFs are typically found when dealing with molecules containing covalent bonds. Although it is possible for organic compounds to have ionic IMFs, most are governed by dipole IMF's. In a covalent bond, electrons are being shared by two atoms, although they are rarely shared equally. *Electronegativity* is a measure of an atom's ability to pull on the electrons in a covalent bond. The elements of the periodic table typically become more electronegative as we travel up and to the right of the table, with fluorine being the most electronegative element. There are many exceptions to this rule, but it is a general trend. In the molecule ICl, the electrons in the I—Cl covalent bond are drawn closer to the chlorine because it is more electronegative than iodine. This gives the chlorine a partial negative charge, which is represented by the symbol δ^-. This also gives the iodine a partial positive charge represented by the symbol δ^+. The ICl molecule is called *polar* because one side of a molecule has a slight positive charge and the other side has a slight negative charge. A dipole IMF is created when the positive side on one molecule is attracted to the negative side of an adjacent molecule. The molecule ICl has only one covalent bond; however, molecules with multiple covalent bonds can also be polar, depending on the orientations of the bonds and the electronegativity of the atoms involved. Note! The dipole IMF is weaker compared to the ionic IMF because only partial charges are being used (Figure 1.10).

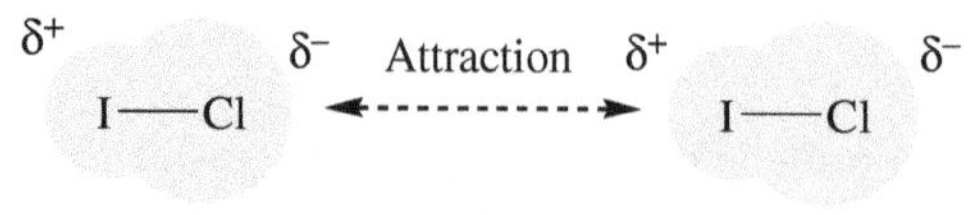

δ^+ δ^- Attraction δ^+ δ^-

I—Cl   ◄------------►   I—Cl

Figure 1.10 An example of a dipole intermolecular force.

An important subset of dipole IMFs is called *hydrogen bonding*, and it is reserved for some of the most electronegative elements in the periodic table, like nitrogen, oxygen, and fluorine. When these elements are directly bonded to hydrogen in a molecule, this special IMF comes about. In this case, the charge separation caused by these highly electronegative elements is so extreme that dipole interaction is enhanced. The hydrogens are "loose" and do not remain permanent bonded. They freely move around, and they can be shared by other hydrogen-bonding molecules. This causes a hydrogen-bonding interaction, which is stronger compared to a typical dipole interaction. Many common liquids utilize hydrogen-bonding IMFs, including water and alcohol. In addition, DNA strands are held together using hydrogen bonding.

The last IMF is probably the most difficult to conceive, although it is still extremely important. Covalent bonds involve electrons being shared; therefore, they are affected by the electronegativity of the atoms involved. However, what happens when the two atoms are the same? A prime example of this is an oxygen molecule. The two oxygen atoms have exactly the same electronegativity, therefore, exactly the same pull on the electrons in the covalent bond. There is no separation of charge; thus, the molecule is deemed *nonpolar*. This can also come about when dealing with substances that have no bonds, such as helium, or with molecules that orientate their bonds so that there is no charge separation in the molecule, such as methane. All these molecules are also considered nonpolar; however, the fact that they can form liquids when made cold enough shows that there must be some type of IMF in them. This last and weakest IMF has many names, including *instantaneous dipole forces*, *Van der Walls forces*, and *dispersion forces*. I will call them instantaneous dipoles in this book because I feel this name best describes the effect.

Instantaneous dipole IMFs can best be explained by looking at a typical helium atom. It is made up of a nucleus containing two protons and two neutrons surrounded by two elections. The two negatively charged electrons balance out the two positively charged protons in the nucleus so the overall atom is neutral. Because both electrons are negatively charged, they repel each other and usually stay on the opposite sides of the nucleus. However, these electrons orbit the nucleus at extremely fast speeds, roughly 80% the speed of light; hence, it is possible for these electrons to crowd on one side of the helium atom. When this happens, the nucleus is momentarily exposed, giving the atom a positively charged side (δ^+) while the crowded electrons give the other side of the atom a negative charge (δ^-). This happens only for an instant because the electrons quickly repel each other. This charge separation gives

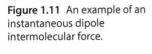

Figure 1.11 An example of an instantaneous dipole intermolecular force.

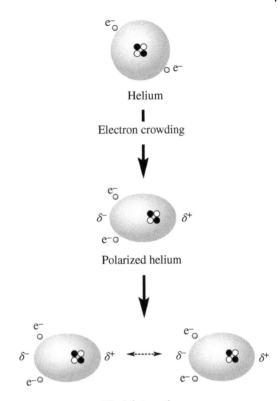

Helium

Electron crowding

Polarized helium

Weak interaction

the atom an "instantaneous dipole" and allows two different helium atoms to have an IMF. This idea is also true for molecules that contain multiple atoms, such as methane. Methane is made up of four hydrogen nuclei and a carbon nucleus, which are all surrounded by a cloud of rapidly moving electrons. These electrons can also momentarily crowd to one side of the atom, giving it an instantaneous dipole IMF (Figure 1.11).

One thing to remember about instantaneous dipole IMFs is that in general, larger nonpolar molecules have more electrons; therefore, they can create a stronger IMF. The hydrocarbons (molecules that only contain hydrogen and carbon) are an excellent example of this. Methane (CH_4) and propane (C_3H_8) are relatively small, and their weak instantaneous dipole IMFs cause them to be gases under normal conditions; however, octane (C_8H_{18}), the major constituent of gasoline, is large enough to allow for a strong enough instantaneous dipole IMF for it to be a liquid. Eventually, larger hydrocarbons with more than about 20 carbons become solids. The polymer polyethylene is made up essentially of large hydrocarbons and is used to create solid commercial products, such as plastic bags and drinking bottles.

2

The Only True Aphrodisiac and Other Chemical Extremes

I think that most of us have looked at the "Guinness World Records" at some point, which is one example of our obsession with extremes: the biggest, the fastest, and so on. Believe it or not, even in chemistry, there are extremes. It is my pleasure to present some of the most intriguing to you.

Death Is Its Withdrawal Symptom!

Though not widely known, alcohol is one of the few drugs for which withdrawal symptoms include death. Most people think drugs such as heroin or methamphetamine would have the worst withdrawal symptoms, since these substances are known to be highly addictive. Dramatic television and movie portrayals of heroin withdrawal symptoms make a great impression on the public as well. Oddly enough, people going through alcohol withdrawal are rarely portrayed in popular media, which is part of the reason most people do not understand the deadly nature of this type of withdrawal.

Why can withdrawal from alcohol be fatal? Consider that alcohol affects virtually all parts of the brain and is involved in a process that activates the nerve cells there. Although the effects of alcohol are complex, its action on *gamma-aminobutyric acid* (GABA) receptors in neurons has long been known to be significant. GABA is the main inhibitory neurotransmitter in the mammalian central nervous system. To put it simply, the presence of GABA prevents the ability of a nerve that contains GABA receptors to fire, thereby deadening it. Alcohol facilitates the inhibitory function of the GABA receptor and keeps the nerves from becoming stimulated, resulting in the euphoric feeling of intoxication (Figure 2.1).

Increased consumption of alcohol causes this numbing of nerve cells to spread, eventually affecting all parts of the brain. Alcohol initially inhibits the most complex brain functions; if consumption continues, it spreads to areas controlling the less complex functions. The first thing to go – the most complex part of the brain – is the cerebral cortex, which controls reasoning.

Strange Chemistry: The Stories Your Chemistry Teacher Wouldn't Tell You, First Edition. Steven Farmer.
© 2017 John Wiley & Sons, Inc. Published 2017 by John Wiley & Sons, Inc.

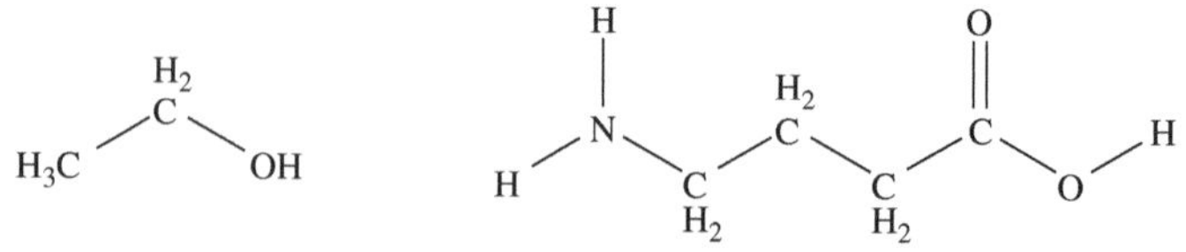

Grain alcohol (ethanol)   Gamma-aminobutyric acid (GABA)

Figure 2.1 The structures of ethanol and GABA.

Therefore, judgment is impaired. With more alcohol, the speech centers of the brain are slowed, resulting in a person slurring their words. Next, the brain area controlling movement and coordination is impaired, and the inebriated individual begins to stagger. If one keeps drinking, the neurons involved in controlling involuntary bodily functions such as breathing and heartbeat are affected, and death can occur.

For people who chronically drink large quantities of alcohol, over time the brain will try to compensate for this continual desensitization of neurons. In a process called *homeostasis*, the nerves of the brain become hyperactive to make up for the long-term deadening effect of habitual alcohol consumption. As a result of homeostasis, chronic alcoholics show reduced brain GABA levels and a decrease in GABA receptor sensitivity as an adaptation to the persistent presence of alcohol. If such a person stops drinking, brain GABA levels fall below normal and GABA activity declines during withdrawal. The resulting decrease in inhibitory function may contribute to symptoms of nervous system hyperactivity, which is associated with alcohol withdrawal. This change in brain chemistry takes some time, though, so here, we are talking about chronic alcohol abusers. In fact, it has been estimated that it would take a continual alcohol intake of roughly 100 g (equivalent to about 1 pt of liquor or 96 ounces of beer) per day for this effect to take place.

The abrupt cessation of long-term alcohol consumption produces symptoms associated with having a hyperactive brain, and this is called acute alcohol withdrawal. Mild withdrawal symptoms start as quickly as 6 hours after the initial decline from peak intoxication, and they include feelings of nervousness, anxiety, irritability, and emotional volatility. One to four days of not drinking alcohol can result in severe withdrawal symptoms including hallucinations called *delirium tremens* (DTs). These are horrible hallucinations, surprisingly not often portrayed in television and movies, and are sometimes described as the feeling of having bugs crawling all over one's skin. Those of you who have watched old cartoons may know the common joke where the alcoholic swears off drinking because they have seen something strange, such as a pink elephant. This is all related to the idea that people who are heavily addicted to alcohol tend to have hallucinations when undergoing withdrawals.

In addition to hallucinations, severe withdrawal symptoms can include grand mal seizures. One can literally have the equivalent of an epileptic seizure from alcohol withdrawal, and it is these seizures that can cause death. The mortality rate among people who are actively having DTs is from 5% to 25%, which means that someone who is having hallucinations from alcohol withdrawal is in serious danger of dying. In addition, clinical data has shown that the likelihood and severity of these seizures increase with the number of past withdrawals.

What circumstances might lead to withdrawal-related death? Typically, death from alcohol withdrawal occurs when alcoholics are arrested or admitted to a hospital and they experience an abrupt removal from alcohol.[1] There are innumerable examples of this reported in the media. In 2009, a county in Montana was ordered to pay $1.35 million for a case in which an 18-year-old died from alcohol withdrawal after being found shivering and unresponsive in a jail cell after being arrested. In another case, a 37-year-old woman went through severe alcohol withdrawal for 3 days after being incarcerated, and she did not receive treatment until she had already lost consciousness. She later died at a hospital. These cases are representative of the public's general lack of knowledge about the deadly nature of alcohol withdrawal. The uninformed are more likely apathetic toward severe symptoms. What would *you* do with someone who is coming off alcohol?

Besides alcohol, a class of sedatives called *benzodiazepines* (Valium® is a common example) work on GABA receptors in neurons, explaining the similarity of their nerve-numbing effect. In fact, benzodiazepines show cross-tolerance and cross-dependence with alcohol. Recent research suggests that benzodiazepines are likely to be the most effective agent for treating alcohol withdrawal symptoms. However, like alcohol, withdrawal from benzodiazepines can be lethal.

Benzodiazepines are commonly used for alcohol detoxification because the dose given can be easily controlled. Ones that may be utilized include the shorter-acting benzodiazepines, such as Serax® and Ativan®, and the longer-acting benzodiazepines, such as Librium® and Valium®. During a controlled withdrawal process, doctors typically employ a progressive decrease in benzodiazepines dosage over the time span of the entire withdrawal process.

To address the danger of grand mal seizures during alcohol withdrawal, doctors may employ antiseizure medications such as Carbamazepine. Furthermore, physicians may prescribe disulfiram (Antabuse®) or naltrexone (ReViaT®) to help prevent people from returning to drinking. Antabuse® combined with alcohol triggers very unpleasant effects including flushing, nausea, vomiting, and dizziness. Naltrexone reduces the alcoholic's craving for

---

1 During the writing of this book, an acquaintance ended up having a seizure during the recovery from a minor surgery, most likely due to alcohol withdrawal. Anyone who regularly abuses alcohol should have a frank discussion with their doctor prior to an extended hospital stay. It could save their life!

alcohol by targeting the brain's reward circuits. Both these drugs have proven to be strong deterrents to drinking, even in chronic alcoholics.

DTs have occasionally been depicted in popular culture. In the 2010 movie *Everything Must Go*, the main character played by Will Ferrell suffers from DTs. In the 2003 movie *The Last Samurai*, the main character played by Tom Cruise suffers from DTs after being captured in a battle. The main character in the 1995 movie *Leaving Las Vegas*, played by Nicholas Cage, suffered from the DTs after a drinking binge.

What Is the Number One Cause of Liver Failure in the United States?

The answer is acetaminophen, the active ingredient found in Tylenol® and many other pain relief formulations. The answer to this question points out the common assumption that over-the-counter (OTC) pharmaceuticals are completely safe. Another fact commonly not considered is that many drugs can interact with each other to sometimes produce lethal results. Owing to these two misconceptions, acetaminophen, the seemingly harmless active ingredient, is the number one cause of liver failure in the United States. This problem is so common that emergency room doctors and health professionals have even started using the term "Tylenol liver" to refer to the inflammation of this organ due to acetaminophen toxicity. The magnitude of the public's obliviousness to the dangers of acetaminophen is evident in the statistics. In 2003, acetaminophen toxicity accounted for 51% of all acute liver failure cases in the United States. During this same year, roughly 5% (52,995) of the 1,079,683 drug-related emergency room visits were due to acetaminophen toxicity. The most disturbing thing, however, aside from liver failure and emergency room visits, is the fact that acetaminophen toxicity can be lethal. In 2003, it was determined that roughly 250 adolescents and adults, as well as over 300 children, died from acetaminophen poisoning!

Why is acetaminophen poisoning such a problem? Part of the answer to this lies in the high frequency with which it is used. It has been estimated that 22% of adults in the United States take acetaminophen-containing products every week, and over 50 million Americans have used it at some point in their lives. In addition, acetaminophen is commonly given to infants and children because it is the most recommended fever-reducing medication. Also contributing to the problem is the presence of acetaminophen as an ingredient in over 600 OTC and prescription formulations, including Actifed®, Benadryl®, Contac®, Dayquil®, Dimetapp®, Dristan®, Formula 44®, Midol®, Nyquil®, Robitussin®, Sinutab®, Sudafed®, Theraflu®, Triaminic®, and Vicks®. Acetaminophen is added to many prescription painkillers as well, including medicines such as

Hydrocodone, Percocet®, and Vicodin®. Because acetaminophen is found in so many formulations, too much can be ingested from multiple sources, resulting in an unintentional overdose and possibly death.

To help combat this problem, the parent company for the Tylenol® brand, McNeil laboratories, has launched the "Get Relief Responsibly" campaign, which involves making drug labels easier to read and lowering the dosage of each individual medicine. They also lowered the recommended maximum number of doses per day and increased the recommended time between doses.

Why does acetaminophen harm your liver if taken in high doses? When acetaminophen enters the liver, it is converted to a highly toxic metabolite called *N-acetyl-p-benzoquinone imine* (NAPQI) via a natural process involving a liver enzyme called cytochrome P450. The P450 enzyme typically changes chemicals to make them more water soluble and plays a role in the process by which chemicals are excreted in our urine. Although the highly reactive and toxic NAPQI molecule is produced in the liver, it is rendered harmless when it reacts with a "suicide molecule" called *glutathione*. In particular, NAPQI reacts with the thiol (SH) moiety present in glutathione. However, if large amounts of acetaminophen are present, the body's glutathione reserves are used up and NAPQI remains. Because proteins often contain thiols due to the presence of the amino acid cysteine, any remaining NAPQI is then free to react with liver tissue causing damage and interfering with DNA replication. Over time, the cells of the liver are unable to reproduce, and in a process called *hepatoxicity*, the liver slowly dies (Figure 2.2).

How is an acetaminophen overdose treated? The first step is to get rid of any undigested acetaminophen by inducing vomiting or by gastric lavage (stomach pumping). Any acetaminophen remaining in the stomach lining is then removed by ingesting activated charcoal. Next, *N-acetylcysteine* (NAC) is administered. NAC is considered the antidote for acetaminophen overdose because it augments the body's glutathione reserves and helps to deactivate any NAPQI still present. However, in order for NAC to be effective, it needs to be given within 8 hours of an individual's overingestion of acetaminophen. Therefore, it is vital for acetaminophen poisoning to be diagnosed and treated as soon as possible (Figure 2.3; Schemes 2.1 and 2.2).

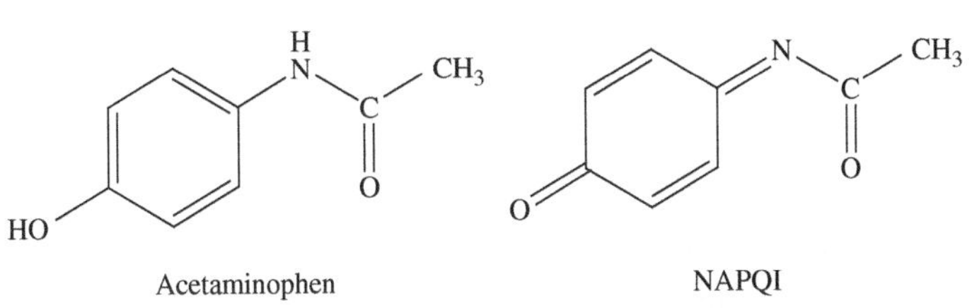

Figure 2.2 The structure of acetaminophen and its toxic metabolite, NAPQI.

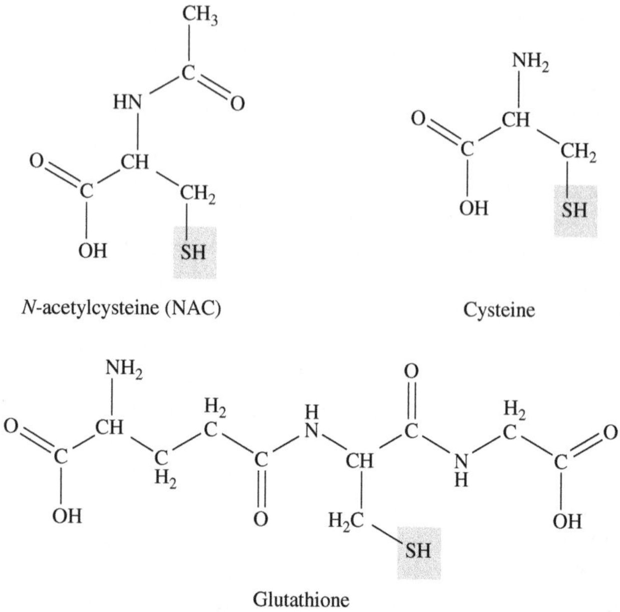

Figure 2.3 Thiol containing compounds that react with NAPQI. The gray box highlights the thiol moiety.

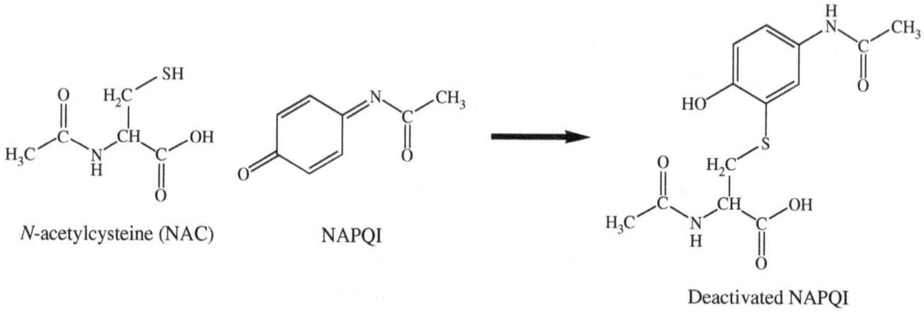

Scheme 2.1 The reaction of NAPQI with the thiol in NAC.

Some of the symptoms of acetaminophen poisoning are nausea, vomiting, and abdominal pain. If tissue damage is severe, the liver may fail and a liver transplant might be required. As a side note, although most of the deaths from acetaminophen poisoning are due to unintentional overdosing, some cases are due to an intentional overdose as a means of suicide. Fortunately, acetaminophen suicide attempts often do not lead to death because the need for NAC is clear, and the antidote is usually given in time to save the patient.

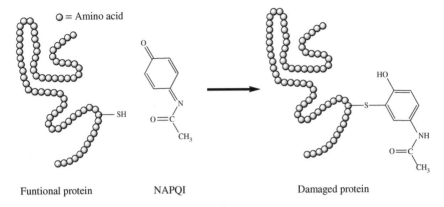

Scheme 2.2 The reaction of NAPQI with the thiol in a protein.

One last thing to be aware of is the effect of acetaminophen and alcohol together. There is currently a very clear warning on bottles of Tylenol® that the medication should not be taken after two or more alcoholic beverages have been consumed. What is the reason for this? People who drink regularly or are currently inebriated are more susceptible to liver damage and failure due to acetaminophen. Alcohol actually amplifies the P450 reaction that produces NAPQI and thereby increases the liver-damaging effects of acetaminophen. It is not surprising that in 14,619 (30%) of the 52,995 emergency room visits caused by acetaminophen in 2003, alcohol was also involved. If people are wondering what they should take for a headache due to a hangover, one thing is clear: *not* acetaminophen!

The Most Addictive Substance Known

Nicotine, the psychoactive substance in tobacco products, is responsible for the most common form of chemical dependence in the United States. In 2010, an estimated 19% (45.3 million) of US adults were cigarette smokers and 78% of them smoked every day. Although addictiveness is difficult to quantify, at least one study has shown nicotine to be among the most addictive substances known, alongside heroin, cocaine, and alcohol (Figure 2.4).

Similar to many addictive substances, nicotine increases the levels of the neurotransmitter dopamine in the region of the brain regulating sensation of pleasure, accounting for the feelings of wellbeing reported by many smokers. In addition, the fact that brain nicotine levels peak within 10 seconds of inhalation adds to its addictive effect. By the same token, the effects of nicotine dissipate quickly, forcing the smoker to continue smoking to maintain the pleasurable feelings and prevent withdrawal.

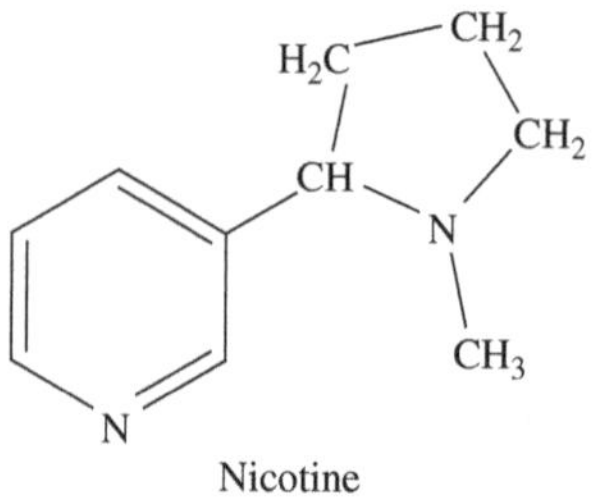

Figure 2.4 The structure of nicotine.

Nicotine

Most regular smokers are addicted to nicotine despite knowing it is harmful to their health. Although many smokers say they want to reduce or stop using tobacco, the intense symptoms of withdrawal – including irritability, anxiety, difficulty concentrating, and increased appetite – often prevent their success. In 2010, 68.8% of adult smokers said they wanted to stop smoking, and 52.4% had attempted to quit in the past year. The end result was that 85% of smokers who tried to quit on their own relapsed, a true demonstration of nicotine's addictiveness. Quitting smoking is difficult, often requiring multiple attempts, and support through counseling and/or cessation medications is needed 31.7% of the time.

Nicotine gum became the first nicotine replacement therapy (NRT) in 1984, and by 1996, it was available OTC to the public. Some smokers gained control over their nicotine habit with use of the gum, but others could not tolerate the taste or had trouble chewing. Therefore, in the early 1990s, the US Food and Drug Administration (FDA) approved four transdermal nicotine patches, two of which were available OTC in 1996. In addition, during the 1990s, a nicotine nasal spray and nicotine inhaler became available by prescription, providing further NRT help for tobacco users wanting to quit.

In addition to NRT, cessation medications were found effective in treating tobacco dependence, including prescription non-nicotine drugs. Examples include the antidepressant bupropion (Zyban®), approved for smoking cessation by the FDA in 1997, and varenicline tartrate (Chantix®), which received FDA approval for this use in 2006. Chantix® works by reducing nicotine's ability to elevate dopamine levels, thereby both easing withdrawal symptoms and blocking the addicting effects of nicotine if the person resumes smoking.

Besides the pharmacological effects of nicotine, behavioral factors contribute to a smoker's urge to light up. For many, the feel, smell, and sight of a cigarette – even the ritual of lighting it – become associated with the pleasure they derive from nicotine, and this worsens cravings. Unfortunately, nicotine replacement therapies do not address environmental cues, so psychological therapies are needed to help smokers with these triggers.

Statistics grimly underline the effects of widespread nicotine abuse in the United States. Tobacco is the leading preventable cause of death in this country, with cigarette smoking killing an estimated 440,000 American citizens each year. This is more than alcohol, illegal drug use, homicide, suicide, car accidents, and AIDS combined! Smoking tobacco, a known cause of lung cancer, is responsible for approximately 90% of lung cancer deaths among men and 80% among women. Furthermore, smokers die from cancer twice as quickly as nonsmokers, and heavy smokers four times as fast. Cigarette smoking harms nearly every organ in the body and has been conclusively linked to cataracts and pneumonia. To top it all off, residential fires caused by lit cigarettes kill approximately 1000 people per year.

With nicotine's estimated toxicity comparable to that of sodium cyanide, it has been estimated that only 50–60 mg of nicotine could kill a person. However, smoking a typical cigarette results in only about 0.1 mg of nicotine being absorbed, so a human would have to smoke more than 500 cigarettes quickly, one after the other, to reach that lethal level.

Interestingly, nicotine extracted from tobacco leaves with water has been used as an insecticide for centuries. One of the most commonly used insecticides in the world, imidacloprid, is similar in chemical structure to nicotine. Although the amount of nicotine ingested via smoking is usually too small to cause extreme toxicity, exposure to the high doses found in some insecticide sprays can cause vomiting, tremors, convulsions, and even death (Figure 2.5).

E-cigarettes ("Vapes") might be a new source of nicotine poisoning as well. Although they are considered safe by their manufacturers, ingesting highly concentrated nicotine e-liquid could be lethal. This is particularly dangerous for children, who might be attracted to the flavored substance. This new form of nicotine exposure is a growing problem; in 2013, there were 1414 reported exposures to e-cigarette liquids, a 300% increase over 2012. The next time you see someone using an e-cigarette, you might want to mention that they are vaping an insecticide (Figure 2.6).

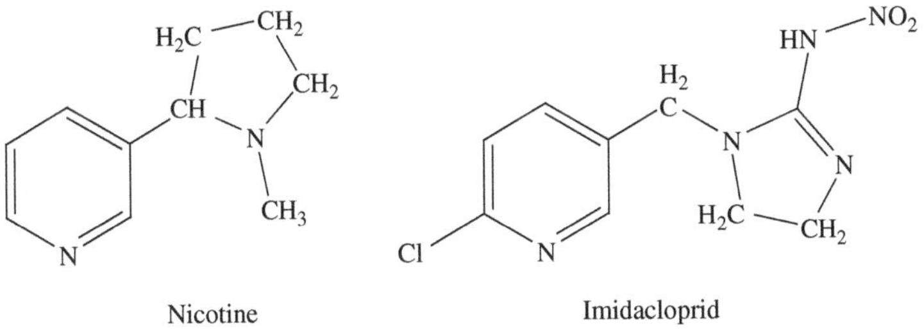

Figure 2.5 A comparison of the structures of nicotine and imidacloprid.

Figure 2.6 Apparently cockroaches do not care that nicotine is used as an insecticide.

40 Million Times Deadlier Than Cyanide

Botulinum neurotoxin type A is a neurotoxic protein produced by the bacterium *Clostridium botulinum* and produces the condition commonly referred to as botulism when ingested. The botulinum toxin is the most toxic substance known with a typical lethal dose being as little as 0.00000002 g! This means that if evenly dispersed, 1 g of crystalline Botulinum toxin could kill more than 1 million people. Another way to consider its toxicity is to realize that it is 40 million times more powerful than cyanide.

When asked what they think is the most toxic substance, people give a huge variety of responses, ranging from odd ones – such as heroin and cocaine (probably because the lethal nature of these drugs is discussed frequently in popular literature) – to more realistic guesses such as polonium-210 or ricin. Polonium-210 is highly toxic when ingested due to its radioactivity, and it has

Table 2.1 A comparison of the lethal doses of toxic compounds.

Toxin	Typical lethal dose (g)
Sodium cyanide	0.0064
Strychnine	0.0001
Ricin	0.000022
Sarin	0.0000172
Polonium-210	0.0000001
Botulinum neurotoxin type A	0.00000002

been linked to many assassinations. Ricin, made from processing castor beans, has recently been made popular by the T.V. series "Breaking Bad." Sarin is another deadly chemical, used in the infamous gas attack on the Tokyo subway by the Aum Shinrikyo cult (all of these poisons will be discussed later in this book). Cyanide and strychnine are also well known among the general public as highly poisonous, but when comparing typical lethal dosages, we can see that botulinum is in a class by itself (although polonium-210 is quite close) (Table 2.1).

The botulinum toxin works by blocking neurons, rendering the muscles unable to contract for a period of 4–6 months. Death is generally caused by suffocation due to the paralysis of the respiratory muscles. Oddly enough, the muscle paralyzing property of the botulinum toxin has been utilized in medicine to treat conditions including muscle spasms, migraines, excessive salivation, and uncontrollable blinking. Perhaps more familiar is the use of botulinum toxin in the most common cosmetic procedure in the United States – Botox® injections. In 2010, there were 2,437,165 reported Botox® injections performed in the United States. The potential for cosmetic use of the botulinum toxin was discovered during its use to treat eye muscle disorders, when it was noticed among patients that frown lines between the eyebrows were softened. This side effect was first published in 1992, but botulinum toxin was not approved for medical use by the FDA until 2002. In fact, Botox® is one of the world's most widely researched medicines, with approximately 2300 related publications in scientific journals. Botox® is currently approved in approximately 85 countries for 25 different medical purposes. In case you are wondering how the most toxic substance known can be used in medical procedures, Botox® injections involve a very dilute solution of the botulinum toxin.

In the United States, Botox® is manufactured by Allergan, Inc. through fermentation of the bacteria *C. botulinum* type A. The neurotoxin is purified from the culture solution by dialysis and a series of acid precipitations and then dissolved in a sterile sodium chloride solution also containing human albumin serum, which is the most abundant protein in human blood plasma. A total treatment dose of Botox® solution corresponds to roughly 1 ng (1×10^{-9} g) of the toxin.

How dangerous is using Botox®? As you can imagine, using the most toxic substance known for medical applications has its concerns. If the FDA-approved guidelines are adhered to, little more than minor problems are associated with using Botox®. However, if FDA directives are not followed, serious problems can arise. In November of 2004, four people became paralyzed after purportedly being injected with potent, unapproved botulinum toxin at an Oakland Park, Florida medical clinic. The victims were hospitalized with severe botulism poisoning; fortunately, the paralysis was temporary. On February 8, 2008, the US FDA notified the public that various types of Botox® have been linked in some cases to adverse reactions, including respiratory failure and death. The most severe adverse effects were found in children treated for spasticity in their limbs associated with cerebral palsy. Treatment of spasticity is not an FDA-approved use of botulinum toxin in children or adults. The adverse reactions appeared to be related to the toxin spreading from the injection site, resulting in symptoms similar to those of botulism.

As far as the danger of botulinum toxin, a total of 140 confirmed cases of botulism were reported to the U.S. Center for Disease Control in 2011, with two of these being fatal. There are five main kinds of botulism. Eating food contaminated with the botulinum toxin causes foodborne botulism. Home-canned foods are often implicated with outbreaks of this type of botulism, and usually two or more persons are affected. Wound botulism occurs when a wound is infected with *C. botulinum* bacteria, which subsequently produces the toxin. Most wound botulism cases are associated with intravenous drug use, especially black tar heroin injections. Infant botulism is caused when an infant consumes spores of the botulinum bacteria from the environment. Because of an infant's weak immune system, the bacteria grow in the intestines and release the toxin. Adult intestinal toxemia is an adult version of infant botulism and is very rare. Lastly, iatrogenic (this term basically means "caused by a doctor!") botulism can occur from accidental overdose of botulinum toxin through administration by a physician. Of the 145 cases of botulism reported in the United States each year, approximately 15% are foodborne, 65% are infant botulism, and 20% are wound. Adult intestinal colonization and iatrogenic botulism also occur but rarely.

The mechanism of botulinum toxin's effect is that it binds to nerves and prevents the release of the neurotransmitter acetylcholine, which in turn causes a localized reduction of muscle activity. In foodborne botulism, symptoms typically occur 18–36 h after eating a contaminated food. The symptoms of botulism are related to muscle paralysis and include blurred vision, drooping eyelids, slurred speech, difficulty swallowing, and muscle weakness. Infants with botulism appear lethargic, have poor muscle tone, feed poorly, are constipated, and have a weak cry. If untreated, symptoms may progress and result in paralysis of the respiratory muscles, which may require a patient to be placed on a breathing machine (ventilator) for weeks or months. Botulism can also be treated with an antitoxin that blocks the action of the toxin circulating in the blood.

The Most Abused Drug in the United States

Most people know that alcohol use and abuse is common, but very few understand exactly *how* common. The fact is, more Americans abuse alcohol than all other drugs combined. In 2013, 86.8% of Americans aged 18 or older reported that they had consumed alcohol at some point; 70.7% reported that they had had an alcoholic drink in the past year; and 56.4% in the past month. In 2011, approximately 16.7 million Americans – roughly 6.5% of the population – were dependent on alcohol or had a history of alcohol abuse.

Why is alcohol abuse so prevalent? Most significant is its availability. Virtually, anyone over the age of 21 can legally purchase alcohol in most states (although there are many places in the United States where the sale of alcohol remains illegal). There is, however, more to the alcohol problem than just availability, as shown by the failure of Prohibition to stop the country's alcohol consumption.

Historically speaking, wine was produced as far back as 6000 BC. Since then, alcohol consumption has become a traditional part of many holidays, such as Memorial Day, New Year's Eve, and St. Patrick's Day. Several religious ceremonies, for example, Holy Communion or wedding festivities, involve drinking alcohol. These are special occasions, but the fact that an overwhelming number of everyday social gatherings involve alcohol is confirmation of the intense social pressure to drink.[2]

The consequences of alcohol abuse in the United States are monumental. In 2014, there were approximately 88,000 deaths due to alcohol abuse, making it the fourth-leading preventable cause of death. Thirty-one percent of driving fatalities (9967 deaths) involved alcohol-impaired drivers.

---

2 I myself had an experience of being pressured to drink alcohol when I attended a friend's wedding. A round of champagne was passed around for the traditional toast to the bride and groom. Since I do not like champagne, I asked for something else. When I was told that champagne was all that was available, I respectfully declined. An inebriated guest overheard me passing on the champagne and made it very clear that he was not going to let me disrespect the wedding couple by not drinking a toast. He actually stood next to me to make sure I partook, and in order to avoid a scene, I begrudgingly had a sip. Any readers planning a wedding might do well to offer sparkling apple cider as an alternative to alcohol. Some of your guests may be battling an alcohol addiction.

In 2006, there were more than 1.2 million emergency room visits and 2.7 million physician visits due to excessive alcohol consumption, resulting in approximated costs of $22.35 billion. In 2013, it was estimated that 1.3 million adults received treatment at a specialized facility for an alcohol use disorder. Unfortunately, this amounts to slightly less than 8% of adults who needed treatment.

Ongoing alcohol abuse causes other kinds of physical damage. Over time, alcohol abuse can lead to dementia, stroke, heart attacks, hypertension, depression, anxiety, suicide, cancer, hepatitis, and cirrhosis. In 2013, a shocking 46.4% of the 71,713 liver disease deaths among people aged 12 and older were attributed to alcohol. In 2009, alcohol-related liver disease was the cause of almost one in three liver transplants in the United States.

Why is alcohol so easy to abuse? As discussed earlier in this chapter, alcohol is extremely addictive and produces severe withdrawal symptoms. Furthermore, the intoxicating effects of alcohol are inherently addicting. The Ernest Gallo Clinic and Research Center at the University of California, San Francisco found that alcohol consumption results in the release of endorphins – which are responsible for the feelings of pleasure and reward – in the *nucleus accumbens* and the *orbitofrontal cortex* areas of the brain. The nucleus accumbens is linked to addictive behavior, and the orbitofrontal cortex is involved with decision-making. The effect of endorphins on these two brain regions can lead to alcohol abuse. In addition, the report suggests that the brains of heavy drinkers have changed to derive unusually high sensations of pleasure from alcohol consumption.

Lastly, there are physiological factors contributing to alcohol abuse. The fact that alcoholism runs in family histories shows that the presence of genes affecting alcohol metabolism can increase the risk of alcohol abuse.

What Is the Only Known Aphrodisiac?

I love asking students this question, because I get such great – albeit some a bit disturbing – answers such as alcohol, cocaine, chili peppers, and oysters. My favorite response, though, came from a student who shouted, "Money!" This is too bad considering chemists are not known for being rich.

An aphrodisiac is any food or drug that increases sexual desire, pleasure, and performance in both men and women. The word "aphrodisiac" gets its name from Aphrodite, the Greek goddess of love and beauty. The search for aphrodisiacs is almost as old as civilization. Some Hindu poems written 3000 or 4000 years ago describe the search for substances that could enhance sexual intercourse, which shows that our modern day obsession with sex is not modern at all.

As a scientist, I enjoy being able to look past speculation, popular beliefs, and myths, to concentrate on evidence-based scientific research. Many supposed aphrodisiacs have been scientifically tested, such as the skin and glands of the Bufo toad, Spanish fly, ginseng, chocolate, alcohol, and yohimbine. However, these substances are usually not tested on humans (unfortunately) but on rats and other small laboratory animals.

The only substance that has been scientifically proven to act as an aphrodisiac in humans is testosterone. *Testosterone* is primarily a male sex hormone, but its aphrodisiac effect is seen in both men and women. It works by increasing reactivity of the sympathetic nervous system, which is involved in the regulation of human sexual responses. Although human sexuality is complicated by psychological and emotional factors, it has been suggested that testosterone level is one of the main determinants in the sexual behavior of men and women. Some of the evidence for this is that the loss of testosterone after castration or ovariectomy is generally accompanied by an almost complete loss of libido. Sexual desire and behavior resume in these cases when testosterone is administered. In fact, scientists use testosterone levels to measure possible aphrodisiac effects of other substances. Although many possible natural aphrodisiacs may have an influence, it is usually because they raise testosterone levels.

In 1889, Dr Charles E. Brown-Séquard, Professor of Experimental Medicine at the Collège de France, reported that he had increased his physical strength, mental abilities, and sexual vigor by injecting himself with a liquid *testiculaire* extract derived from the testicles of dogs and guinea pigs (Yes, scientists used to do things like this!). By the end of that year, more than 12,000 physicians were administering Brown-Séquard's formula to their patients, and many businesses across the world were making fortunes selling the new "Elixir of Life."

In 1935, a group of scientists from the Organon Company in The Netherlands published the now-classic paper, "On Crystalline Male Hormone from Testicles (Testosterone)," which reported on the isolation of testosterone. The article was the first to use the term "testosterone" for the newly identified hormone (*testo* = testes, *ster* = sterol, *one* = ketone). A sterol is any of a group of solid, mostly polycyclic (many rings) alcohols derived from plants or animals. A ketone is represented in testosterone by the oxygen doubly bonded to carbon (Figure 2.7).

The artificial synthesis of testosterone was accomplished virtually at the same time by two different research groups led by Leopold Ruzicka (Yugoslavia) and Adolf Butenandt (Germany) later that year. They were both awarded the 1939 Nobel Prize in Chemistry, although the Nazi government initially forced Butenandt to decline the honor.

In case you are wondering, Viagra® does not meet the criteria for an aphrodisiac because it only increases performance, not desire. However, testosterone is also widely used to treat erectile dysfunction.

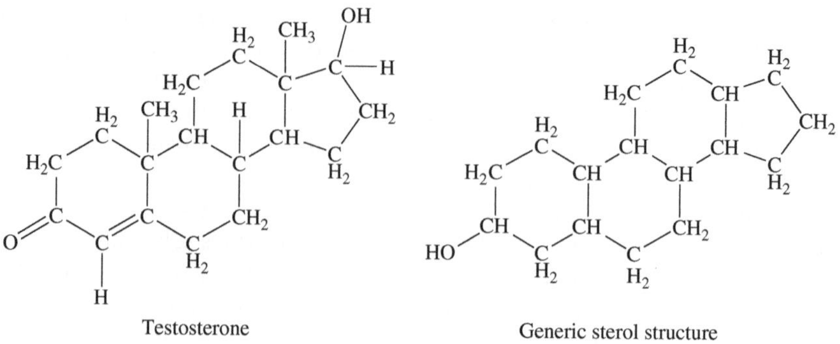

Testosterone   Generic sterol structure

Figure 2.7 A comparison of testosterone and a generic sterol structure.

The Most Consumed Psychoactive Substance

To understand this question, one must know the meaning of the word "psychoactive." The technical definition of a psychoactive substance is one that acts on the central nervous system and affects brain function, resulting in changes in perception, behavior, and mood. Once this is understood, most people immediately think of illicit drugs such as marijuana, methamphetamine, and cocaine as being the most consumed psychoactive substance. After it is explained that many commonly consumed legal products are, in fact, psychoactive and actually change the way one's brain works, then people quickly come up with the top three most consumed psychoactive substances: caffeine, alcohol, and tobacco.

Of these three, caffeine is the most consumed psychoactive substance in the world. About 90% of the people in North Americans use caffeine every day. Caffeine consumption from all sources has been estimated to around 70–76 mg/person/day worldwide but reaches a high of 210–238 mg/day in the United States. This equals about a half cup of coffee per day for every person in the world, and one-and-one-half cups per day for people in the United States. Coffee represents 75% of all the caffeine consumed in the United States. Americans consume 400 million cups of coffee per day, which works out to 146,000,000,000 (146 billion) cups of coffee per year. For this reason, it is easy to see why the United States ranks as the number one consumer of coffee in the world (Figure 2.8).

In 1980, the FDA actually tried to eliminate caffeine from soft drinks because of its psychoactive nature. The soft drink manufacturers, however, were able to keep caffeine in their products by justifying it as a flavor enhancer. How strange to think that if caffeine was not regarded as a flavor enhancer, soft drinks might be regulated by the FDA as a drug. Imagine needing a prescription to buy a bottle of Coke®! Despite caffeine being regarded as a psychoactive substance,

Figure 2.8 The structure of caffeine.

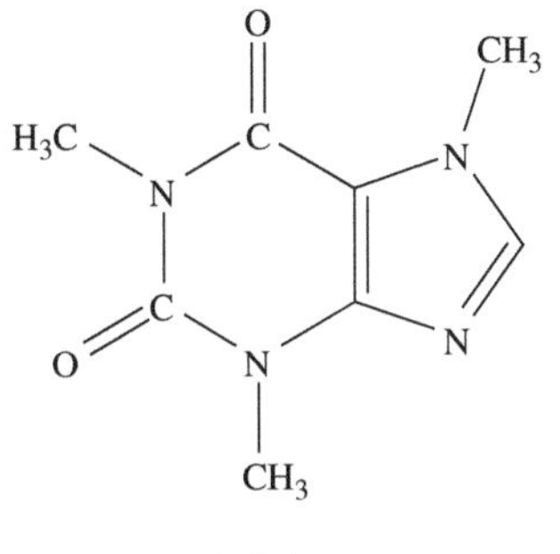

Caffeine

the FDA approved it in soft drinks but limits the maximum caffeine content to 0.02% or 71 mg/12 fl oz.

The advent of energy drinks has drastically changed the caffeine marketplace. By claiming these products are dietary supplements, due to their ingredients being derived from herbs and natural sources, energy drink manufacturers were able to get around the caffeine content restrictions for soft drinks. Currently, at least 130 energy drinks exceed the 0.02% caffeine restriction and at least one contains 505 mg (the equivalent to several cups of coffee) in a 24 fl oz can.

More serious problems with caffeine have started, in fact, because of energy drinks. 5 Hour Energy®, labeled as a dietary supplement, has been implicated in 33 hospitalizations and 13 deaths and is currently under investigation by the FDA. The FDA is also taking a good look at Monster Energy Drink® due to possible connections with five deaths and one heart attack, but with a name such as "Monster," the danger should already be apparent. In one case, a 28-year-old motorcycle athlete's heart stopped during a competition because he had consumed eight cans of Red Bull® over a 5-hour period.

The Substance Abuse and Mental Health Services Administration (SAMHSA) reported that emergency room visits related to energy drinks increased 10-fold from 2005 to 2009. In 2011, there were 20,783 reported emergency room visits related to these drinks. Half of these visits involved the products mixed with alcohol or drugs. In fact, alcohol-laced energy drinks were banned in 2010 after reports of deaths and various illnesses. One study showed that ingestion of a caffeinated energy drink (Red Bull®) with vodka reduced the participants' perception of impairment of motor coordination, leading to a tendency for them to drink more than if drinking vodka alone.

In an interesting side note, caffeine shows up in the Psychiatric Association's Diagnostic and Statistical Manual of Mental Disorders (DSM) in a general section called caffeine-related disorders. These include a condition caused by excessive intake called *caffeine intoxication.* You qualify for caffeine intoxication if you have had more than 250 mg of caffeine, which is roughly two to three cups of brewed coffee, and you experienced five or more of the following symptoms: restlessness, nervousness, excitement, insomnia, flushed face, diuresis, gastrointestinal disturbance, muscle twitching, rambling flow of thought, rapid heartbeat, and periods of inexhaustibility. The newest edition, DSM-5, even includes a new entry – caffeine withdrawal! According to the DSM-5, withdrawal from caffeine includes symptoms such as fatigue, headache, and difficulty focusing. It all goes to show something important: It is best not to mess with someone until after they have had their morning cup of coffee!

The fact that caffeine withdrawal appears in the DSM-5 reflects the growing problem with this substance. A recent survey showed that 30% of a sample of 162 caffeine users met the diagnostic criteria for caffeine dependence. Studies with adult twins have shown that lifetime caffeine intake, caffeine toxicity, and caffeine dependence are associated with various psychiatric disorders including depression, panic disorder, alcohol dependence, and cannabis and cocaine abuse.

Is it possible to die from a caffeine overdose? Of course it is possible, but quite rare and involves a large quantity of the drug. Generally, one would need to consume more than 5 g of caffeine (about 30 cups of coffee) for it to be lethal. In the United States, as few as two deaths per year on average can be attributed to caffeine (Figure 2.9).

" I'm prescribing you a cup of coffee first thing in the morning! "

Figure 2.9 A treatment for caffeine withdrawal.

40,000 Tons of Aspirin

Aspirin is the most used drug by humans. Taking into account consumption over the entire world, the amount adds up to 40,000 tons or 120 billion aspirin tablets per year. Most of you have taken aspirin at least once in your life…either that, or you just do not realize you have taken it at some point (Figure 2.10).

Why is aspirin so popular? First of all, aspirin has been around for a long time, so most people are accustomed to taking it. Aspirin was first created in 1897 by a scientist named Felix Hoffman, who was working for Bayer, a German company. Think of Bayer® aspirin; this is the same company. Another reason for aspirin's popularity is that it is one of the safest medicines known, and very few harmful side effects are associated with it. Furthermore, aspirin is one of the least expensive painkillers on the market, and it is actually quite easy to make. Aspirin was one of the first molecules made from coal tar derivatives, which will be discussed later in this book and is even commonly made as part of college general chemistry laboratory courses.

Another quality of aspirin that makes it such a go-to medicine is its ability to fight a large number of symptoms, most commonly general pain or fever. It also treats many diseases such as rheumatic fever, gout, and rheumatoid arthritis. Aspirin has even been shown to help avoid heart attacks. The scientist Sir John Vane actually won the Nobel Prize for suggesting that aspirin could be used to reduce the occurrence of heart attacks because of its ability to thin blood. Lastly, aspirin is one of the ingredients in many different OTC products such as Alka Seltzer®, Anacin®, or Pamprin® and is sometimes included in prescription-strength, opioid-based painkillers such as Darvon® and Percodan®.

Can someone overdose with aspirin? Absolutely! Even something as harmless-sounding as aspirin can be lethal under the right conditions. In 2004, U.S. poison centers reported that there were roughly 40,000 exposures to salicylate-containing products, aspirin belonging in this group. In 2011, there were 60 reported deaths due to individuals who overdosed aspirin alone (not in conjunction with other drugs). Many of the people who have overdosed

Figure 2.10 The structure of aspirin.

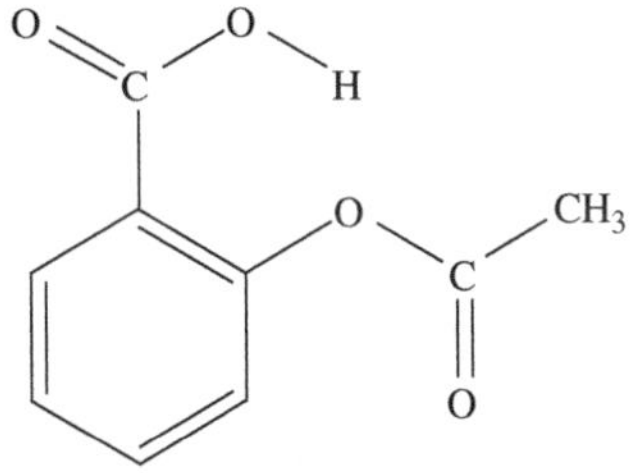

Aspirin

or died used aspirin as a method for self-harm or committing suicide. In the movie, *Girl Interrupted*, the main character, played by Winona Ryder, checks into a psychiatric hospital after ingesting a bottle of aspirin in an attempt to commit suicide.

The more frequent accidental cases of aspirin overdosing are caused in part by the fact that it is an ingredient in so many OTC products, as mentioned above. Suppose that you take some aspirin and then take something else such as Alka Seltzer® in combination with it. Now you are in danger of overdosing. In fact, there is a whole class of drugs called salicylates, of which aspirin is one. Because salicylates all have a very similar effect, combining them can cause an overdose. In mild cases, aspirin overdose can result in symptoms including lethargy, nausea, vomiting, ringing in the ears, and dizziness. In moderate poisoning cases, there may be additional symptoms such as rapid breathing, fever, sweating, loss of coordination, and restlessness. Finally, severe poisoning cases can give rise to hallucinations, convulsions, kidney failure, heart failure, coma, and death.

For emergency room patients suffering an overdose, medical treatment varies depending on the symptoms they are exhibiting, as well as the length of time since consumption of the aspirin. Remedies include ingesting activated charcoal to soak up the aspirin in the stomach and ingesting laxatives to promote bowel movements to help rid the intestines of aspirin and the activated charcoal. If the overdose is extremely severe, dialysis may be used to remove aspirin from the blood. In addition, other medicines such as potassium salt and sodium bicarbonate can be given intravenously to assist in removing aspirin from a person's system.

Part of why people die from aspirin overdosing is it generally is not considered dangerous. The issue arises among some medical and nursing home staff who underestimate the ability of aspirin to poison, which causes a delay in diagnosis. When this happens, there can be failure to administer sufficiently vigorous treatments early in the overdosing period, which are needed to prevent the death of the patient. Death of patients exhibiting severe symptoms occurs about 5% of the time, and delay in diagnosis corresponds to an increased mortality rate of 15%. Simply stated, if people fail to act quickly, the chances of someone dying from aspirin poisoning are quite a bit higher. Death is usually due to either cardiopulmonary arrest after a prolonged coma or abnormal bleeding. Aspirin is known to be a blood thinner that causes impairment of a clotting factor in the blood, and this effect may lead to abnormal bleeding and death.

How Bitter Is the Bitterest?

After reading thus far, you may realize that there are numerous poisons commonly found in the household, some of which are not even usually considered dangerous. Wouldn't it be great if children or pets could be totally prevented from unknowingly ingesting these compounds and becoming sick? In some cases, because many household chemicals contain ethanol, kids looking to experiment are tempted to drink them in order to get drunk. Wouldn't it be nice to be able to stop this, too? Problems such as this have led to the idea of adding a substance to ethanol, which renders it so bitter tasting that it is undrinkable, a process called *denaturing*. Denaturing alcohol is also done so that whatever product containing it is not treated as an alcoholic beverage with respect to taxation and sales restrictions.

A wide variety of substances have been used to denature alcohol: ammonia, benzene, chloroform, gasoline, isopropyl alcohol, kerosene, nicotine solution, and vinegar. However, many of these were relatively ineffective or toxic in themselves. The answer was found in 1958 when scientists discovered the most bitter substance known, denatonium benzoate (DB), in the course of looking for a new local anesthetic similar to lidocaine. Lidocaine and DB have similar structures but vastly different effects. In solution, DB was found to be more potent than the standard alcohol denaturant at the time, brucine. It was found that DB solutions containing as little at 50 ppb (parts per billion) could be detected by humans, and solutions with as little as 10 ppm (parts per million) of the compound were unbearably bitter. This means that one teaspoon of DB in an Olympic-sized swimming pool would make the water noticeably bitter! In fact, DB is officially recognized by the Guinness Book of World Records as the bitterest substance in the world. If you are interested, you can actually get a sample of DB from the company that makes it under the trade name Bitrex® (Figure 2.11).[3]

Depending on regulations specific to each country, DB is added in amounts ranging from 6 to 50 ppm for the purpose of denaturing alcohol. The fact that only a small about of DB is needed is important because the small amount added does not change the properties of the substance, only its taste. After the

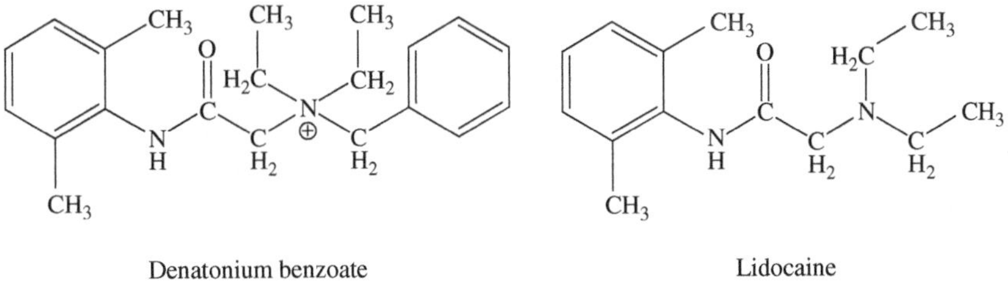

Denatonium benzoate   Lidocaine

Figure 2.11 A comparison of the structures of Denatonium benzoate and lidocaine.

---

3 I actually tried a sample of Bitrex® out of curiosity and I can honestly report that it was not as bad as I expected. I would describe it as being much, much worse than tonic water and more like broccoli without the leafy taste to it... Anyway, if you do end up trying Bitrex®, make sure to follow the manufacturer's advice and eat chocolate right afterward!

ethanol has been denatured using DB, the designation SD-40B is applied to the product.

Under the trade names of Bitex® or Aversion®, DB has been added to numerous household products to prevent accidental poisoning. In particular, it has been used in denatured alcohol, antifreeze, liquid laundry detergents, liquid soaps, shampoos, disinfectants, windshield wiper fluid, and pesticides. In addition, it is added to nail polish to prevent nail biting and used to coat children's thumbs to prevent thumb sucking. Since 2012, most major U.S. antifreeze marketers have been adding bittering agents to antifreeze, and in fact, the state of Oregon requires that DB is added to antifreeze and windshield washer fluid. Some countries in the world have also made DB mandatory in antifreeze.

Why is accidental poisoning from commercial products such a problem? It has been suggested that young children's sense of taste is not as well developed as adults', making it easier for them to ingest larger quantities of unpalatable products. Combine with this the fact that many household products contain alcohol – food flavorings, colognes, and mouthwash are examples – and you have a problem. Mouthwashes are particularly dangerous because they can contain up to 75% ethanol. In looking at the numbers relating to causes of human poisoning for children under the age of 5, cosmetics/personal care products were ranked first, with 166,246 reported cases (13.95% of all cases), and household cleaning substances were ranked third, with 109,442 reported cases (9%) in the year 2011. For children under the age of 5, there were four reported deaths due to cleaning substances in 2011 and one reported death from a cosmetics/personal care product.

Our ability to detect the property of bitterness lies in the physiology of the tongue, which is covered with receptors activated by specific kinds of molecules. The activation of different receptors triggers the sensation of different flavors. Bitter compounds are detected by a specific subset of taste receptor cells called TASTE 2 Receptor (TAS2R or T2R). In fact, there are 25 different bitter taste receptors on the human tongue. DB in humans is recognized by no less than eight distinct bitter taste receptors. Although there is no clear connection between bitterness and toxicity, it is generally believed that bitter taste helps mammals avoid potentially harmful food constituents. There are many different compounds that taste bitter, and their chemical structures are quite diverse.

$62.5 Trillion per Gram

Right now, antimatter – with a price tag of about $62.5 trillion per gram – is the most expensive substance on the Earth. When groups of people are asked to name the most expensive substance, the variety of answers is hilarious. Often, the responses are droll ones such as love, cocaine, cash money, computer ink, or even sex. More serious guesses include rare substances such as diamonds, uranium, and platinum. Some particularly astute people realize that the most expensive substance probably has to be created artificially by some sort of man-made process. Some good guesses in this area are the artificial elements such as ununoctium or unobtanium. Unlike the artificially made substances, however, antimatter has been the target of serious attempts to capture and contain it, which greatly increases its cost of production.

What is antimatter? Most people know that the basic nuclear particles that make up matter are protons, electrons, and neutrons. In 1930, Paul Dirac developed a description of the electron, which also predicted that an antiparticle of the electron should exist. This antielectron (also called a positron) was predicted to have the same mass as the electron but an opposite electric charge. Later, it was discovered that the other basic atomic particles had antimatter counterparts, the antiproton and the antineutron. When a particle and its antimatter counterpart meet, they are both annihilated, which means that the two particles disappear and their mass is converted to energy, following the principle embodied in Einstein's famous equation $E = mc^2$. As you may well know, "c" in this equation is the speed of light, which is a large number. Because this number is squared, this means that a small amount of mass can be converted to an enormous amount of energy. To give you an idea of how much energy is evolved during a matter/antimatter annihilation, this reaction is considered to be 100,000,000,000 times more powerful than a typical chemical explosion such as trinitrotoluene (TNT) and 10,000 times more powerful than a nuclear explosion.

To actually create antimatter, scientists focused on the simplest form of matter, hydrogen. A hydrogen atom consists of just one electron and one proton. This means that the simplest form of antimatter, an *antihydrogen*, is made up of an antiproton and a positron. The positron is attracted to an antiproton in the same way an electron is attracted to a proton.

The first antihydrogen was made in 1995 at the CERN (European Organization for Nuclear Research) super collider by colliding antiprotons with xenon atoms. This collision produces a positron, which is electrically attracted to another antiproton, subsequently forming antihydrogen. Unfortunately, it only takes a few millionths of a second for antimatter particles to come into contact with their matter counterparts, annihilating themselves and giving off energy. Because of this, scientists have worked on ways to make antimatter stable enough to be contained. The key was to slow the antihydrogens to keep them from colliding, and this was achieved by containing the antimatter in a bottle at just half a degree above absolute zero, the theoretically coldest achievable temperature. In 2011, scientists were able to hold produced antihydrogen for over 15 minutes using this method.

The reason for antimatter's tremendous expense is easy to understand when you realize the technology involved in creating it. To make antihydrogen, the required antiprotons must be literally made one atom at a time using a particle accelerator. The CERN super collider is the most complex piece of machinery every constructed by humans. It took about a decade to construct, at a cost of about \$4.75 billion. It is roughly 10 miles across and contains 9300 magnets, all of which must be super cooled to −456.25°F using liquid helium. For the collision to occur, the particles are accelerated to 99.99% of the speed of light, which requires an incredible 120 MW of electric power, enough to power a large city. The collider has a total operating budget of about \$1 billion per year, with electricity costs alone running at \$23.5 million per year. When you also consider the fact that it has been estimated to take 100 billion years to produce 1 g of antihydrogen, you begin to see why the costs are so high.

Do you think there is no practical application for matter/antimatter annihilation? Think again! The well-known positron emission tomography (PET) scan uses matter/antimatter annihilation as a method to image biological tissue, typically cancer metastasis. This is accomplished by injecting the patient with a biologically active molecule that contains a radioactive isotope. A typical example is fludeoxyglucose (FDG), which is a glucose molecule with one of the hydroxyl groups (an oxygen atom bonded to a hydrogen atom, expressed as OH) replaced with the positron-emitting radioactive isotope ^{18}F. Because cancer has an increased metabolism compared to regular tissue, it absorbs glucose more quickly. With FDG being analogous to glucose, it is also absorbed more quickly in cancer cells and tends to accumulate there. Once the FDG is absorbed into biological tissues, the radioactive ^{18}F undergoes positron emission decay. The emitted positron travels through tissue a short distance before eventually colliding with an electron. The collision annihilates both the electron and positron and produces a pair of gamma rays. These gamma rays are detected by the PET scanner and are eventually converted into a three-dimensional image of the body (Figures 2.12 and 2.13).

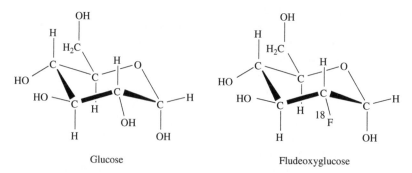

Glucose   Fludeoxyglucose

Figure 2.12 A comparison of the structures of glucose and fludeoxyglucose.

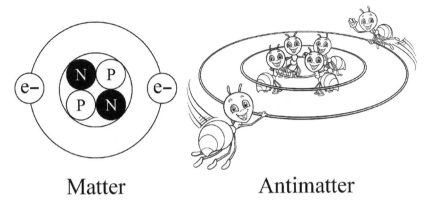

Matter   **Antimatter**

Figure 2.13 The true structure of antimatter.

What Is the Most Abundant Source of Air Pollution?

In 1981, President Ronald Reagan famously stated that "approximately 80% of our air pollution stems from hydrocarbons released by vegetation." At the time, he was using this as a rationale to limit air pollution control. As odd as this may sound, he was actually correct to an extent…but the story behind air pollution is a bit more complicated.

When temperatures are high, many plants emit an organic molecule called *isoprene*. Although the mechanism is not exactly known, it is believed that isoprene protects the plants from potential damage from heat stress. Trees, especially oaks and aspens, are some of the greatest sources of isoprene because leaves at the tops of trees are exposed to intense sunlight, which is variable depending on cloud cover. In addition, very large trees can create an insulating boundary, causing leaves to exceed the air temperature by up to 10 °C. In consideration of this, for plants to develop some type of protection against temperature change makes perfect sense.

Isoprene emission is the predominant biological source of hydrocarbons in the atmosphere, with roughly 600 Tg (600×10^{12} g) being emitted globally each year. Although isoprene is the single most common biogenic volatile organic compound (BVOC) given off by plants, the majority of BVOCs emitted by vegetation as a whole consist of a larger group of plant-produced chemicals called *terpenoids*. Terpenes, one type of terpenoids, are often strong smelling and usually the primary constituents of fragrant essential oils derived from plants and flowers. Two of the most familiar are α-pinene (associated with pines) and D-limonene (associated with lemons). Terpenes are believed to play a role in pollination as well as in attracting predators and parasites of herbivores to keep herbivores from ingesting the plants (Figure 2.14).

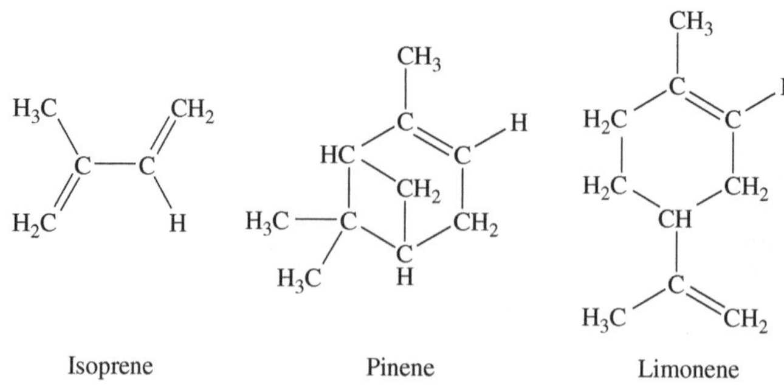

Figure 2.14 The structure of common biogenic volatile organic compounds.

Altogether, plants give off 1150 Tg of carbon per year of BVOC, of which roughly 35% is from isoprene. Human beings contribute roughly 1/10th of this amount in volatile organic compounds (VOCs), such as gasoline and formaldehyde, from vehicles, fossil-fueled power plants, fuel storage, solvent usage, and emissions from landfills.

Both terpenes and isoprenes are linked to *photochemical smog*, which is created when sunlight causes reactions between nitrogen oxides and VOCs in the atmosphere. The end products are airborne particles and ozone – the main components of smog. *Nitrogen oxides*, such as nitric oxide (NO) and nitrogen dioxide (NO_2), are gases produced when nitrogen-containing compounds are combusted. The major natural sources of nitrogen oxides are lightning and forest fires. Unfortunately, humans contribute nitrogen oxides through motor vehicle exhaust, tobacco smoke, and the burning of coal, oil, or natural gas. Furthermore, many industrial processes, including welding, electroplating, engraving, and blasting, produce nitrogen oxides. In the absence of VOCs, the level of nitrogen oxides determines the amount of ozone generated in the atmosphere. However, the presence of VOCs in the atmosphere causes reactions that produce more NO_2, therefore more ozone and smog.

The other part of smog, particulate pollution, is caused by terpenes reacting with oxygen in the atmosphere and gaining an – OH group. This group allows the resulting molecules to hydrogen bond to each other. When this happens, the increase in intermolecular forces enables the molecules to bind together and form minute airborne particles collectively called an *aerosol*. These particles are roughly 2.5 μm in diameter, about 1/30th that of the average human hair. Aerosol particles are small and airborne, so they can be easily breathed into our lungs and cause serious health problems. These particulate pollutants can even get into the bloodstream, with exposure leading to an assortment of issues including asthma, decreased lung function, coughing, irregular heartbeat, heart attack, and premature death for people with heart or lung disease.

Aerosol particles are the main cause of the haze we commonly associate with smog. When sunlight hits the particles, some of the light is absorbed and the rest dispersed, resulting in the haze we see. The overall effect reduces visual clarity, casting a blur that obstructs distant objects. It is shocking and unfortunate to consider that air pollution has become so concentrated in the United States, it has reduced the human average visual range from 90 miles to about 15–20 miles in the eastern part of the country, and from 140 miles to about 60–90 miles in the western states. In fact, the haze caused by natural pollution accounts for the names of the Great Smoky Mountains, Blue Ridge Mountains, and Australia's Blue Mountains. In these examples, aerosol particles scatter blue light to produce this famous effect. Despite some air pollution being natural, the combustion emissions generated by human beings have exacerbated the problem.

Terpenes, α-pinene and D-limonene, are ingredients in products such as room fresheners, pine cleaners, wood polishing products, and wood-based furniture coatings. Despite their widespread use, these cleaning products do not generate a significant amount of outdoor pollution. Indoor pollution, however, was found to be an issue when a recent study showed that if a source of terpenes was present in an office building, there was a higher concentration of aerosol particles in the workplace air. Aerosols are formed by the terpenes reacting with ozone from the outdoors. The same situation occurs in homes where ozone-generating air purifiers are used. In addition, terpenes exposure in households, where other chemicals are also present, is reportedly the most frequent cause of non-work-related hospital admissions for chemically related respiratory disease in the United States.

The outlook for smog concentrations in our air is not cheery. Because terpenes emission is directly related to temperature, as carbon dioxide levels rise in the atmosphere, the Earth will warm, and more terpenes will be given off. This, of course, means more smog.

Where Did That Rash Come From?

Recent studies done by the North American Contact Dermatitis Group (NACDG) showed that the most common contact allergen was nickel sulfate, with 19.0% of tested patients having a reaction to this compound. This means that roughly one in five people will develop dermatitis (a rash) after prolonged contact with items containing the metal nickel. Other symptoms of nickel allergy include headache, malaise, diarrhea, and fever.

Why is nickel such a common allergen? Part of the answer comes from the fact that nickel is one of the most widely used metals, present in alloys, plating, and stainless steel. Many coins are made from nickel, and it is also a common material used to make cheap jewelry. You can see from these facts that we are constantly exposed to nickel. This frequent exposure is likely the greatest contributor to the high allergic response, since repeated exposures can cause an increased sensitization. A good piece of evidence supporting this is the fact that nickel allergy is roughly five times more common in women than men, possibly due to women being exposed to more nickel-containing items – for example, when they have their ears pierced and wear inexpensive earrings and jewelry – when they are young. In fact, nickel in jewelry is such a problem that the European Union actually has passed the Nickel Directive, which limits the amounts of nickel present in the post assemblies of piercing and other products that have prolonged contact with the skin.

What is happening during the allergic reaction, anyway? A recent report showed that nickel activates toll-like receptors 4 (TLR4s), generating a signal that promotes inflammation and skin rash. Basically, the immune system views nickel salts as a foreign body, responding in much the same way it would to a germ. The activation of the immune system causes immune cells, such as T cells, to migrate to the area, which causes swelling.

One of the most significant sources of nickel exposure is occupational. Eczema from contact with nickel was first noted in the late 1880s in the nickel plating industry. Those who commonly work with nickel – including metal platers, mechanics, construction workers, electronics workers, and hairdressers – all have a higher risk of getting nickel dermatitis. Even cashers are exposed to nickel due to repeated handling of nickel-containing coinage.

Although nickel in jewelry is present in metal form, it is the nickel salts produced when perspiration comes in contact with the metal that usually causes the allergy. Think you might be suffering from nickel dermatitis? Look on your hands, feet, and elbows for the rash, as these are the places the dermatitis tends to be located. There is no known cure for nickel allergy; avoiding contact with nickel is the only real solution. Nickel allergy is so common and problematic that commercial nickel test kits are available to determine if an item contains nickel. There are even products that provide a barrier allowing people with nickel allergies to wear nickel-containing jewelry.

Another important source of nickel ingestion is smoking. The use of tobacco is one of the most significant sources of nickel for people who are not exposed to it as part of their job. Most tobacco plants contain nickel, likely due to their absorbing it from the soil or pesticides. A study of cigarettes showed a high content of nickel (2.32–4.20 mg/kg) regardless of the origin of the tobacco. Smoking tobacco causes nickel to enter the body as an inhaled gas. As can be expected, a recent study disclosed that previous and current smokers have a higher chance of being allergic to nickel than nonsmokers. Although some people may believe that they are avoiding this exposure by using e-cigarettes, this is unfortunately not the case. Because many of the heating and other components of e-cigarettes are actually made out of nickel, they have also been shown to be a source of inhaled nickel.

One surprising source of nickel dermatitis is cell phones. In a recent study, over half of the investigated cell phones tested positive for nickel, with buttons, logos, and metallic frames being the most common places where it was found. When you consider the fact that cell phones are held against the face, often for long periods of time, you can see why a problem might occur.

Some men have noticed that after switching to an electric razor, they sometimes get a red rash on their neck. Although you may think this is just simple shaving irritation, it actually may be an allergic reaction to the nickel contained in the stainless steel shaving heads. Perhaps it is time to go back to regular razors.

Multiple studies have shown that allergic reactions, such as hand dermatitis or generalized eczematous breakouts, can also occur due to dietary nickel ingestion. Some common foods that contain relatively high amounts of nickel are chocolate, all nuts, all seeds, commercial salad dressing, green beans, broccoli, peas, and black tea. In addition, nickel can also infuse canned foods via leaching from the metal can. Tap water sitting overnight can contain nickel due to leaching from faucet fixtures. Stainless steel cookware can also leach nickel into food, especially when used to cook acidic things such as tomato, vinegar, or lemon.

Some things you can do to avoid ingesting nickel are avoiding foods high in nickel, canned foods, and stainless steel cookware. Run tap water for a few seconds before using it. In addition, taking vitamin C and/or iron supplements with meals has been shown to decrease absorption of dietary nickel.

It Would Take an Elephant on a Pencil

Graphene is a form of carbon, just as diamond or charcoal. There are many different ways that carbon atoms can bond together to form various materials. How they bond is what determines the properties of the substance. In other words, diamond and graphene are both made of carbon, but each compound has a different arrangement of their carbon atoms. In a diamond, each carbon atom is single bonded to four other carbon atoms; however, in graphene, each carbon atom is bonded to three other carbon atoms, and some are double bonded. In this case, the differences between the materials are quite drastic (Figure 2.15).

Graphene actually got its name from the fact that it is similar in structure to graphite. Graphite is made up of many sheets of carbon atoms arranged in a

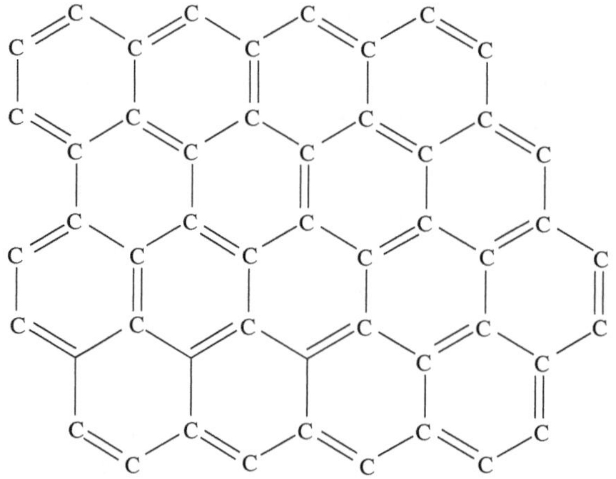

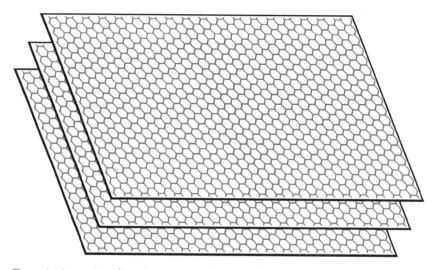

Figure 2.15 The basic structure of graphene.

A graphene fragment

Figure 2.16 Layering of graphene sheets to form graphite.

hexagonal honeycomb structure. The individual sheets in graphite are actually graphene. Graphite works as a writing material or a lubricant because these sheets are loosely bonded and therefore slip easily over each other. In the case of a pencil, the graphene sheets are pulled off by the paper, leaving behind the characteristic black line. In fact, the first characterized sample of graphene was obtained from graphite (Figure 2.16).

Graphene was first identified in 2004 by two University of Manchester scientists, Andre Geim and Konstantin Novoselov, who used a piece of adhesive tape to pull off a sheet of graphene from a sample of ordinary graphite. For this and their subsequent work on graphene, both Geim and Novoselov received the 2010 Nobel Prize for Physics.

Graphene consists of a single atomic layer of carbon, arranged in a honeycomb lattice. Studies of the material led to the discovery of its many interesting properties, including being the strongest material ever studied. The strength of graphene was tested with an instrument called an atomic force microscope, which showed that graphene has a breaking strength more than 100 times greater than that of steel and resists loads of 150,000,000 pounds per square inch. To give you an idea of the strength of graphene, it has been estimated that it would take an elephant standing on a pencil to break through a single sheet of graphene with the thickness of plastic wrap!

The fact that a graphene sheet is only 1 atom thick also makes it the thinnest material ever made. How thin? If you were to stack 3 million sheets of graphene on top of each other, they would be only 1 mm high. In addition to being the strongest and thinnest material ever tested, graphene is an excellent conductor of electricity. In 2008, Geim and Novoselov created the world's smallest transistor using graphene. The transistor was 1 atom thick and 10 atoms wide.

In 2012, graphene was used to make an aerogel determined to be the lightest material ever made. An *aerogel* is a material made from a gel where the liquid component has been replaced by a gas. The density of the graphene aerogel was found to be 0.16 mg per cubic centimeter. Given that air is 1.2 mg per cubic centimeter, this new material was actually 7.5 times lighter than air. To illustrate, a cubic inch of graphene aerogel can be balanced on a blade of grass or on the fluffy seed head of a dandelion.

Still wondering about the practical applications of graphene? How about this! In 2013, a research team from the University of Manchester received a $100,000 grant from the Bill and Melinda Gates Foundation to use graphene to make the next generation of condoms. The Gates Foundation was seeking to develop new technology, which would make condoms more desirable to use. The new condom would be composed of graphene and an elastic polymer, such as latex, creating a stronger, thinner condom, which would encourage greater use by increased sensation.

Because of its almost science fiction qualities, graphene has started to appear in popular culture. In the 2014 movie, Robocop, the main character was given a more modern, nonmetallic look because his armor was made out of graphene (Figure 2.17).

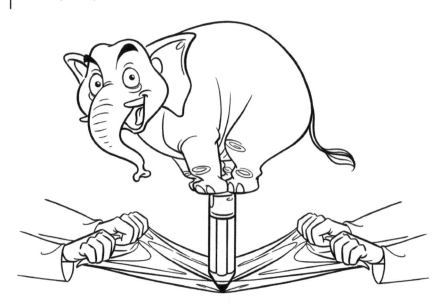

Figure 2.17 A representation of graphene's strength.

The Largest Industrial Accident in World History

Phosgene ($COCl_2$) is a highly toxic compound that was first synthesized in 1812. At room temperature (70 °F), phosgene is a poisonous gas that may appear either colorless or as a white to pale yellow cloud that can have a pleasant odor similar to that of newly mown hay or green corn. Phosgene is a major industrial chemical used to make plastics, pharmaceutical agents, synthetic foam, dyes, and pesticides with the worldwide chemical industry annually producing more than 2–3 million tons of phosgene (Figure 2.18).

Phosgene first gained its deadly reputation during World War I, when it was used in chemical warfare. Phosgene was used extensively as a choking (pulmonary) agent and was responsible for a large majority of chemical warfare deaths. It has been estimated that phosgene accounted for 80,000 of the 100,000 deaths from chemical gas exposure in World War I.

Exposure to dangerous concentrations of phosgene may cause the following symptoms to develop quickly: coughing, burning sensation in the throat, watery

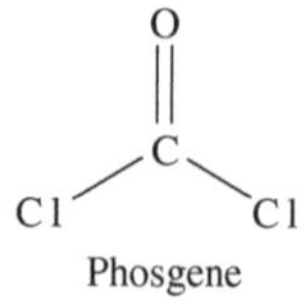

Figure 2.18 The structure of phosgene.

O
‖
C
Cl   Cl

Phosgene

eyes, difficulty breathing, nausea, and vomiting. Direct skin contact with phosgene can result in lesions similar to those from burns.

Phosgene causes damage to biological molecules in two ways. It can react with water (hydrolysis) to form hydrochloric acid ($COCl_2 + H_2O \rightarrow CO_2 + 2\,HCl$). When considering the fact that water is present in the lungs and on the skin, it is easy to see how exposure to phosgene can cause significant damage. This first reaction contributes far less to the typical symptoms of phosgene exposure but is likely responsible for the irritant effects.

The second reaction is called an acylation. Phosgene attaches to reactive groups on biological molecules, such as proteins and phospholipids. These reactions can result in structural changes in membranes and proteins and stop them from functioning properly. Inhaled phosgene attacks the major constituents of surfactants and tissue membranes in the lungs causing irreversible acute lung injury and life-threatening fluid accumulation in the lungs (pulmonary edema) (Scheme 2.3).

As a consequence of its toxic action, phosgene may produce symptoms that may not be apparent for up to 48 hours after exposure, even if the person appears well following removal from exposure. Delayed effects of phosgene exposure include difficulty breathing, coughing up white- to pink-tinged fluid (a sign of pulmonary edema), low blood pressure, heart failure, and death. Unfortunately, there is no specific antidote for phosgene exposure and medical treatment is mainly supportive.[4]

Phosgene at a concentration of 1 ppm (parts per million) in the air causes little effect in the short term but can cause serious damage if exposure continues for

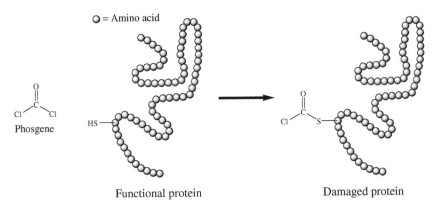

Scheme 2.3 The reaction of phosgene with a protein.

---

4 In one of the most chilling stories I have heard as an academic, a renowned researcher received a significant exposure to phosgene gas in a laboratory accident. Knowing his fate was sealed he went back to his office, wrote out his will and then went home. He died the next day from the phosgene exposure.

several hours. Parts per million (ppm) means parts of a substance in 1 million parts of another substance and is used to express very dilute concentrations. One part per million is like 4 drops of ink in a 55-gallon drum, 1 inch in 16 miles, or 1 second in 11.5 days. Phosgene concentrations of 4–10 ppm in the air can cause immediate irritation of the respiratory tract and eyes. A lethal dose of phosgene in humans is approximately 500 ppm/min of exposure. This means a 2-minute exposure to phosgene concentrations of 250 ppm can be lethal as is a 10-minute exposure at 50 ppm.

Because phosgene is such a commonly used industrial chemical, there have been multiple cases when it was accidentally released. On the night of December 2, 1984, a breakdown occurred at Union Carbide India Limited plant pesticide plant in Bhopal, Madhya Pradesh. A runaway reaction in a tank caused the pressure relief system to vent large amounts of poisonous gas into the atmosphere. An estimated 40 tons of phosgene mixed with methyl isocyanate (also highly toxic) were released into the atmosphere and spread through towns located near the plant. The Bhopal disaster is considered the worst industrial disaster in history. Over 500,000 people were exposed to the gases and between 3000 and 10,000 people died within the first week. In 1989, Union Carbide paid $470 million in compensation to the Indian government. Moreover, seven Union Carbide employees were convicted of "death by negligence" for their role in the Bhopal tragedy.

Warren Anderson, the chairman and CEO of Union Carbide never faced trial over the deadly industrial accident. Shortly after the incident, Anderson visited Bhopal and was arrested but was released after paying a $2000 bail and fled the country. Since 1993, the Indian government tried several times to extradite him but never succeeded. Anderson escaped all attempts to bring him to trial and recently died in a Florida nursing home on September 29, 2014, at the age of 92.

More recently, on January 23, 2010, there was a release of phosgene at a DuPont facility in Belle, West Virginia. An employee was exposed when a braided steel hose connected to a 1-ton capacity phosgene tank suddenly ruptured. He was hospitalized and died a day later.

Think you couldn't possibly get exposed to phosgene? Think again! Certain chlorinated organic compounds found in many household solvents, paint removers, and dry cleaning fluids, such as tetrachloroethylene (PERC) and methylene chloride, can be chemically converted to phosgene when brought into contact with contact with flames and/or other intense heat sources. Chlorinated polymers such as polyvinyl chloride can form phosgene under similar conditions. These dangers are particularly high for people who work around open flames (firemen, welders) and chlorinated solvents, such as painters. Due to a similarity in structure, these chlorinated molecules can react with oxygen in the air and under intense heat can form phosgene (Figure 2.19).

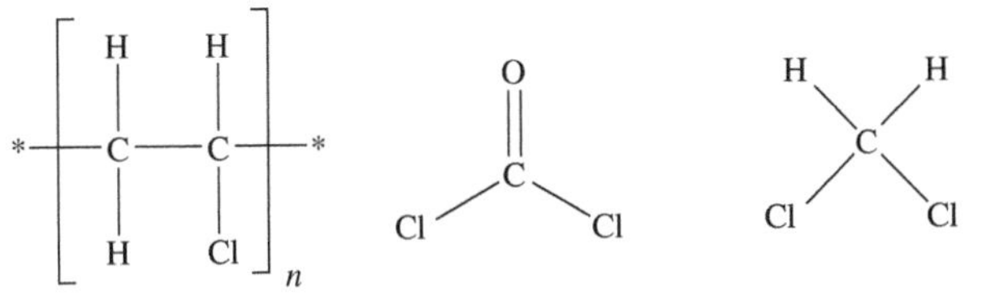

Figure 2.19 A comparison of the structures of PVC, methylene chloride, and phosgene.

What Is the Most Important Chemical Reaction?

The field of organic chemistry encompasses thousands of reactions, each of which has its subtle importance. Often, I am asked which of these reactions I consider to be the most important. As mentioned in the preface of this book, I am usually shocked by the results of my investigations into simple questions such as this. How could something so important not be common knowledge to the public, let alone a chemist? After some study, I think that the most important reaction in chemistry is also the most important invention of the twentieth century: the industrial synthesis of ammonia.

Although carbon is the atom that makes life possible, nitrogen is also crucial. Without nitrogen, life as we know it would not exist. Nitrogen is present in protein, DNA, and a multitude of biologically important molecules, so the ability of living things to incorporate nitrogen is vital to their existence. One would not think that coming up with a source of usable nitrogen would be a problem because we are literally surrounded by it. Earth's atmosphere is roughly 80% molecular nitrogen (N_2); however, this nitrogen cannot be used by more than 99% of living organisms. The triple bond between the nitrogens is one of the strongest chemical bonds known. This makes the molecule very stable and unreactive. Converting molecular nitrogen to a form that plants and animals can use – *nitrogen fixation* – requires a tremendous amount of energy. The process can be performed by only a few species, including a small number of specialized nitrogen-fixing microbes. Certain plants, including legumes such as peas, beans, and clover, nurture nitrogen-fixing rhizobium bacteria, in nodules attached to their roots. As these plants secrete sugars to feed the bacteria, the bacteria supply the plants with a ready supply of fixed nitrogen. When the plants die, the remaining nitrogen is returned to the soil where it nourishes other crops. Farmers 8000 years ago understood how growing these crops in rotation with other crops like cereals helped kept the soil nitrogen-rich. This coupled with recycling nitrogen-rich crop residues and animal manures were the main methods of fertilization. However, in Europe 150 years ago, farmers still had trouble with infertile land, as the soil being used for cultivating crops

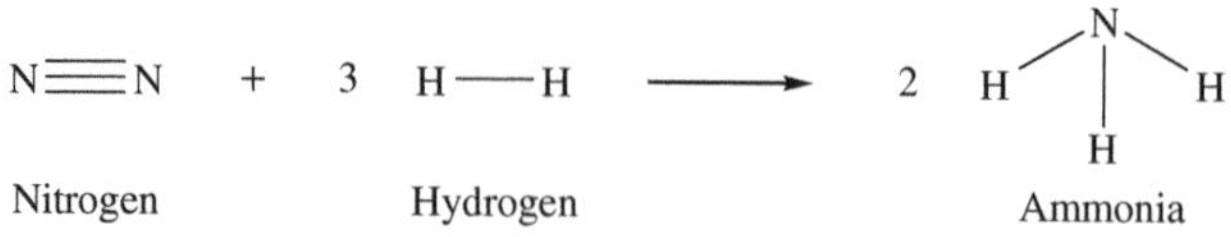

$$N\equiv N \quad + \quad 3 \quad H\!-\!H \quad \longrightarrow \quad 2 \quad H\!\!\diagup\!\!\overset{\displaystyle N}{\underset{\displaystyle H}{|}}\!\!\diagdown\!\! H$$

Nitrogen   Hydrogen   Ammonia

Scheme 2.4 The Haber process.

became depleted of nitrogen, and they are less and less productive. It was not enough to help feed Europe's increasing population; famines were prevalent, and people were dying of starvation.

The solution came in 1913, when the German chemist Fritz Haber found that it was possible to produce ammonia from molecular nitrogen and hydrogen by using a simple reaction (Scheme 2.4).

The secret to the Haber process was using an iron catalyst along with a small amount of aluminum. The Haber process operates at high pressure ($\sim$3000 psi) so as to force the reaction to form more ammonia and high temperatures ($\sim$400–450 °C) to increase the rates of the reaction. The hydrogen gas used is obtained in a process called *steam reforming*, in which methane reacts with high-temperature steam to form hydrogen gas and carbon monoxide. The nitrogen used in the process is obtained from air. Air is liquefied by cooling it down to $-$200 °C. Then trace gasses and water vapor are removed by filtering. The liquefied air is then allowed to warm up and the major gases are separated because of their different boiling points. Nitrogen, with a boiling point of $-$196 °C, becomes a gas at a lower temperature than oxygen, with a boiling point of $-$183 °C. Fortunately, this provides a virtually limitless supply of nitrogen in the Earth's atmosphere.

Although this process would eventually have a massive impact on civilization, it did not originally have a peaceful purpose. Ammonia is a crucial molecule in the creation of the nitro-based munitions used in World War I such as TNT, Dunnite, and Lyddite. The generated ammonia NH_3 is oxidized to make nitric acid (HNO_3). Nitric acid is the main reagent in the reaction that forms the nitro group (NO_2) in these molecules (Figure 2.20).

Prior to the Haber process, the only other source of nitric acid was nitrate mineral deposits found in Chile. Chile is unique as a source of natural nitrates due to specific geological factors. The 9 km-deep Peru–Chile ocean trench, which lies roughly 100 miles off the coast, hosts marine algae. Seawater uplift caused by the trench feeds plant nutrients to the algae. The algae fix the nitrogen, and after they die, they release ammonium (NH_4^+) into the water. Sea spray brought inland from high seas and winds causes ammonium to accumulate on the coast. The ammonium is then exposed to the sun and undergoes photochemical oxidation to become nitrate (NO_3^-). Although the deposits in Chile were abundant during World War I, they still needed to be shipped to Germany by tanker load across the Atlantic and past patrolling British warships.

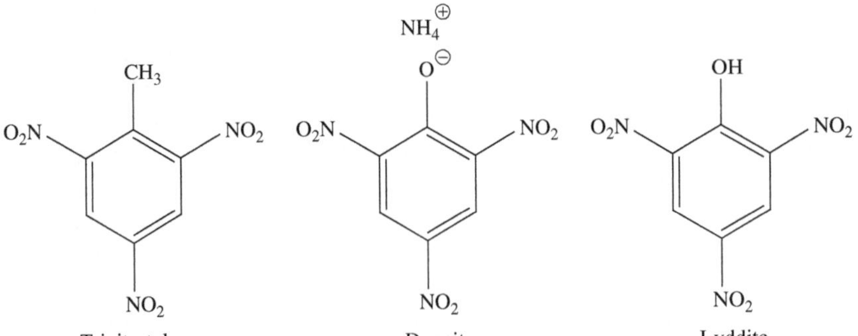

Figure 2.20 Explosives created using nitric acid.

This lack of availability of nitrates put a stranglehold on munition generation for Germany during World War I. Haber's great reaction allowed Germany to continue to make munitions.

Despite its rather dubious start, we still have not discussed the reason why the Haber process is considered one of the most important inventions of the twentieth century. You see, nitrates made from the Haber process can also be used to make fertilizers. Traditionally, agriculture production was limited by the amounts of nitrogen gained by recycling organic wastes and by planting leguminous crops. The Haber synthesis of ammonia removed this key constraint on crop productivity, and in combination with pesticides, the abundant fertilizers have quadrupled the productivity of agricultural land. If crop yields remained at the 1900 level, the crop harvest in 2000 would have required nearly half of the land on all ice-free continents, rather than under 15% of the total land area that it actually required.

Ammonia synthesis took off only after World War II, its global output rising from 3.7 million tons in 1950 to about 133 million tons in 2010, with about 75% of this being used to create fertilizer. The rate of change is astounding, and half of all of the synthetic nitrogen fertilizers ever used on the Earth has been produced since 1985. Molecular nitrogen is now the second-most produced chemical in the United States. So massive is ammonia production that it utilizes about 1–2% of the world's annual energy supply!

This discovery fueled a revolution and allowed a massive expansion of global agriculture, with concomitant decrease in hunger and malnutrition. This increase in crop production caused by the Haber process has been called the detonator of the population explosion, enabling the global population to increase from 1.6 billion in 1900 to today's 7 billion. Without this extra crop production, almost two-fifths of the world's population would not be here. More than half of the world's food production depends on Haber ammonia synthesis, and this share will keep rising for decades. Without the use of nitrogen fertilizers, we could not secure enough food for the nearly 45% of the world's population or 3 billion people. Use of this man-made fertilizer is so prevalent that it has been estimated that nearly 80% of the nitrogen found in human tissues originated from the Haber process! By the way, the fertilizer created by the Haber process cannot be utilized in organic farming, where only plant, animal, or manure-based fertilizers can be used.

Unfortunately, there is a downside to the Haber process. Nature maintains a delicate balance of fixed nitrogen in the environment. All the extra nitrogen that humans are extracting from the air has to end up somewhere. Due to inefficiencies of nitrogen uptake by plants and animals, only about 10–15% of the fixed nitrogen in fertilizer ever produces food. The rest is lost to the environment causing ecosystems to become overloaded with nitrogen, a process known as *"eutrophication."* This extra fertilizer is washed straight off the fields, leaching the nitrates into ground water, rivers, ponds, lakes, and eventually coastal ocean waters. The excess of nitrogen supply causes great blooms of algae and bacteria, which feed on the surplus nitrogen. In the process, they suck all the oxygen from the water, killing fish and other organisms. Finally, the excess nitrogen can also be put back into the atmosphere as nitrous oxide (N_2O), now the third most important greenhouse gas following carbon dioxide and methane. In addition, N_2O is currently the largest stratospheric ozone-depleting substance. Emissions from agricultural soils constitutes ~50% of global anthropogenic (man-made) N_2O produced and is predominantly made by the microbial processes of nitrification (oxidation of ammonium) and denitrification (reduction of nitrate).

So why have you never heard of Fritz Haber? Haber was a fanatical patriot and did everything in his power to help Germany during World War I. Haber had the idea to use poison gases to kill troops in the trenches and has been called the Father of Chemical warfare. In 1915, he personally oversaw the use of 498 tons of chlorine gas at the battle of Ypres, killing over 10,000 French troops in a few minutes. Over the next few years of the war, he helped develop more infamous chemical warfare agents, such as phosgene (see the previous story) and finally mustard gas, all of which were used against Allied troops.

Although Haber received the Nobel Prize in 1918 for his ammonia process, there were many objections because of his association with chemical weapons. In fact, he was branded as a war criminal by the Allies in 1920 but was never prosecuted. This was probably due to the political embarrassment it would have created to prosecute a Nobel Prize winner for war crimes.

After the war, he continued to make poison gases for the German government under the cover of projects to make insecticides. One gas he invented was a cyanide-based gas named Zyklon B, which was used in Auschwitz during World War II. Admittedly, its initial invention was for use as a pesticide, but it is difficult not to blame Haber for the use to which the poison would be put.

In the end, Haber paid a dear price for his patriotism toward Germany. Haber's wife, also a chemist, deplored his use of poison gas during the World War I and his preoccupation with work. She took her own life by shooting herself in his living room. When the Nazis came to power in 1933, they treated Haber with suspicion because of his Jewish heritage. When the Nazis started deporting his Jewish technicians, Haber fled to Switzerland. He remained there in exile, abandoned by his own country, until his death in 1934.

Further Reading

Death is its Withdrawal Symptom!

Ballenger, JC & Post, RM 1978, 'Kindling as a model for alcohol withdrawal syndromes', *British Journal of Psychiatry*, vol. 133, pp. 1–14.

Brown, ME, Anton, RF, Malcom, R & Ballenger, JC 1988, 'Alcohol detoxification and withdrawal seizures: Clinical support for a kindling hypothesis', *Biological Psychiatry*, vol. 23, pp. 507–514.

Espinoza, M 2011, *Records show man who died on hospital grounds treated for alcohol withdrawal*, The Press Democrat, Viewed 2 October 2016, http://www .pressdemocrat.com/article/20111005/articles/111009791#page=.

Miller, N & Gold, M 1998, 'Management of withdrawal syndromes and relapse prevention in drug and alcohol dependence', *American Family Physician*, vol. 58, pp. 139–146.

Paul, SM 2006, 'Alcohol-sensitive GABA receptors and alcohol antagonists', *Proceedings of the National Academy of Sciences of the United States of America*, vol. 103, pp. 8307–8308.

Talwan, S 2013, *Blaine counties liable for $1.35M in inmate's alcohol withdrawal death*, Billings Gazette, Viewed 2 October 2016, http://billingsgazette.com/ news/state-and-regional/montana/hill-blaine-counties-liable-for-m-in-inmate-s-alcohol/article_482ae554-71fb-550b-b268-d331ef62e5a8.html.

Trevisan, LA, Boutros, N, Petrakis, IL & Krystal, JH 1998, 'Complications of alcohol withdrawal', *Alcohol Health and Research World*, vol. 22, pp. 61–66.

Yoshida, K, Funahashi, M, Masui, M, Ogura, Y & Wakasugi, C 1990, '[Sudden death of alcohol withdrawal syndrome–report of a case]', *Nihon Hōigaku Zasshi*, vol. 44, pp. 243–247.

What is the Number One Cause of Liver Failure in the United States?

Drugs.com 2016, *Acetaminophen overdose*, Drugs.com, Viewed 2 October 2016, https://www.drugs.com/cg/acetaminophen-overdose.html.

Dart, RC, Erdma, AR, Olson, KR, Christianson G, Manoguerra AS, Chyka PA, Caravati EM, Wax PM, Keyes DC, Woolf AD, Scharman EJ, Booze LL, Troutman WG & American Association of Poison Control Centers 2006, 'Acetaminophen poisoning: An evidence-based consensus guideline for out-of-hospital management', *Clinical Toxicology (Philadelphia, Pa.)*, vol. 4, pp. 1–18.

Getting Relief Responsibly 2016, *Getting Relief Responsibly*, Johnson & Johnson Consumer Inc., Viewed 2 October 2016, http://www.getreliefresponsibly.com/.

Larson, AM, Polson, J, Fontana, RJ, Davern, TJ, Lalani, E, Hynan, LS, Reisch, JS, Schiødt, FV, Ostapowicz, G, Shakil, AO & Lee, WM 2005, 'Acetaminophen-induced acute liver failure: Results of a United States multicenter, prospective study', *Hepatology*, vol. 42, pp. 1364–1372.

Moore, M, Thor, H, Moore, G, Nelson, S, Moldéus, P & Orrenius, S 1985, 'The toxicity of acetaminophen and N-acetyl-p-benzoquinone imine in isolated hepatocytes is associated with thiol depletion and increased cytosolic Ca^{2+}', *Journal of Biological Chemistry*, vol. 260, pp. 13035–13040.

Pal, AK, Ghosh, S, Bera, AK, Bhattacharya, S, Chakraborty, S, Ghatak, KL & Banerjee, A 2000, 'Structural basis of inactivation of thiol protease by N-acetyl-p-benzoquinone imine (NAPQI). A knowledge-based molecular modeling of the adduct of NAPQI with thiol protease of the Papain family', *Molecular Modeling Annual*, vol. 6, pp. 648–653.

The Drug Abuse Warning Network 2004, *Highlights of the 2003 Drug Abuse Warning Network (DAWN) Findings on Drug-Related Emergency Department Visits*, Substance Abuse and Mental Health Service Administration.

The Most Addictive Substance Known

Centers for Disease Control and Prevention 2015, *Quitting smoking, Centers for Disease Control and Prevention*, Viewed 2 October 2016, http://www.cdc.gov/tobacco/data_statistics/fact_sheets/cessation/quitting/index.htm.

National Institute on Drug Abuse 2012, *Tobacco/nicotine*, National Institute on Drug Abuse, Viewed 2 October 2016, http://www.drugabuse.gov/publications/research-reports/tobacco-addiction.

Centers for Disease Control and Prevention 2011, *Morbidity and mortality weekly report (MMWR)*, Centers for Disease Control and Prevention, Viewed 2 October 2016, http://www.cdc.gov/mmwr/preview/mmwrhtml/mm6035a5.htm?s_cid=mm6035a5_w.

Centers for Disease Control and Prevention 2011, *Quitting smoking among adults – United States, 2001–2010*, Centers for Disease Control and

Prevention, Viewed 2 October 2016, http://www.cdc.gov/mmwr/preview/ mmwrhtml/mm6044a2.htm?s_cid=%20mm6044a2.htm_w.

National Cancer Institute 2015, *Tobacco*, National Cancer Institute, Viewed 2 October 2016, http://www.cancer.gov/cancertopics/tobacco/smoking.

Centers for Disease Control and Prevention 2014, *Nicotine*, Centers for Disease Control and Prevention, Viewed 2 October 2016, http://www.cdc.gov/niosh/ idlh/54115.html.

Med-Health.net 2016, *How many milligrams of nicotine in a cigarette?*, Med-Health.net, Viewed 2 October 2016, http://www.med-health.net/How-Many-Milligrams-Of-Nicotine-In-A-Cigarette.html.

National Institute on Drug Abuse 2012, *Is nicotine addictive?*, National Institute on Drug Abuse 2012, Viewed 2 October 2016, http://www.drugabuse.gov/ publications/research-reports/tobacco/nicotine-addictive.

40 Million Times Deadlier Than Cyanide

Allergan 2016, *Botox*, Botox.com, Viewed 3 October 2016, http://www.botox .com/.

Arnon, SS, Schechter, R & Inglesby, TV 2001, 'Botulinum toxin as a biological weapon', *Journal of the American Medical Association*, vol. 285, pp. 1059–1070.

Carruthers, JD & Carruthers, JA 1992 'Treatment of glabellar frown lines with *C. botulinum* – A exotoxin', *The Journal of Dermatologic Surgery and Oncology*, vol. 18, pp. 17–21.

Centers for Disease Control 2011, *Botulism annual summary, 2011*, Centers for Disease Control, Viewed 3 October 2016, http://www.cdc.gov/ nationalsurveillance/PDFs/Botulism_CSTE_2011.pdf.

Centers for Disease Control and Prevention 2016, *Botulism*, Viewed 3 October 2016, http://www.cdc.gov/botulism/.

Gill, MD, 1982, 'Bacterial toxins: A table of lethal amounts', *Microbiological Reviews*, vol. 46, pp. 86–94.

Plastic Surgery Research 2015, *2015 statistics*, Plastic Surgery Research, Viewed 3 October 2016, http://www.cosmeticplasticsurgerystatistics.com/statistics.html.

US Food and Drug Administration 2002, *Botulinum toxin type A product approval information – licensing action 4/12/02*, US Food and Drug Administration, Viewed 2 October 2016, http://www.fda.gov/Drugs/ DevelopmentApprovalProcess/HowDrugsareDevelopedandApproved/ ApprovalApplications/TherapeuticBiologicApplications/ucm080509.htm.

US Food and Drug Administration 2008, *FDA notifies public of adverse reactions linked to botox use ongoing safety review of botox, botox cosmetic and myobloc taking place*, US Food and Drug Administration, Viewed 2 October 2016, http://www.fda.gov/newsevents/newsroom/pressannouncements/2008/ucm116857.htm.

US Food and Drug Administration 2015, *FDA law enforcers crack down on illegal botox scammers*, US Food and Drug Administration, Viewed 2 October 2016, http://www.fda.gov/forconsumers/consumerupdates/ucm048377.htm.

The Most Abused Drug in the United States

Centers for Disease Control and Prevention 2010, *Summary health statistics for U.S. adults: National health interview survey, 2010*, Centers for Disease Control and Prevention, Viewed 2 October 2016, http://www.cdc.gov/nchs/data/series/sr_10/sr10_252.pdf.

Centers for Disease Control and Prevention 2011, *Deaths: Final data for 2009*, Centers for Disease Control and Prevention, Viewed 2 October 2016, http://www.cdc.gov/nchs/data/nvsr/nvsr60/nvsr60_03.pdf.

Centers for Disease Control and Prevention 2015, *Alcohol-related disease impact (ARDI)*, Centers for Disease Control and Prevention, Viewed 2 October 2016, http://www.cdc.gov/alcohol/onlinetools.htm.

Chan, A 2012, *Why alcohol is so addictive*, The Huffington Post, Viewed 2 October 2016, http://www.huffingtonpost.com/2012/01/13/alcohol-addictive-endorphins-_n_1202406.html.

Mokdad, AH, Marks, JS, Stroup, DF & Gerberding, JL 2004, 'Actual causes of death in the United States', *JAMA* vol. 291, pp. 1238–1245.

National Institute of Drug Abuse 2011, *Drug facts: Drug-related hospital emergency room visits*, National Institute of Drug Abuse, Viewed 2 October 2016, http://www.drugabuse.gov/publications/drugfacts/drug-related-hospital-emergency-room-visits.

National Institute of Drug Abuse 2015, *Drug facts: Nationwide trends*, National Institute of Drug Abuse, Viewed 2 October 2016, http://www.drugabuse.gov/publications/drugfacts/nationwide-trends.

Raymond, J 2012, *Study explains the science behind your beer buzz*, NBC News, Viewed 2 October 2016, http://bodyodd.nbcnews.com/_news/2012/01/11/10120223-study-explains-the-science-behind-your-beer-buzz.

Rehm, J, Gmel, G, Sepos, CT & Trevisan, M 2003, 'Alcohol-related morbidity and mortality', *Alcohol Research and Health*, vol. 27, pp. 39–51

Ryan, C 2004, *Why alcohol is addictive*, BBC News Online, Viewed 2 October 2016, http://news.bbc.co.uk/2/hi/health/3537387.stm.

The Only Known Aphrodisiac

Freeman, ER, Bloom, DA & McGuire, EJ 2001, 'A brief history of testosterone', *The Journal of Urology*, vol. 165, pp. 371–373.

Greenblatt, RB, Karpas, A 1983, 'Hormone therapy for sexual dysfunction. The only true aphrodisiac', *Postgraduate Medicine*, vol. 74, pp. 78–80.

Kingsberg, SA, Simon, JA & Goldstein, I 2008, 'The current outlook for testosterone in the management of hypoactive sexual desire disorder in postmenopausal women', *The Journal of Sexual Medicine*, vol. 5, pp. 182-193.

Morales, A 2013, 'The long and tortuous history of the discovery of testosterone and its clinical application', *The Journal of Sexual Medicine*, vol. 10, pp. 1178–1183.

Shamloul, R 2010, 'Natural aphrodisiacs', *The Journal of Sexual Medicine*, vol.7, pp. 39–49.

Sherwin, BB 1988, 'A comparative analysis of the role of androgen in the human male and female sexual behavior: Behavioral specificity, critical thresholds, and sensitivity', *Psychobiology*, vol. 16, pp. 416-425.

Tuiten, A, Van Honk, J, Koppeschaar, H, Bernaards, C, Thijssen, J & Verbaten, R 2000, 'Time course of effect of testosterone administration on sexual arousal in women', *Archives of General Psychiatry*, vol. 57, pp. 149–153.

The Most Consumed Psychoactive Substance

Castillo, M 2013, FDA *investigating 13 deaths tied to 5-hour Energy*, CBS News, Viewed 2 October 2016, http://www.cbsnews.com/news/fda-investigating-13-deaths-tied-to-5-hour-energy/.

Centers for Disease Control and Prevention 2015, *Caffeine and alcohol*, Centers for Disease Control and Prevention, Viewed 2 October 2016, http://www.cdc.gov/alcohol/fact-sheets/cab.htm.

Gilbert, RM 1981, *Nutrition and Behavior*, Franklin Institute Press, Philadelphia, PA.

Gilbert, RM 1984, *Caffeine Consumption. In The Methylxanthine Beverages and Foods: Chemistry, Consumption, and Health Effects*, Alan R. Liss, Inc., New York, NY.

Kerrigan, S & Lindsey, T 2005, 'Fatal caffeine overdose: two case reports', *Forensic Science International*, vol. 153, pp. 67–69.

Mednick, SC, Cai, DJ, Kanady, J & Drummond, SP 2008, 'Comparing the benefits of caffeine, naps and placebo on verbal, motor and perceptual memory', *Behavioural Brain Research*, vol. 193, pp. 79–86.

O'Shea, K 2013, *Caffeine withdrawal is now a considered mental disorder*, Philly.com.

Reissig, CH, Strain, EC & Griffiths, RR 2009, 'Caffeinated energy drinks – a growing problem', *Drug and Alcohol Dependence*, vol. 99, pp. 1–10.

40,000 Tons of Aspirin

Bronstein, AC, Spyker, DA, Cantilena, LR, Rumack, BH & Dart, RC 2011, '2011 annual report of the American Association of Poison Control Centers' National Poison Data System (NPDS): 29th annual report', *Clinical Toxicology*, vol. 50, pp. 911–1164

Centers for Disease Control and Prevention 2010, *National hospital ambulatory medical care survey: 2010 emergency department summary tables*, Centers for Disease Control and Prevention, Viewed 2 October 2016, http://www.cdc.gov/nchs/data/ahcd/nhamcs_emergency/2010_ed_web_tables.pdf.

Chyka, PA, Erdman, AR, Christianson, G, Wax, PM, Booze, LL, Manoguerra, AS, Caravati, EM, Nelson, LS, Olson, KR, Cobaugh, DJ, Scharman, EJ, Woolf, AD & Troutman WG 2007, 'Salicylate poisoning: An evidence-based consensus guideline for out-of-hospital management', *Clinical Toxicology (Philadelphia, Pa.)*, vol. 45, pp. 95–131.

Dargan, PI, Wallace, CI & Jones, AL 2016, *An evidence based flowchart to guide the management of acute salicylate (aspirin) overdose*, Emergency Medicine Journal, Viewed 2 October 2016, http://emj.bmj.com/content/19/3/206.full.pdf+html.

MedlinePlus 2016, *Aspirin overdose*, MedlinePlus, Viewed 2 October 2016, http://www.nlm.nih.gov/medlineplus/ency/article/002542.htm

National Reye's Syndrome Foundation Inc. 2008, *Medications containing aspirin (acetylsalicylate) and aspirin-like products*, National Reye's Syndrome Foundation Inc., Viewed 2 October 2016, http://reyessyndrome.org/pdfs/medicationscontainingaspirin.pdf.

US Food & Drug Administration 2015, *Assessment of safety of aspirin and other nonsteroidal anti-inflammatory drugS (NSAIDs)*, US Food & Drug Administration, Viewed 2 October 2016, http://www.fda.gov/ohrms/dockets/ac/02/briefing/3882B2_02_McNeil-NSAID.htm.

How Bitter is the Bitterest?

Alcohol and Tobacco Tax and Trade Bureau 2015, *Laws and regulations – completely denatured alcohol (CDA)*, Alcohol and Tobacco Tax and Trade Bureau, Viewed 2 October 2016, http://www.ttb.gov/industrial/sda_regs_laws2.shtml.

Aversion Technologies, Inc 2016, *Denatonium benzoate*, Aversion Technologies, Viewed 2 October 2016, http://denatonium-benzoate.com/.

Bronstein, AC, Spyker, DA, Cantilena, LR, Rumack, BH & Dart, RC 2011, '2011 annual report of the American Association of Poison Control Centers' National Poison Data System (NPDS): 29th annual report', *Clinical Toxicology*, vol. 50, pp. 911–1164.

Johnson Matthey Chemicals 2016, *BITREX®*, Johnson Matthey Chemicals, Viewed 2 October 2016, http://www.jmfinechemicals.com/bitrex/.

Macfarlan Smith 2016, *Bitrex*, Macfarlan Smith, Viewed 2 October 2016, www .bitrex.com/.

Meyerhof, W, Batram, C, Kuhn, C, Brockhoff, A, Chudoba, E, Bufe, B, Appendino, G & Behrens, M 2009, 'The molecular receptive ranges of human TAS2R bitter taste receptors', *Chemical Senses*, vol. 35, pp. 157–170.

Wilerson, R, Northington, L & Fisher, W 2005, 'Ingestion of toxic substances by infants and children, what we don't know can hurt', *Critical Care Nurse*, vol. 25, pp. 35–44.

$62.5 Trillion per Gram

Amos, J 2011, *Antimatter atoms are corralled even longer*, BBC News, Viewed 2 October 2016, http://www.bbc.co.uk/news/science-environment-13666892.

Berkeley Lab 2004, *Antimatter*, US Department of Energy, Viewed 2 October 2016, http://www.lbl.gov/abc/Antimatter.html.

Freedman, DH 1997, *Antiatoms: Here today …*, Discover Magazine, Viewed 2 October 2016, http://discovermagazine.com/1997/jan/antiatomsheretod1029.

Knapp, A 2012, *How much does it cost to find a Higgs Boson?*, Forbes, Viewed 2 October 2016, http://www.forbes.com/sites/alexknapp/2012/07/05/how-much-does-it-cost-to-find-a-higgs-boson/.

NASA 1999, *Antimatter*, NASA, Viewed 2 October 2016, http://science.nasa.gov/science-news/science-at-nasa/1999/prop12apr99_1/.

What is the Most Abundant Source of Air Pollution?

Grace, F 2003, *Here we go again!*, CBS News, Viewed 2 October 2016, http://www .cbsnews.com/news/here-we-go-again-544188/.

Guenther, A, Karl, T, Harley, P, Wiedinmyer, C, Palmer, PI & Geron C 2006, 'Estimates of global terrestrial isoprene emissions using MEGAN (model of emissions of gases and aerosols from nature', *Atmospheric Chemistry and Physics*, vol. 6, pp. 3181–3210.

Harling, A, Milton, M, Brookes, C, Corbel, M 2015, *Measuring the impact of trace emissions from trees on air quality and climate*, National Physical Laboratory,

Viewed 2 October 2016, http://www.npl.co.uk/upload/pdf/measuring_harling
.pdf.

Jenkin, ME & Clemitshaw, KC 2000, 'Ozone and other secondary photochemical
pollutants: chemical processes governing their formation in the planetary
boundary layer', *Atmospheric Environment*, vol. 34, pp. 2499–2527.

Pio, CA & Valente, AA 1998, 'Atmospheric fluxes and concentrations of
monoterpenes in resin-tapped pine forests', *Atmospheric Environment*, vol. 32,
pp. 683–691.

Sharkey, TD, Wiberley, AE & Donohue, AR 2008, 'Isoprene emission from plants:
why and how', *Annals of Botany*, vol. 101, pp. 5–18.

US Environmental Protection Agency 2006, *How air pollution affects the view*,
US Environmental Protection Agency, Viewed 2 October 2016, https://www
.epa.gov/sites/production/files/2015-05/documents/haze_brochure_20060426
.pdf.

Where Did That Rash Come From?

Ani 2010, *Why nickel causes allergies in humans*, The Gaea Times, Viewed 2
October 2016, http://science.gaeatimes.com/2010/08/16/why-nickel-causes-
allergies-in-humans-20732/.

Bercovitch, L & Luo, J 2008, 'Cellphone contact dermatitis with nickel allergy',
CMAJ, vol. 178, pp. 23–24.

Darlenski, R, Kazandjieva, J & Pramatato, K 2012, 'The many faces of nickel
allergy', *International Journal of Dermatology*, vol. 51, pp. 501–629.

European Council Directive 1976, *On the approximation of the laws of the
Member States relating to cosmetic products*, 76/768/EEC. Official Journal of
the European Communities.

Stojanović, D, Nikić, D & Lazarević, K 2004, 'The level of nickel in smoker's blood
and urine', *Central European Journal of Public Health*, vol. 12,
pp. 187–189.

Williams, M, Villarreal, A, Bozhilov, K, Lin, S, & Talbot, P 2013, 'Metal and silicate
particles including nanoparticles are present in electronic cigarette cartomizer
fluid and aerosol', *PLoS One*, vol. 8, pp. e57987.

Zirwas, MJ & Molenda, MA 2009, 'Dietary nickel as a cause of systemic contact',
Journal of Clinical and Aesthetic Dermatology, vol. 2, pp. 39–43.

Zug, KA, Warshaw, EM, Fowler, JF Jr,, Maibach, HI, Belsito, DL, Pratt, MD,
Sasseville, D, Storrs, FJ, Taylor, JS, Mathias, CG, Deleo, VA, Rietschel, RL &
Marks J 2015, 'Patch-test results of the North American contact dermatitis
group 2005-2006', *Dermatitis*, vol. 20, pp. 149–160.

It Would Take an Elephant on a Pencil

BBC News 2013, *Bill Gates condom challenge 'to be met' by graphene scientists*, BBC News, Viewed 2 October 2016, http://www.bbc.com/news/uk-england-manchester-25016994.

Geim, AK 2009, 'Graphene: Status and prospects', *Science*, vol. 324, pp. 1530–1534.

Hudson, A 2011, *Is graphene a miracle material?*, BBC News, Viewed 2 October 2016, http://news.bbc.co.uk/2/hi/programmes/click_online/9491789.stm.

Lee, C, Wei, X, Kysar, JW & Hone, J 2008, 'Measurement of the elastic properties and intrinsic strength of monolayer graphene', *Science*, vol. 321, pp. 385.

Nature Research Highlights 2013, *Solid carbon, springy and light*, Nature, Viewed 2 October 2016, http://www.nature.com/nature/journal/v494/n7438/full/494404a.html.

PhysOrg.com 2008, *Engineers prove graphene is the strongest material*, PhysOrg.com, Viewed 2 October 2016, http://phys.org/news135959004.html#jCp.

The Royal Swedish Academy of Sciences 2010, *Graphene*, The Royal Swedish Academy of Sciences, Viewed 2 October 2016, http://www.nobelprize.org/nobel_prizes/physics/laureates/2010/advanced-physicsprize2010.pdf.

The Largest Industrial Accident in World History

Bogdanović, M 2009, 'Widely known chemical accidents', *Working and Living Environmental Protection*, vol. 6, pp. 65–71.

Centers for Disease Control and Prevention 2011, *PHOSGENE (CG): Lung damaging agent*, Centers for Disease Control and Prevention http://www.cdc.gov/niosh/ershdb/EmergencyResponseCard_29750023.html

Eckerman, I 2004, *The Bhopal Saga – Causes and Consequences of the World's Largest Industrial Disaster*, Universities Press (India) Private Limited.

Glass, D, McClanahan, M, Koller, L & Adeshina, F 2009, 'Provisional advisory levels (PALs) for phosgene (CG)', *Inhalation Toxicology*, vol. 21, pp. 73–94.

India Together 2012, *Bhopal timeline*, India Together, Viewed 2 October 2016, http://indiatogether.org/campaigns/bhopal/timeline.htm#sthash.wogwjmGz.dpuf.

International Programme on Chemical Safety 1998, *Phosgene health and safety guide*, International Programme on Chemical Safety, Viewed 2 October 2016, http://www.inchem.org/documents/hsg/hsg/hsg106.htm.

McKeown, NJ & Burton, BT 2012, 'Acute lung injury following refrigeration coil deicing', *Clinical Toxicology (Philadelphia, Pa.)*, vol. 50, pp. 218–220.

Raina, P 2014, *Indians angry Anderson never tried over Bhopal disaster*, Yahoo News, Viewed 2 October 2016, http://news.yahoo.com/indians-angry-anderson-never-tried-over-bhopal-disaster-143048387.html;_ ylt=AwrSyCMoAlxUTWEACFbQtDMD.

Rega, PP 2012, *Phosgene toxicity*, Medscape, Viewed 2 October 2016, http://misc .medscape.com/pi/android/medscapeapp/html/A832454-business.html.

US Chemical Safety Board 2010, *DuPont corporation toxic chemical releases*, US Chemical Safety Board, Viewed 2 October 2016, http://www.csb.gov/dupont-corporation-toxic-chemical-releases/.

Vaish, AK, Consul, S, Agrawal, A, Chaudhary, SC, Gutch, M, Jain, N & Singh, MM 2013, 'Accidental phosgene gas exposure: A review with background study of 10 cases', *Journal of Emergencies, Trauma, and Shock*, vol. 6, pp. 271–275.

Wijte, D, Alblas, MJ, Noort, D, Langenberg, JP & van Helden HPM 2011, 'Toxic effects following phosgene exposure of human epithelial lung cells in vitro using a CULTEX_ system', *Toxicology in Vitro*, vol. 25, pp. 2080–2087.

What is the Most Important Chemical Reaction?

Arias, J 2003, *On the origin of saltpeter, northern Chile Coast*, XVI INQUA Congress, Viewed 2 October 2016, https://gsa.confex.com/gsa/inqu/finalprogram/abstract_55601.htm.

Eurekalert 2008, *Addressing the 'nitrogen cascade'*, Eurekalert, Viewed 2 October 2016, http://www.eurekalert.org/pub_releases/2008-05/uov-at051208.php.

Fritz, S 2012, 'Fritz Haber: Flawed greatness of person and country', *Angewandte Chemie-International Edition*, vol. 51, pp. 50–56.

Galloway, JN, Aber, JD, Willem Erisman, J, Seitzinger, SP, Howarth, RW, Cowling EB & Cosby, BJ 2003, 'The nitrogen cascade', *BioScience*, vol. 4, pp. 341–356.

Howarth, RW 2008, 'Coastal nitrogen pollution: A review of sources and trends globally and regionally', *Harmful Algae*, vol. 8, pp. 14–20.

Smil, Vaclav 1999, 'Detonator of the population explosion', *Nature*, vol. 400, pp. 415.

Smil, Vaclav 2011, 'Nitrogen cycle and world food production', *World Agriculture*, vol. 2, pp. 9–10.

Szinicz, L 2005, 'History of chemical and biological warfare agents', *Toxicology*, vol. 214, pp. 167–181.

3

The Poisons in Everyday Things

If you follow the news, you know that chemical poisoning from readily available household products is quite common. This might be related to the false assumption that products are completely safe if they can be purchased at a drugstore or found in certain foods. In general, it is not safe to consume a huge amount of anything in a short time. In fact, one reason this book has been written is to educate people so they will be careful with *all* chemicals.

Toxicity is arguably one of the most fascinating aspects of chemistry. You may hear that someone ingested cyanide and died. But how, exactly? What actually caused death? Similar to many other things, toxicity can be understood at the molecular level. The body is a complex machine, with many chemical cogs and wheels turning to keep it alive. All a molecule has to do is stop one of these components from functioning, and death can quickly follow.

Why Is Antifreeze Lethal?

Most antifreeze used in engines is a solution of ethylene glycol and water, and in the United States, more than 9 billion pounds of ethylene glycol is produced each year. The structure of ethylene glycol is similar to some types of sugars, for example, xylitol, which gives it a sweet taste that can sometimes be attractive to children and animals. Once inside the body, ethylene glycol is converted to glycolic acid in the liver by the enzyme alcohol dehydrogenase. Although weakly acidic, glycolic acid causes blood to become acidified in a process called *acidosis*. Human blood normally has a pH between 7.36 and 7.42, and even slight deviations can cause serious health problems. Acidosis reduces the blood's ability to absorb oxygen because the hydrogen from the acid competes with oxygen to bond with hemoglobin, the protein that carries oxygen in the blood. Acidosis is an early warning sign of ethylene glycol poisoning, causing symptoms that include confusion, lethargy, rapid breathing, shortness of breath, and blue lips and fingernails (Figure 3.1 and Scheme 3.1).

Strange Chemistry: The Stories Your Chemistry Teacher Wouldn't Tell You, First Edition. Steven Farmer.
© 2017 John Wiley & Sons, Inc. Published 2017 by John Wiley & Sons, Inc.

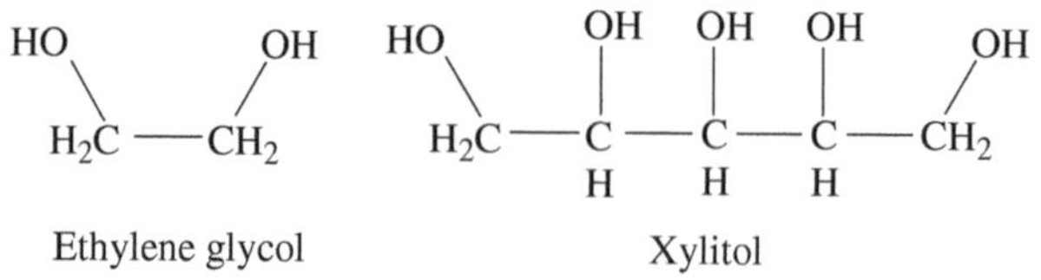

Figure 3.1 A comparison of the structures of ethylene glycol and xylitol.

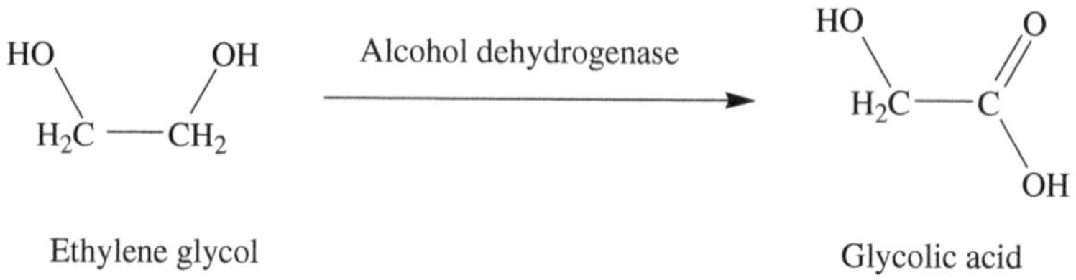

Scheme 3.1 The conversion of ethylene glycol to glycolic acid.

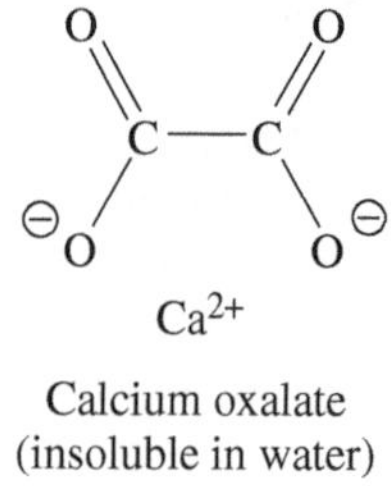

Figure 3.2 The structure of calcium oxalate.

If this is not bad enough, an even greater peril awaits the patient who survives acidosis. The glycolic acid is further metabolized to oxalic acid, again by the enzyme alcohol dehydrogenase. The oxalic acid travels to the kidneys, where it combines with calcium to form calcium oxalate, which precipitates out of blood. Calcium oxalate is the major component of kidney stones. Once the calcium oxalate crystals are produced, they block critical parts of the kidneys – the nephrons – causing kidney failure. At this stage of antifreeze poisoning, the unfortunate victim suffers symptoms including blood in the urine, no urine output, and blurry vision. This is followed by blindness, convulsions, coma, and death (Figure 3.2 and Scheme 3.2).

Because of differences in the alcohol dehydrogenase enzyme among species, the effect of ethylene glycol varies with the animal that has ingested the antifreeze. For example, 2 ounces of ethylene glycol can kill a dog, and only one teaspoon can be lethal to a cat. As little as 4 ounces (118 ml) can kill an adult human.

One of the treatments for ethylene glycol poisoning is large doses of ethanol (grain alcohol). The same enzyme, alcohol dehydrogenase, that converts ethylene glycol to glycolic acid also converts ethanol to acetic acid as part of

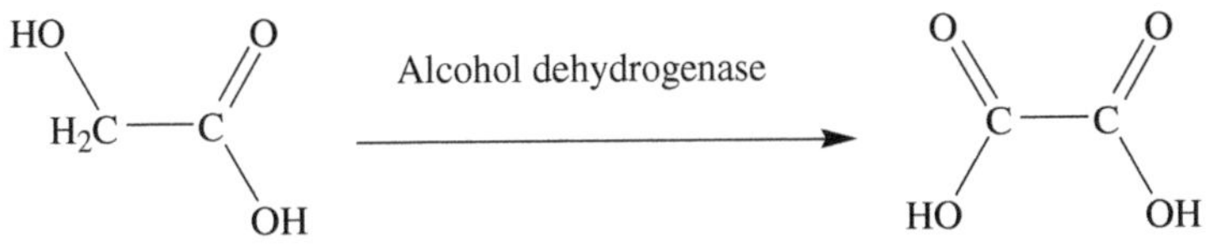

Scheme 3.2 The conversion of glycolic acid to oxalic acid.

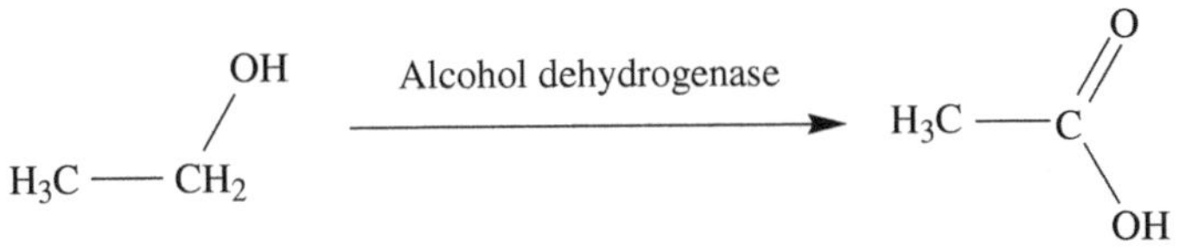

Scheme 3.3 The conversion of ethanol to acetic acid.

the human body's ethanol metabolism. The two molecules, ethylene glycol and ethanol, compete for the enzyme, but the enzyme prefers ethanol. This allows the ethylene glycol to pass through unchanged and be expelled in the urine. It is interesting to note the acetate formed by ethanol does not form insoluble crystals with calcium and therefore does not have the lethal effects of ethylene glycol (Scheme 3.3).

The American Association of Poison Control Centers reported that in 2011, there were 5694 cases of ethylene glycol poisoning in humans, of which 727 were intentional cases of poisoning that resulted in seven deaths. The Humane Society estimates that up to 10,000 cats and dogs die each year from ethylene glycol poisoning. To help with this problem, bittering agents such as denatonium benzoate, the bitterest substance known, has been added to antifreeze by major marketers since 2012 (see the section later in this book about the "most bitter substance").

Because antifreeze is odorless and sweet-tasting, it is a popular poison for the criminal-minded. One of the most famous people who used antifreeze with intent to kill is Stacey Castor, who was convicted in 2009 of murdering her second husband by poisoning him with antifreeze, and then attempting to murder one of her daughters and frame her for the crime. Castor is believed to have murdered her first husband the same way. She was sentenced to serve 51.3 years to life in a New York State prison, although she was found dead in her cell on June 16, 2016.

In 2014, the actor Daniel Radcliffe, of *Harry Potter* fame, became ill after accidentally ingesting antifreeze on the set of the movie *Horns*. Because filming was taking place in snowy Canada, antifreeze had been added to the water pipes supplying his trailer to keep the water from freezing. Not knowing that the water was unsafe, Radcliffe drank from this source and fell ill. Fortunately, he was not seriously hurt and recovered fully.

Aqua Dots: What a Difference a Carbon Makes!

Aqua Dots was a children's arts and crafts toy made up of brightly colored beads, which could be arranged into designs and fused when sprayed with water. In 2007, a recall was issued for 4.2 million units of this toy when five children in the United States and Australia were hospitalized after swallowing the beads. To their credit, the makers of Aqua Dots were aware that children might mistake the vibrant little balls for candy, so they used safe ingredients in manufacturing them.

The problem occurred with the plasticizer used in the toy. The *plasticizer* – small organic molecules added to plastic products, generally to soften them – should have been the nontoxic 1,5-pentanediol molecule. However, a Chinese manufacturer decided to substitute 1,4-butanediol, a toxic compound as a substitute plasticizer. When comparing these two molecules, the only difference between them is one carbon atom. But that one carbon can make a huge difference inside the human body! (Figure 3.3)

In light of the fact that the substituted molecule, 1,4-butanediol, costs about $2000 per metric ton, while the original molecule, 1,5-pentanediol, costs more than four times as much, $9700 a ton, it was suspected that the Chinese manufacturer made the change to cut costs. However, the consequences of this offense in China are not that prohibitive; the company found guilty of this material substitution was fined a mere $12,877.

What is so bad about 1,4-butanediol? It has everything to do with this chemical's metabolism, that is, the way most drugs are changed inside the body prior to excretion. Sometimes, the metabolic process can turn a harmless molecule into a more dangerous one. Using the enzyme alcohol dehydrogenase, our body converts 1,4-butanediol to 4-hydroxy butanoic acid, also called *gamma-hydroxybutyric acid* (GHB) or colloquially, a "date rape drug." Therefore, ingesting 1,4-butanediol has effects similar to the ingestion of

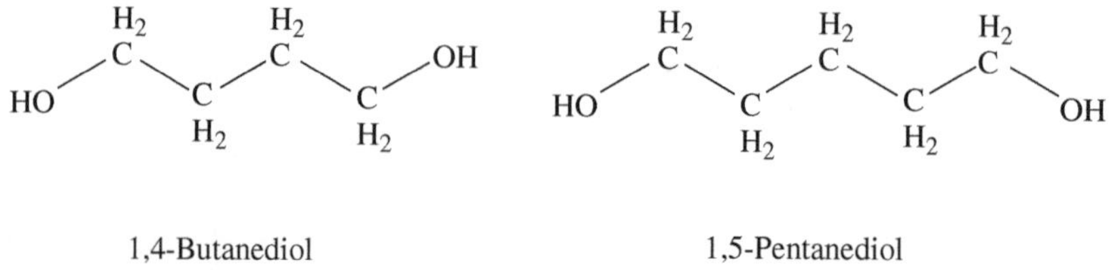

1,4-Butanediol   1,5-Pentanediol

Figure 3.3 A comparison of the structures of butanediol and pentanediol.

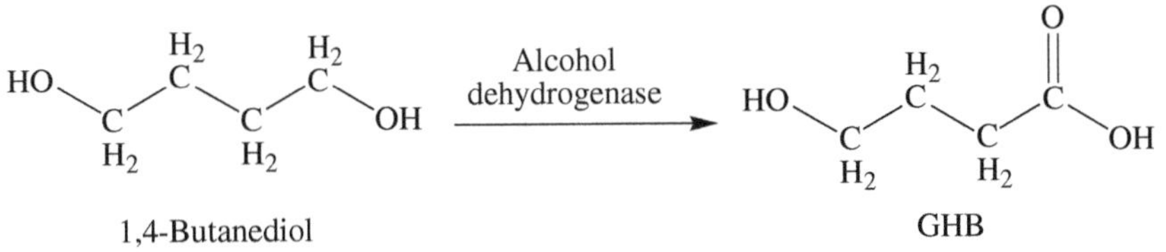

Scheme 3.4 The conversion of butanediol to GHB.

GHB. Consequently, alcohol dehydrogenase was the same enzyme that made antifreeze lethal, as discussed in the previous story (Scheme 3.4).

Scientific data suggest that GHB can function as a neurotransmitter or neuromodulator in the brain. It produces depressant effects similar to those of the barbiturates and can intensify the effects of alcohol. GHB has been used as a general anesthetic and to treat insomnia, clinical depression, narcolepsy, and alcoholism. Because GHB was purported to act as a fat burner and a growth hormone promoter, in the late 1980s, it was used as food supplement for body builders and dieters. Low doses of GHB produce drowsiness, nausea, and visual distortion. At high doses, GHB overdose can result in unconsciousness, seizures, slowed heart rate, severe respiratory depression, decreased body temperature, vomiting, nausea, coma, or death.

Infamous as a date rape drug, GHB is colorless and odorless and can be easily added to a person's drink without their knowledge. It can cause unconsciousness within 20 minutes, and people who have consumed it, frequently have no memory of what happened for several hours afterwards. In addition, GHB usually leaves the body within 24 hours, which makes it difficult to detect in date rape victims. The US Food and Drug Administration (FDA) prohibited the sale and manufacture of GHB in 1990, and it has been listed as a Schedule I controlled substance since 2000. Schedule I controlled substances have a high potential for abuse and no currently accepted use for medical treatment in the United States. Since 1990, GHB has been linked to at least 58 deaths and more than 5700 recorded overdoses.

While 1,4-butanediol is not scheduled federally in the United States, several states have classified 1,4-butanediol as a controlled substance, so 1,4-butanediol falls between the legal cracks. Technically, it is not illegal to use, but it is illegal to purchase or sell online.

GHB gets its effect by mimicking the structure of the neurotransmitter *gamma-aminobutyric acid* (GABA). Because of the structural similarity, the body cannot tell the two molecules apart. As previously discussed, GABA is the main inhibitory neurotransmitter in the body. The presence of GABA or GHB causes nerves to become inactive, which depresses the nervous system. It is easy to see from Figure 3.4 that the only difference between these two molecules is that GHB has a hydroxyl group (OH) where GABA has an amino group (NH_2).

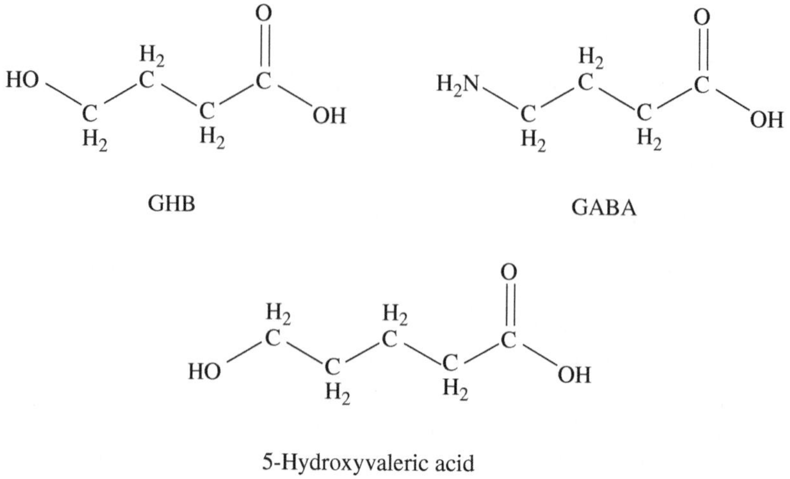

GHB

GABA

5-Hydroxyvaleric acid

Figure 3.4 A comparison of the structures of GHB, GABA, and 5-hydroxyvaleric acid.

The extra carbon in 1,5-pentanediol keeps this molecule from being converted to GHB, which has only four carbons. The five-carbon product resulting from 1,5-pentanediol in the presence of alcohol dehydrogenase, 5-hydroxyvaleric acid, does not mimic GABA and thus does not have the same dangerous effects in the body.

How Can Visine® Kill You?

The well-known company Johnson & Johnson produces Visine®, a brand of eye drops containing a formulation of potassium chloride and tetrahydrozoline hydrochloride. Tetrahydrozoline, patented in 1956, is a vasoconstrictor that causes the superficial blood vessels in the eye to contract, diminishing the redness. Tetrahydrozoline is found in many over-the-counter eye drops and some nasal sprays (Figure 3.5).

The problems with tetrahydrozoline started after the popular movie *Wedding Crashers* came out in 2005. In the movie, Owen Wilson's character made a rival ill by putting eye drops into his glass of wine, saying, "Put a few drops of these in his drink, and he'll be going down on a toilet seat for the next 24 hours." Unfortunately, the scene caused people to believe that this prank would result in nothing worse than an embarrassing case of diarrhea. They couldn't be more wrong! This urban myth went viral. One blog about the worst drinks made with alcohol listed the "Visine Surprise," a shot of any kind of liquor mixed with two or three drops of Visine®. In this case, the "Surprise" would be a trip to the hospital and some possible jail time. The story, written by a bartender, was that

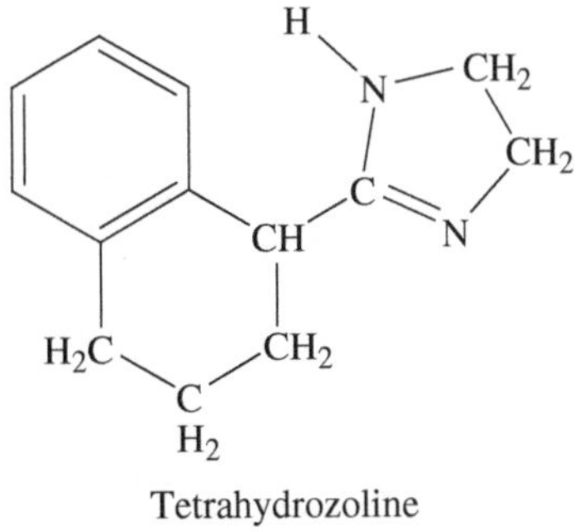

Figure 3.5 The structure of tetrahydrozoline.

Tetrahydrozoline

after a customer berated one of his coworkers, the humiliated employee told the bartender to put some eye drops in his drink the next time he came in. Fortunately, the bartender did not because he was worried that it might kill the customer (however, this did not stop him from publishing the drink recipe).

In actuality, tetrahydrozoline eye drops are known to be fatal when ingested. A college student in Wisconsin poisoned her roommate with Visine® as a "joke." When interrogated by the police, she admitted that she got the idea from *Wedding Crashers*. She was sentenced to 90 days in prison and 30 months of probation. A California man poisoned his girlfriend with Visine® to get even with her. The eye drops made her ill and she contacted police after seeing a suspicious text on her boyfriend's phone: "If she's going to be talking crap, then she's going to be crapping." In July of 2014, a Pennsylvania woman was charged with aggravated assault and child endangerment after she confessed to adding over a bottle of Visine® to her 3-year-old son's drink. The child had to be airlifted to a hospital after his heartbeat dropped to 40 beats per minute. The woman claimed that she did not want to harm her son, but just wanted her boyfriend, the boy's father, to pay more attention to her.

The toxicity of Visine® is clearly stated on its packaging: "If swallowed, get medical help or contact a Poison Control Center right away." Nonetheless, the American Association of Poison Control Centers reported 1238 cases of tetrahydrozoline poisoning in 2012, 36 of which were intentional. Accidental poisoning has become such a problem that in 2012, the U.S. Consumer Product Safety Commission proposed requiring child-resistant closures on all tetrahydrozoline-containing products.

Applied to the eyes, tetrahydrozoline in eye drops constricts dilated microvessels. However, this same chemical has a completely different effect when swallowed, causing vasodilation, decrease in blood pressure, and depression of the central nervous system. Symptoms of tetrahydrozoline poisoning include nausea, vomiting, diarrhea, dizziness, drowsiness, hypothermia, tremors, blurred vision, muscle flaccidity, difficulty breathing, slowed heartbeat, seizures, coma, and eventually death.

Figure 3.6 A helpful warning.

The lethal dose of tetrahydrozoline is not yet known. However, the cases seen so far show that even a fraction of a bottle of a typical tetrahydrozoline-containing product can cause serious harm. The FDA, however, has reported that ingestion of only 1–2 ml could be harmful to young children (Figure 3.6).

Death by BENGAY®

In 2007, the New York City Office of the Chief Medical Examiner ruled that a 17-year-old cross-country runner died from an accidental overdose of methyl salicylate, an anti-inflammatory ingredient found in popular sports balms. It was discovered that she had been rubbing BENGAY® on her legs between running sessions and at the same time using painkilling adhesive pads that also contained methyl salicylate. Her death brought to light the fallacy of assuming that all over-the-counter products are completely safe. In fact, topical preparations such as BENGAY® are the seventh most common substance in human poisonings, with 109,831 incidents involving topical substances reported in 2005 alone.

The active ingredient in BENGAY®, methyl salicylate, is oil of wintergreen (or wintergreen oil), an organic ester produced by many plant species, some of which are called *wintergreens*. Methyl salicylate gives BENGAY® its characteristic strong, minty smell. Methyl salicylate is also used to flavor candies, such as wintergreen Life Savers® and Altoids®.

Native Americans used wintergreens to treat back pain, rheumatism, fever, headaches, and sore throats. In topical form, it is generally considered safe; however, in large amounts, it can be lethal. When swallowed, one teaspoon (5 ml) of wintergreen oil is equivalent to approximately 21.5 adult aspirin tablets. Ingestion of as little as 4 ml by a child and 6 ml by an adult has been shown to be fatal. For this reason, food products must contain less than 0.05% oil of wintergreen.

The high risk of overdose is partly related to the fact that methyl salicylate is just one member of a larger class of drugs called salicylates. This class of drugs includes acetylsalicylic acid (aspirin) and bismuth subsalicylate (commonly found in Pepto-Bismol®). Considering the fact that salicylates are found in more than 200 formulations, such as Alka-Seltzer®, Dristan®, Pepto-Bismol®, Sine-Aid®, and Percodan®, patients and physicians commonly overlook them as a source of toxicity. In one reported case, an 82-year-old woman died after taking 66 Pepto-Bismol® tablets.

Because salicylates are anticlotting agents, very high doses can cause internal bleeding, heart arrhythmias, and liver problems. These conditions can result in shortness of breath, hyperventilation, rapid heartbeat, fever, hemorrhage, increased body temperature, and low blood sugar – all of which can lead to altered mental status, seizures, and death.

It is estimated that 6.9 g of methyl salicylate can be a lethal dose. Considering that Ultra Strength BENGAY® Cream contains 30% methyl salicylate, a typical adult dose of aspirin contains 0.3–1.0 g of acetylsalicylic acid, and Pepto-Bismol® contains 0.262 g of bismuth subsalicylate per 15 ml dose, it is easy to see how an overdose is possible (Figure 3.7).

This discussion begs the question of why the toxicity of such a common substance has not been made public. In the early part of the twentieth century, oil of wintergreen was a common means of suicide. It was available in its pure form in any drugstore, and a sip was sufficient to cause death. Currently, the pure form is no longer commercially available, and virtually all discussion of its toxicity has been eliminated in the popular media. Although fewer people now take their own lives by ingesting oil of wintergreen, the number of accidental overdoses has increased because there is little information made available about its potential danger.

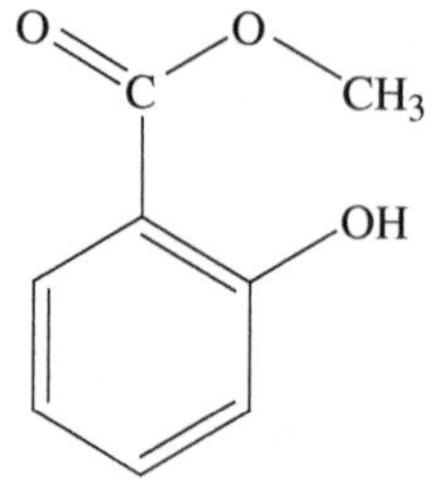

Methyl salicylate
(oil of wintergreen)

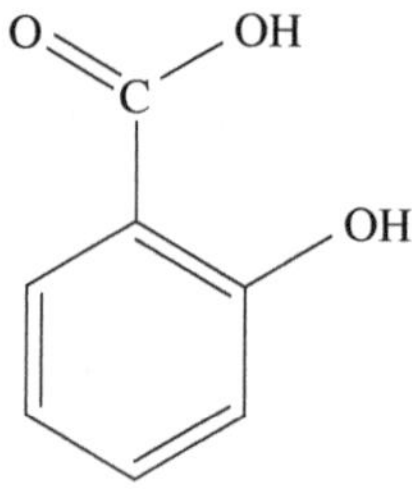

Salicylic acid

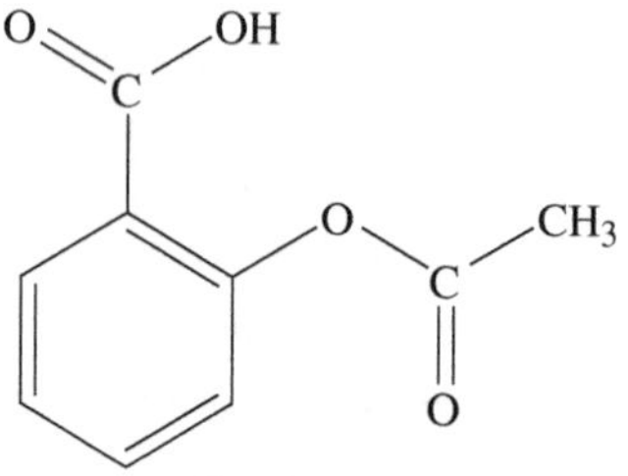

Acetylsalicylic acid
(aspirin)

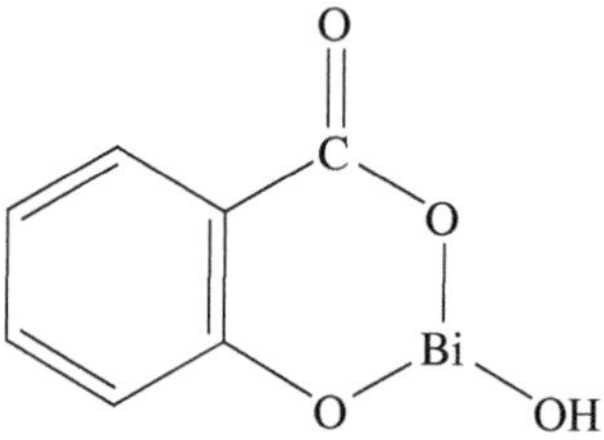

Bismuth subsalicylate

Figure 3.7 The structures of various salicylates.

It Is in 93% of People in the United States

First of all, what *is* bisphenol A? Abbreviated BPA, this organic chemical is one of the highest volume chemicals produced in the world, with over 6 billion pounds made each year, including the reported 2.3 billion pounds produced in the United States. BPA is typically used to create a hard, clear plastic called *polycarbonate* through a reaction with phosgene, and since the 1960s, polycarbonate has been used in countless consumer products such as plastic bottles, lining for food cans, lining for beverage cans, compact discs, digital versatile disc (DVD)s, household electronics, and eyeglass lenses. BPA is also used as a color developer in thermal fax paper and many printed paper receipts (Scheme 3.5).

Because BPA is used in the synthesis of polycarbonates, some BPA remains as a trace contaminant. When polycarbonates are used as a lining for canned food and water bottles, for example, trace amounts of BPA can leach from the containers and be ingested. It has also been shown that handling thermal paper, such as printed receipts, can transfer BPA to the skin, where it is absorbed into

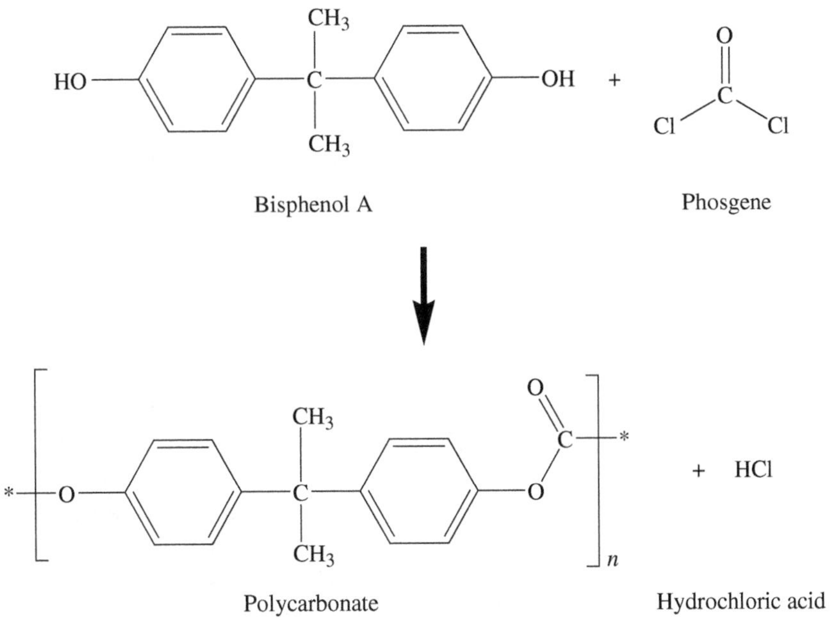

Scheme 3.5 The synthesis of polycarbonate using bisphenol A.

the body. As much as 3% of a receipt's weight can be BPA. In the United States and Canada combined, receipts add 33.5 tons of BPA to the environment each year. The Environmental Protection Agency estimates that more than a million pounds of BPA are released in the environment every year from all sources.

BPA shows up in many unexpected places. Because roughly 30% of thermal paper is recycled, BPA can show up in toilet paper, napkins, and food packaging made from recycled paper. A study has shown that 95% of dollar bills have trace amounts of BPA, presumably by coming into contact with receipts. Traces of BPA can be found in water, soil, food, and sewage. A study conducted in 2003 and 2004 by the Center for Disease Control and Prevention found BPA in the tissues of 93% of people in the United States.

BPA mimics the function of the female estrogen hormone, estradiol, because of similarity in these two compounds' chemical structure. Thus, BPA binds to the same estrogen receptors in the body as estradiol. This is a matter of concern because some animal studies show that BPA affects fetuses and newborns. There are very few studies of possible health effects of BPA exposure in humans. However, there are many published studies regarding the effects of BPA in animals, showing a correlation between elevated levels of this chemical and the incidence of prostate cancer, mammary gland cancer, earlier puberty, body weight increase, and genital malformations (Figure 3.8).

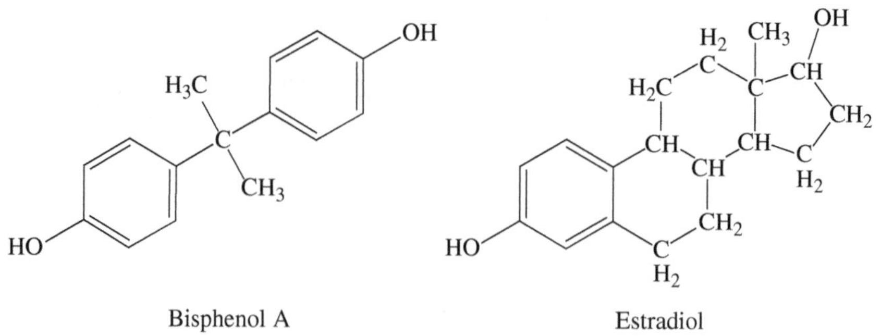

Figure 3.8 A comparison of the structures of bisphenol A and estradiol.

In 2010, Canada became the first country to declare BPA a toxic substance and begin regulating it. Because of its known effects on newborn animals, the European Union, Canada, and the United States have banned its use in baby bottles. In June of 2009, Connecticut became the first state to ban the use of BPA in reusable beverage containers and baby food containers. People who want to limit their exposure to BPA should reduce their use of canned foods and avoid microwaving polycarbonate plastic food containers, handling receipts, and using home coffeemakers that have polycarbonate-based water tanks.

Do you think you are safe if something is labeled "BPA Free"? Think again! After BPA was linked to cancer, plastic manufacturers switched to bisphenol S (BPS). Unfortunately, BPS is just as pervasive in the environment, in paper products, and in the human body. BPS has been shown to speed up embryonic development and disrupt the reproductive system. A 2012 study found BPS in 81% of Americans' urine (Figure 3.9).

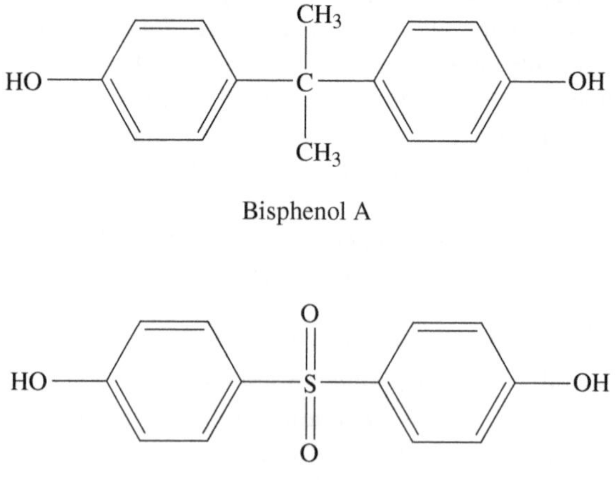

Figure 3.9 A comparison of the structures of bisphenol A and bisphenol S.

The Dreaded…Apricot Pits?

In order to protect themselves against herbivores and insects, some plants contain cyanides, usually bound to sugar molecules and called *cyanogenic glycosides*. Under certain conditions, this inert form of cyanide changes to become a sort of natural pesticide. In the plant, one type of cell contains the cyanogenic glycosides, while stored in a different type of cell is an enzyme that, in the presence of cyanogenic glycosides, causes a reaction producing extremely poisonous hydrogen cyanide (HCN). When plant material is chewed by an insect or animal, the compartments break open, allowing the cyanogenic glycoside and enzyme to mix, resulting in HCN release.

Cyanogenic glycosides can be found in small amounts in seeds from apples, mangos, peaches, and bitter almonds. Another plant that contains cyanogenic glycosides is the cassava, also called *tapioca*. Fortunately, an edible plant that contains cyanides can have the poison removed by crushing the plant and then washing the mash with water. Mashing releases the cyanide, which dissolves in the water and is carried away.

The cyanogenic glycoside found in bitter almonds is called *amygdalin*. When amygdalin decomposes, it becomes benzaldehyde, the sugar glucose, and HCN. One almond can produce as much as 4–9 mg of HCN. It is estimated that 20 bitter almonds will kill an adult. Eating foods that contain HCN may result in some or all of the following symptoms: dizziness, headache, nausea and vomiting, rapid breathing, rapid heart rate, restlessness, and weakness (Scheme 3.6).

Ingestion of higher quantities of food containing HCN may cause other more serious symptoms including convulsions, loss of consciousness, low blood pressure, lung injury, slow heart rate, and death. In 2011, there were 2470 total reported exposures to foods containing cyanogenic glycosides, according to the American Association of Poison Control Center.

Cyanide is probably the most notorious of all chemical poisons. HCN (prussic acid) under the trade name of Zyklon B was believed to have caused more than 1 million deaths in Nazi gas chambers during World War II. In 1978, cyanide along with grape flavoring was used to kill more than 900 people in a mass suicide led by Jim Jones of the Peoples Temple in Guyana. In 1982, seven people died after an unknown person in the Chicago area added cyanide to Tylenol® capsules. In fact, the public outrage over this event led to tamper-resistant packaging for over-the-counter drugs in the United States.

Cyanide interferes with respiration, the process by which our cells use oxygen molecules we take in through breathing, to free up energy from the sugars we eat. Cyanide irreversibly binds to the iron component of an enzyme called *cytochrome c oxidase*, inhibiting its action. Normally, this enzyme would bind to the oxygen in our blood and use it to create energy. However, when cyanide is present, this enzyme binds to the cyanide molecules, preventing it from using

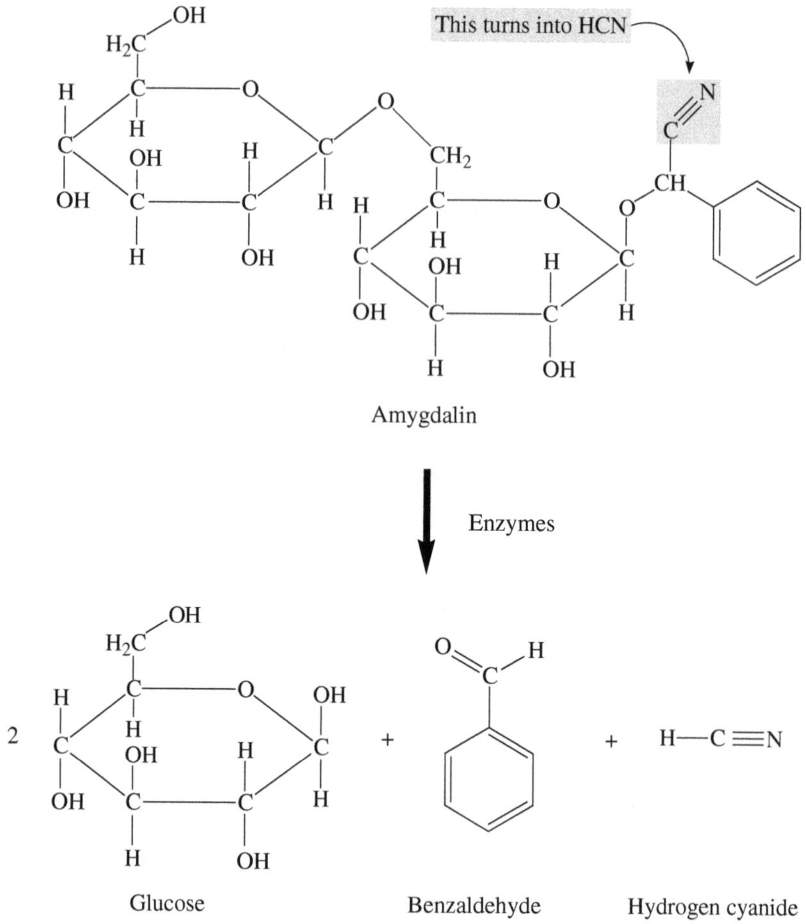

This turns into HCN

Amygdalin

Enzymes

2   Glucose   +   Benzaldehyde   +   H—C≡N

Hydrogen cyanide

Scheme 3.6 The decomposition of amygdalin.

oxygen for the body's needs. The victim suffocates because the heart can no longer beat.

Since oxygen is critical to brain and muscle functioning, death from cyanide poisoning occurs very quickly compared to other poisons, often within minutes. Because the oxygen in the blood is not being used, the skin of victims of cyanide poisoning takes on a characteristic cherry red hue. Plants, on the other hand, contain an alternative pathway for respiration. The enzymes in this alternative pathway are immune to cyanide. It is owing to this that plants are able to generate cyanide as a defense mechanism.

Because cyanide is so fast acting, recovery from an exposure depends on the promptness of antidote administration. One antidote is *hydroxocobalamin*, a vitamin B_{12} precursor called B_{12a}, which binds cyanide to its cobalt moiety to form cyanocobalamin (vitamin B_{12}). Because hydroxocobalamin has a greater affinity for cyanide than cytochrome oxidase, respiration is restored and the patient survives.

From a different perspective, cyanogenic glycosides have been used therapeutically in the treatment of cancer. The cyanogenic glycoside found in bitter almonds, amygdalin, was first isolated in 1830 and used as an anticancer agent in the United States in the 1920s. Unfortunately, amygdalin was found to be too toxic, and work with the compound was stopped.

In the 1950s, a partly synthetic form of amygdalin, called *laetrile*, was patented and used as an anticancer agent. Laetrile is also known as Vitamin B_{17}, even though it is not a vitamin. Laetrile was purported to be a more potent anticancer drug and less toxic than amygdalin. From the 1950s through the 1970s, laetrile grew in popularity in the United States as an alternative treatment for cancer. Despite being distinctly different, amygdalin is also commonly referred to as laetrile. In addition, many products sold as anticancer drugs, such as laetrile, consist mostly of amygdalin. However, amygdalin isolated from apricot pits is the "laetrile," most often used by Mexican cancer clinics, since laetrile is no longer sanctioned for use in the United States (Figure 3.10).

How laetrile works as an anticancer agent is still a matter of conjecture. One popular explanation is that cancer cells contain more of an enzyme that decomposes the laetrile molecule and releases HCN. Subsequently, the cancer cells die from cyanide poisoning. Supposedly, healthy cells do not have as much of this enzyme and instead contain an enzyme that renders the laetrile harmless. Because of this, the healthy cells are not affected by laetrile.

Neither laetrile nor amygdalin has been approved by the FDA. Despite extensive testing, there is no evidence that laetrile is either safe or effective. In fact,

Figure 3.10 The structure of laetrile.

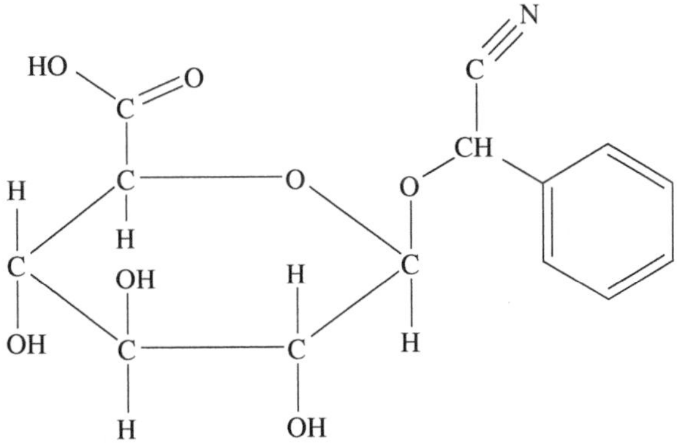

Laetrile

many patients who took laetrile died sooner than expected. Because of the risk of cyanide poisoning, the US government has put penalties in place for the sale of laetrile and banned its transport into the United States or across state lines. In fact, since 2000, several people have been prosecuted for transporting laetrile across state lines. Nonetheless, products advertised as laetrile and amygdalin can still be purchased online. In addition, as mentioned previously, some cancer treatment centers in Mexico offer laetrile as an alternative cancer treatment.

The other molecule released by the breakdown of amygdalin, *benzaldehyde*, is the main molecule responsible for the smell and taste of almonds. "Pure" almond flavoring is made from extracts of the almond nut and distilled to be free of any traces of cyanide. Benzaldehyde is also found in other plants, which are used as alternate, cheaper sources. The so-called natural almond extract is usually flavored with benzaldehyde made from apricot pits, peach pits, or cassia, a relative of cinnamon. Note! Just because a product says it uses "natural" flavors does not mean that the flavor came from that particular source; it just means that it was created from natural ingredients. You may be surprised to find out that Honey Nut Cheerios® actually does not contain any nuts. After 2006, the natural almond flavoring used in the product was made from apricot or peach pits. Even McCormick® almond extract is made from apricot pits. "Imitation" almond extract uses synthetic benzaldehyde made from petrochemical base materials (Scheme 3.7).

Pure benzaldehyde and pure bitter almond oil are both monitored by the Drug Enforcement Administration (DEA) because they can be used in the manufacture of methamphetamine. Using a reaction called a Knoevenagel condensation, benzaldehyde is condensed with nitroethane ($CH_3CH_2NO_2$). The product is then reduced with lithium aluminum hydride to create methamphetamine (Scheme 3.8).

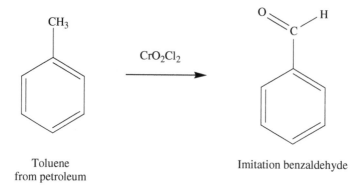

Scheme 3.7 The synthesis of imitation benzaldehyde.

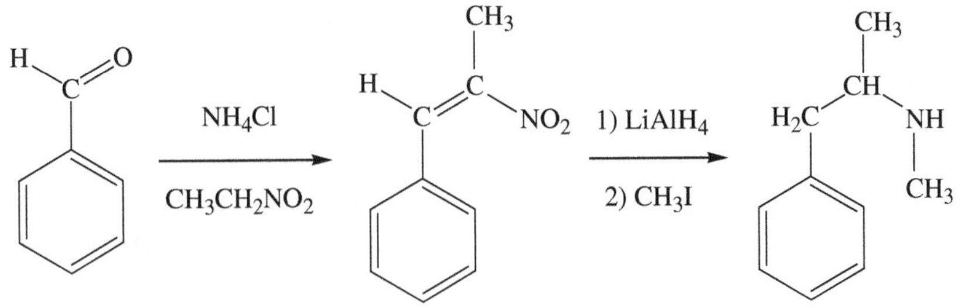

Benzaldehyde Methamphetamine

Scheme 3.8 The synthesis of methamphetamine using benzaldehyde.

Honey Intoxication

Rhododendrons and other plant members of the family *Ericaceae* contain a group of toxic compounds called *grayanotoxins.* When bees use the nectar from these plants to make honey, the grayanotoxins are transferred to create what is called "mad honey." Eating this honey causes rhododendron poisoning or honey intoxication. The effect of mad honey is dependent on the amount consumed, with between 5 and 30 g causing poisoning. The grayanotoxins are neurotoxins that interfere with the transmission of nerve signals and, in very small doses, can induce a hallucinogenic state. However, larger amounts can produce loss of coordination, muscle weakness, slowed heartbeat, low blood pressure, and even death. This poisoning has been associated with *Rhododendron luteum* and *Rhododendron ponticum,* normally found around the Black Sea region of Turkey. Nevertheless, cases occasionally show up in the Pacific Northwest of the United States and in other parts of the world (Figure 3.11).

Accounts of mad honey intoxication were recorded as far back as 401 BC by the Greek warrior/writer Xenophon. In fact, mad honey was used as a weapon

Figure 3.11 The structure of a grayanotoxin.

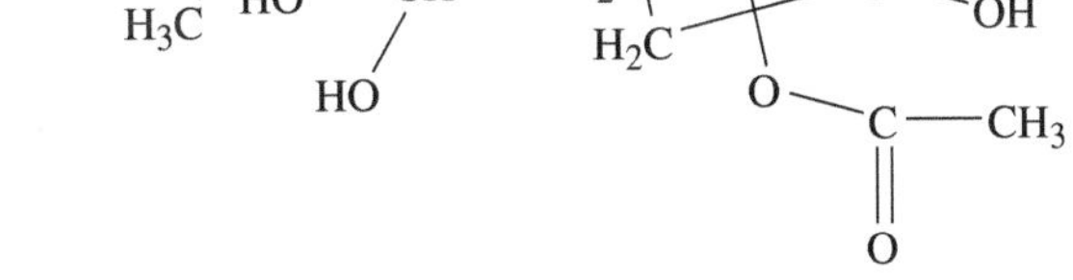

A grayanotoxin

by King Mithradates IV against Pompey the Great in 67 BC. Mithradates IV retreated and left behind mad honey–containing honeycombs in the path of the advancing Roman troops. After consuming the honey, the Roman troops became incapacitated and were easily overcome.

While most cases of mad honey ingestion are believed to be accidental, deliberate intoxication may originate from its use as an alternative medicine for the treatment of ulcers, hypertension, arthritis, and diabetes. In addition, mad honey is used as an aphrodisiac and to intensify the effect of alcoholic drinks.

The toxin from the rhododendron was used as a plot device in the 2009 film *Sherlock Holmes*. In the movie, Holmes suggested the villain, Lord Blackwood, used hydrated *R. ponticum* to fake his death, which some symptoms of consuming grayanotoxins might. Holmes says, "There is a toxin refined from the nectar of *Rhododendron ponticum*. It's quite infamous in the region of Turkey, bordering the Black Sea, for its ability to induce an apparently mortal paralysis. Enough to mislead a medical mind even as tenacious and well trained as your own. It's known locally as... mad honey's disease." Sherlock illustrates his point by using the poison to make his dog, Gladstone, pass out.

Figure 3.12 A bear feeling the effects of honey intoxication.

In addition to the grayanotoxins from rhododendrons, there have been many reports of honey becoming toxic when bees produce it using the nectar of poisonous plants. Honey collected from hives near opium poppy cultivation has been reported to have narcotic properties. A French beekeeper trained his bees to make honey from the marijuana plant. The "cannahoney" reportedly combines the characteristics of cannabis and honey, as if you needed any more reason to save the bees. On the poisonous side, honey made from the nectar of *Gelsemium sempervirens*, commonly called *yellow jessamine* or *jasmine*, contains highly toxic strychnine-related alkaloids, gelsemine and gelseminine.

In March of 2008, there were 22 reported cases of severe illness and hospitalization when people consumed honey containing the toxic molecule tutin. This toxic honey is produced in New Zealand when honeybees feed from the honeydew produced by an insect called the passionvine hopper (*Scolypopa australis*) when this pest feeds on tutu (*Coriaria arborea*), a plant containing the poisonous tutin molecule. The tutin from the plant is transferred to the passionvine hopper and subsequently to the honey by the bees (Figure 3.12).

The DMSO Patient

On February 19, 1994, 31-year-old Gloria Ramirez was brought in the emergency room of Riverside General Hospital in Riverside, California. She had cervical cancer. Her heart was beating too rapidly and her blood pressure was plummeting. After failing to respond to treatment with drugs, the staff tried to defibrillate her heart with electricity. After removing her shirt, several people saw an oily sheen covering her body, and some noticed a garlicky odor. A medical resident noticed unusual manila-colored particles floating in a sample of Ramirez's blood. At this point, things started happening. A nurse in the emergency room passed out, saying later that her face felt like it was burning. Then a medical resident began feeling queasy. Complaining that she was light-headed, she left the trauma room, and ended up passing out. She shook intermittently and would stop breathing for several seconds, take a few breaths, then stop breathing again. Back in the emergency room, a respiratory therapist passed out, and when she woke up, she said she could not control the movements of her limbs. In total, 23 of the 37 emergency room staff members in the hospital that day experienced at least one symptom.

Finding the cause of this siege of symptoms began one of the most extensive investigations in forensic history. Dozens of possible reasons were investigated, including mass hysteria, chemicals from a methamphetamine laboratory, and even poison sewer gas.

The most widely accepted explanation was that Ramirez had rubbed *dimethylsulfoxide* (DMSO) on herself to relieve the pain from her cervical cancer. DMSO is an organic solvent, and although it is not approved for human

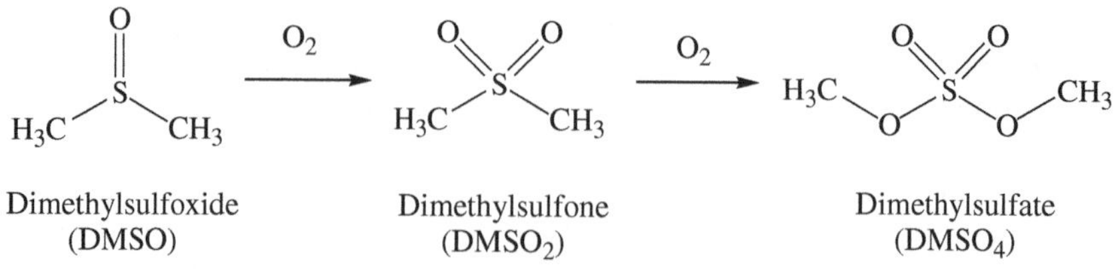

Scheme 3.9 The conversion of DMSO to DMSO₄.

use, it is popular as a painkiller and inflammation fighter. It was theorized that the DMSO in Ramirez's system might have built up due to urinary blockage. Combined with oxygen administered by the paramedics, the DMSO would form dimethyl sulfone ($DMSO_2$). After this, an undetermined process – one possibility is the electric shocks administered during defibrillation – may have then converted the $DMSO_2$ into *dimethyl sulfate* ($DMSO_4$), a poisonous gas. Because $DMSO_4$ is volatile, it would have evaporated without a trace (Scheme 3.9).

The latter explanation made sense, considering that most of the reported symptoms are commonly associated with dimethyl sulfate exposure, and Ramirez's blood and tissues were shown to have a high concentration of $DMSO_2$. Furthermore, the presence of DMSO on her body would account for the oily sheen and garlicky odor observed by the staff, and when blood containing $DMSO_2$ is drawn out in the syringe and cools to room temperature, crystals form like the ones witnessed in Ramirez's blood sample.

$DMSO_4$ was first considered as a potential chemical warfare agent by the Germans during World War I. In 1987, a declassified Department of Defense document reported that a 10-minute exposure to half a gram of dimethyl sulfate dispersed in a cubic meter of air can kill a person. Symptoms of exposure of $DMSO_4$ are irritation of the eyes, dizziness, skin burns, breathing difficulty, vomiting, and delirium much like those suffered by the emergency room workers at Riverside General Hospital.

This event was dramatized in the episode of the television show *Grey's Anatomy* titled "'Wishin' and 'Hopin'.'" In the episode, a cancer patient is operated on in the hospital and found to have toxic blood. After several hospital staff fall ill, the operation is completed by staff using respirators and later by holding their breath. The doctors speculate that an herbal remedy along with the chemotherapy drugs the patient had been taking combined to create a nerve gas.

Deadly Helium Balloons

On February 22, 2012, a girl died after inhaling helium from a pressurized tank. She was at a party where people were inhaling helium to make their voices high pitched. The cause of death? A helium gas bubble formed in her bloodstream, was trapped in a blood vessel in her brain, and caused a stroke.

This is the same thing that happens when a SCUBA diver surfaces too quickly and gets "the bends," otherwise known as *decompression illness*. It is common knowledge that gasses can be dissolved in liquids. Fish survive underwater by virtue of respiratory systems specialized to extract the oxygen gas dissolved in water. How much gas is dissolved follows what is called *Henry's law*, which says that the amount of dissolved gas is directly proportional to the pressure of the gas. This means that at higher gas pressures, more gas is dissolved. Divers' pressured air tanks cause more nitrogen to be dissolved in their blood. If the pressure is released too quickly, the nitrogen becomes less soluble in blood and comes out as bubbles. This is much like how carbon dioxide bubbles form when a bottle of soda is opened. The bottles are under a small amount of pressure. Opening the bottle releases the pressure, and the carbon dioxide fizzes due to the reduced solubility. To combat the bends, divers slowly relieve the pressure, and the excess nitrogen does not form bubbles but is harmlessly released in their breath.

In the case of inhaling helium from a balloon, the increase in helium pressure is negligible, so normally there is no possibility of gas bubbles forming. The girl who died, however, inhaled helium directly from a pressurized tank. This should never be done because these tanks are pressurized to as much as 1500 psi, which is almost 50 times the pressure of a car tire.

Helium can also be lethal as an *asphyxiant*, a material causing severely deficient oxygen supply to the body. There are media reports of people breathing in too much helium from a balloon, passing out, and being in danger of dying of *hypoxia* – lack of sufficient oxygen. Unfortunately, helium's availability and properties as an asphyxiant have made it a popular method of suicide. There are few statistics on deaths from helium inhalation, since the federal government does not collect them. Florida is one of the few states that track helium-caused deaths. In 2010, nine Floridians died from helium inhalation. In 2011, 37 people in the United Kingdom died the same way.

The 2007 Pet Food Recall

One of the largest pet food recalls in the US history occurred in 2007, after numerous pet owners reported their animals becoming ill and/or dying after eating a certain products. It was discovered that the pet food contained an organic compound called *melamine*, which becomes poison when combined with another compound inside the body. Over 60 million units of pet food – about 1% of all pet foods in the United States, spanning 100 brand names and involving retailers across the nation – were recalled. The problem was traced back to one pet food supplier that had received wheat gluten with melamine added to it. Wheat gluten is used to thicken and enrich the gravy in many pet foods. Some of the tainted wheat gluten had recorded melamine concentrations as high as 6.6%, and melamine crystals could literally be seen in wheat gluten samples. Sadly, more than 100 deaths and nearly 500 cases of kidney failure in pets were reported. It is suspected that the actual number of animals affected was much higher, considering the unreported cases.

An even worse incident involving melamine contamination occurred in China the following year, 2008. After being fed milk powder that had been tainted with melamine, 300,000 babies were sickened, and 6 were believed to have died because of the contamination. Because milk is used in many other consumable products, melamine was soon discovered in liquid milk, yogurt, and chocolates, although not in large enough amounts to cause significant harm. Two men were sentenced to death for producing and selling the contaminated food. The head of Sanlu, the milk company at the center of the scandal, was sentenced to life in prison after admitting that the company had known about the problem months before alerting the government.

In both cases, it is believed that the melamine was added to increase the apparent amount of protein in the product in the least costly way. Nitrogen content is commonly used to estimate protein content because proteins contain nitrogen, while most other food constituents do not. Because of melamine's high nitrogen content (66% by mass vs 10–12% for typical protein), this ingredient can be added to increase the nitrogen content of a sample. In one case, 800 tons of artificial "milk" powder were made by only mixing melamine and malt starch.

Melamine can be used to create plastics such as Formica® laminate; however, it has not been approved for human or animal food in the United States. The toxic dose of melamine is on a par with common table salt; an estimated half-pound of melamine is toxic to a 150-pound human. Despite melamine's seemingly nontoxic nature, it becomes deadly when administered with *cyanuric acid*, a molecule classified as "essentially" nontoxic. Insoluble crystals result when both melamine and cyanuric acid are present, leading to the formation of solids similar to kidney stones, which can cause kidney failure and death. Additional signs of melamine poisoning may include irritability, blood in urine, little urine, and high blood pressure (Figure 3.13).

We are constantly exposed to cyanuric acid because the FDA permits a certain amount of cyanuric acid in some nonprotein nitrogen additives used in animal feed. In addition, cyanuric acid is a common ingredient in swimming pools, to protect free chlorine from being destroyed by the sun's UV rays. Moreover, melamine itself has been shown to convert to cyanuric acid in mammals and bacteria. If humans or pets are exposed to a source of melamine, they will also be exposed to cyanuric acid.

Melamine and cyanuric acid form insoluble crystals due to a strong interaction between the two molecules, called *hydrogen bonding*. The hydrogen atoms of melamine interact with the oxygen of cyanuric acid, and the hydrogen atom of cyanuric acid interacts with the ring nitrogen of melamine. This strong

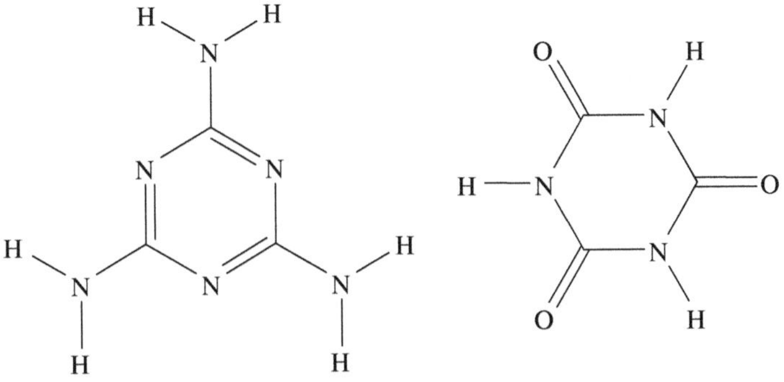

Melamine   Cyanuric acid

Figure 3.13 The structure of melamine and cyanuric acid.

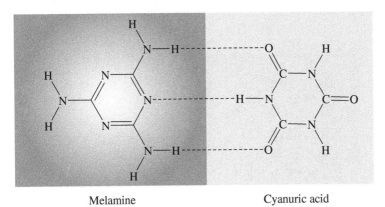

Melamine   Cyanuric acid

Figure 3.14 The interaction of melamine and cyanuric acid. The hydrogen bonds are shown as dashed lines.

reaction prevents the two molecules from being soluble in water. If sufficient quantity of these two molecules is present in the kidneys, they will interact and form a solid that can be lethal to humans (Figure 3.14).

The interaction between melamine and cyanuric acid is very similar to the reaction between DNA base pairs, wherein the strong hydrogen bonding allows for specific base pairs, such as cytosine and guanine, to become coupled with each other (Figure 3.15).

In the United States, melamine-formaldehyde resin is approved for use in the manufacture of some cooking utensils, as well as tableware. In forming these products, a small amount of the melamine does not convert to the plastic and remains. It is possible for the melamine remaining on the plastic to leach out of the plastic into food that comes into contact with the tableware. This is especially true when highly acidic foods, such as tomato sauce, are heated to 160 °F or higher. Because of this, foods and drinks should not be heated on melamine-containing tableware in microwave ovens (Figure 3.16).

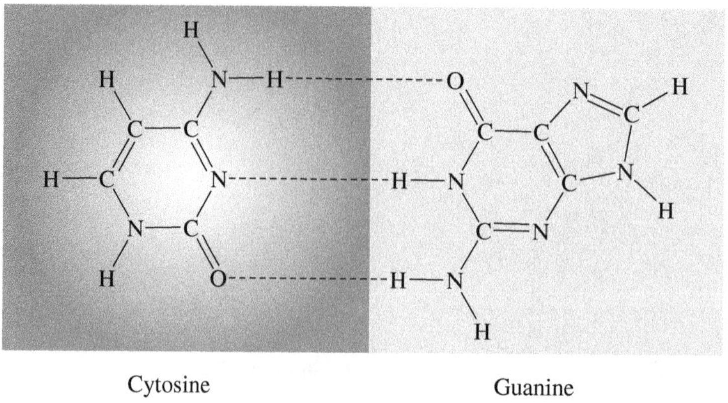

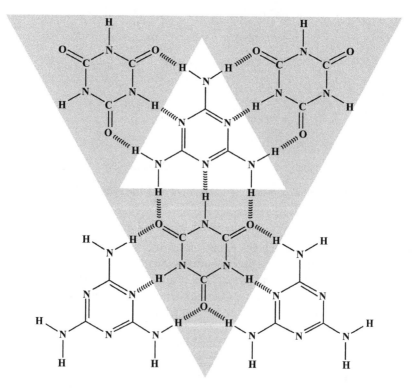

Figure 3.15 The interaction of cytosine and guanine. The hydrogen bonds are shown as dashed lines.

Figure 3.16 How melamine and cyanuric acid can combine to form a solid that causes kidney damage. The hydrogen bonds are shown as dashed lines.

Mercury in Vaccines and Eye Drops?

Thimerosal, also known as ethyl mercury thiosalicylic acid, or its Eli Lilly and Company trade name Merthiolate, is a preservative that has been used since the 1930s in vaccines, antivenins, eye drops, and nasal sprays. In fact, it is the most widely used preservative for vaccines. Having thimerosal or some other preservative in these products is necessary to help prevent potentially life-threatening contamination by microbes. There are several cases that highlight the necessity of preservatives in vaccines. In Queensland, Australia, in 1928, a diphtheria vaccine contaminated with staphylococci caused the death of 12 of the 21 children inoculated with it (Figure 3.17).

Thimerosal is effective at very low concentrations and normally does not reduce the potency of a vaccine. Typically, vaccines contain thimerosal in concentrations of 0.001–0.01%. This means a vaccine containing 0.01% thimerosal contains approximately 25 µg of mercury per 0.5 ml dose. However, doses of thimerosal ranging from approximately 3 mg/kg to several hundred mg/kg have been found to be dangerous. Several incidents of acute mercury poisoning from thimerosal-containing products have occurred. In one case, a formulation was accidently made with 1000 times the safe dose of thimerosal. Another case was a suicide attempt.

Thimerosal use as a vaccine preservative is controversial and is at the heart of the notion that vaccines are linked to autism. Under the FDA Modernization Act (FDAMA) of 1997, the FDA conducted a comprehensive review of the use of thimerosal in childhood vaccines and found no evidence of harm other than local allergic reactions. Despite these conclusions, thimerosal is being phased out from routine childhood vaccines in the United States and the European Union as a precautionary measure. The World Health Organization still advocates continued use of thimerosal-containing vaccines in developing countries because of their effectiveness and low cost. Vaccines that contain thimerosal include those against diphtheria, tetanus, and pertussis (DTP), hepatitis B, *Hemophilus influenzae* type b (Hib), rabies, influenza, and meningococcal diseases.

Figure 3.17 The structure of thiomersal.

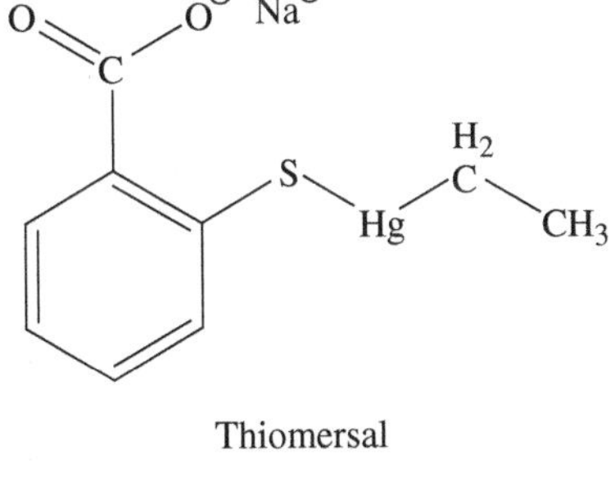

Thiomersal

In addition to vaccines, thimerosal is used in many commercial products such as antibiotics for the eye (Cortisporin® Ophthalmic Suspension and Neosporin® Ophthalmic Solution), nasal preparations (Nasal Moist® AF and Neo-Synephrine®), ear preparations (Coly-mycin® S Otic and Cortisporin® Otic Suspension), and cosmetics (L'Oreal® Miracle Wear Mascara and Stagelight® Mascara).

The World's Deadliest Frog

The golden poison dart frog (*Phyllobates terribilis*) is considered the most toxic vertebrate on earth. Measuring just 2 inches long, this frog has sufficient venom to kill *10* men. The golden poison dart frog inhabits the Pacific Coast of Columbia, South America, where the Emberá people indigenous to this region have used it to tip their blowgun darts when hunting; hence the species' name (Figure 3.18).

The toxic molecule in the frog is called *batrachotoxin*, which comes from the Greek words *batrachos*, meaning "frog," and *toxine*, meaning "poison." When threatened, the frogs release batrachotoxin as milky secretions from secretory canals and glands located on their back and behind their ears. Only 100 μg, the equivalent weight of two grains of table salt, can kill a 150-pound person. This

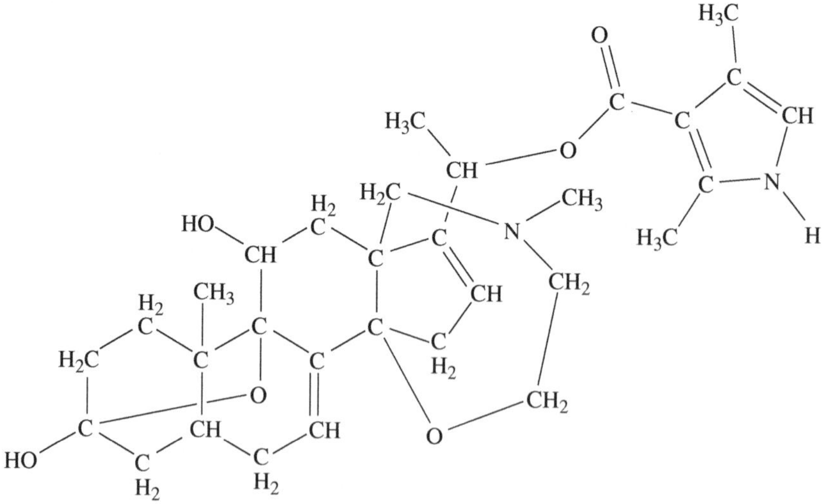

Batrachotoxin
(a complex molecule)

Figure 3.18 The structure of batrachotoxin.

makes batrachotoxin more than 1000 times more poisonous than cyanide and more than 15 times more potent than curare, a plant-derived arrow poison also used by South American Indians.

Batrachotoxin is a neurotoxin that acts on the peripheral nervous system and kills by blocking nerve signal transmission to the muscles, resulting in paralysis. Simply stated, batrachotoxin irreversibly locks neurons into the "on" position and does not allow them to refire. This is similar to hampering the reloading mechanism of a gun. Once you fire the gun, it cannot fire again. Because the neuron is rendered incapable of carrying commands to the muscles, paralysis results. Upon exposure to batrachotoxin, death is almost inevitable. The toxin has a particularly severe effect on the heart, where it causes arrhythmias, cardiac failure, and death in less than 10 minutes. The golden poison dart frog itself is virtually immune to its own batrachotoxin, probably due to minor changes in the structure of its neurons which prevent the toxin from binding.

Oddly, scientists are unsure of the source of this frog's amazing toxicity. Arrow frogs are not poisonous in captivity, so scientists believe that batrachotoxin originates from the frog's natural diet. The current theory is that the frogs produce their poison from ingesting insects such as the melyrid beetle, also native to Colombia. Batrachotoxin has been detected in these insects. Researchers conjecture that the arrow frogs consume large numbers of these insects and accumulate batrachotoxin as a defensive mechanism. Coincidentally, it has been discovered that some bird species in New Guinea also contain the batrachotoxin on their skin and feathers. Because these birds also feed on the melyrid beetle, it is assumed that these birds accumulate the toxin in the same manner as the arrow frog.

Taking the oddness a step further, another theory is that these beetles consumed by the frogs acquire the poison from the plants they eat. This would mean that the poison is transferred from a plant, to the beetle, and finally to the frog. Biologists agree that it would be extremely unlikely for the beetle to be able to biosynthesize a complex molecule like batrachotoxin, and they continue to investigate the possible sources of the production of batrachotoxin in beetles.

Leaded Candy

In 2003, there was a sudden rise in lead-poisoned children in Orange County, California. Upon investigation, California authorities estimated that as many as 15% of children in California (about 1000 per year) who suffered lead poisoning had eaten candy from Mexico. Further examination showed that imported products – especially those containing tamarind, chili powder, or mined salt – tended to show high levels of lead contamination. Most of these candies were from Mexico, although some were from Malaysia, China, and India.

It seems that chilies in the candy were the culprit. A report by the Orange County Register showed that more than 90% of the chili powder samples from Mexico tested high for lead. An estimated 33% of Mexican candy sold in the United States contains chilies. One chili-based candy imported from Mexico, called Chaca-Chaca, was shown to have as much as 0.15 ppm (parts per million) of lead, while levels in excess of 0.10 ppm are known to be unsafe.

How do chilies become contaminated by lead? Investigators found that the chilies started out safe, but by the time chili powder reached the market, it had become tainted with lead. The conclusion drawn was that the lead was being introduced into the chilies through improper handling and processing. It was finally determined that chilies, destined for local markets that make Mexican candy, were not washed, individually inspected, or packed into clean boxes. The dirt clinging to the chilies was determined to be the probable source of the contamination.

Lead exposure is especially dangerous to children because it can harm a child's developing nervous system and brain. This leads to difficulties in learning and performing in school, hyperactivity, irritability, reduced attention span, poor appetite, constipation, and stomach pain. Because children's brains and nervous systems are still developing, lead poison can have long-term effects such as lower intelligence, behavior problems, hyperactivity, impaired speech, slowed growth, and hearing damage.

In adults, high levels of lead in the blood lead can cause kidney damage, nerve damage, seizures, and death to any human. Lead is also especially dangerous to pregnant women because exposure can cause premature delivery and poor fetus growth.

After the candy problem, there was concern about the possibility of lead in chilies and chili pepper products sold in the United States, since Mexico is the source of nearly 98% (~388 million pounds in 2003) of imported fresh peppers. Fortunately, this issue was investigated and the chilies destined for export from Mexico were found to be properly cleaned and packaged, with minimal lead contamination.

Why not Drink "Real" Root Beer?

Root beer was first introduced in 1876 by pharmacist Charles Hires, and was originally made from an extract of a sassafras plant. However, in 1960, sassafras extract was banned by the FDA because tests showed it to be a carcinogen, owing to the presence of a molecule called *safrole*. The FDA and European Commission on Health determined safrole to cause cancer in tests on animals.

After this ban, root beer makers began experimenting with recipes that did not use sassafras extract. It was discovered that the carcinogenic parts of the sassafras could be removed and the rest used to flavor root beer. Therefore, modern root beers contain the same flavoring as the original, except for safrole.

Safrole is also banned for use in soap and perfumes by the International Fragrance Association. The compound has been used for the illicit production of the drug ecstasy and been designated a list I chemical by the US DEA. Ecstasy is sometimes cut with safrole, which can make it carcinogenic.

The carcinogenic effects of safrole come from the biotransformation that occurs with the cytochrome P450 enzyme found in the human liver. This enzyme changes molecules as part of the process by which the body excretes wastes. Typically, this enzyme changes the molecule to increase its water solubility, so it becomes easier to remove via the urine. In the case of safrole, the P450 enzyme changes it into a carcinogenic compound, 1′-hydroxy-safrole. This carcinogen is subsequently converted to a highly reactive sulfonate species by reactions with another liver enzyme called *sulfotransferase*. The reactive sulfonate formed can react with DNA in the liver and become a chemically bound compound called a *DNA adduct*. Once the DNA becomes changed, it may replicate improperly, leading to cancer (Figure 3.19 and Schemes 3.10, 3.11).

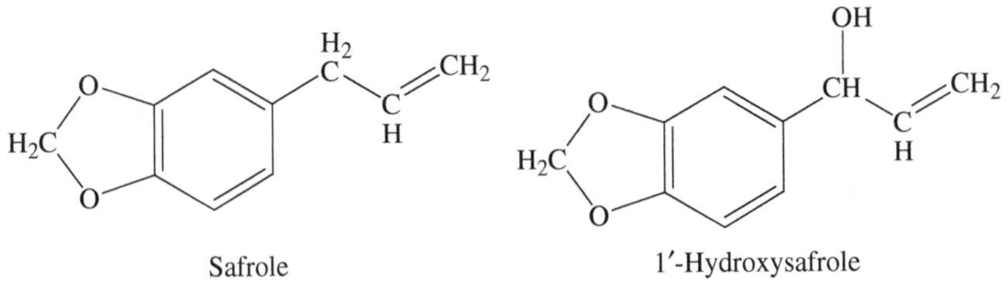

Safrole   1′-Hydroxysafrole

Figure 3.19 The structures of safrole and 1′-hydroxysafrole.

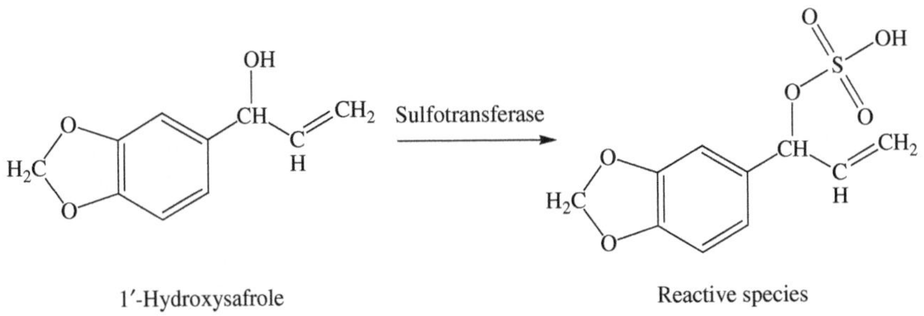

1′-Hydroxysafrole   Reactive species

Scheme 3.10 The conversion of 1′-hydroxysafrole to a reactive species.

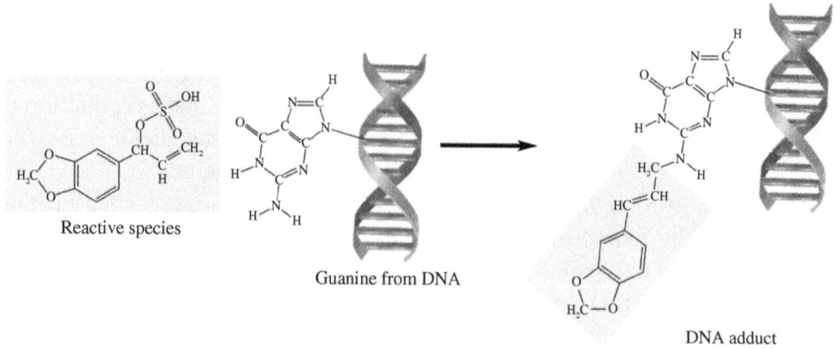

Reactive species

Guanine from DNA

DNA adduct

Scheme 3.11 The formation of a DNA adduct.

The Killer Fog

On the afternoon of December 4, 1952, a thick black fog descended on London, England. The sun was blotted out, and everyone in the city walked around in the dark for the next 4 days. The smog dispersed after a change of weather on December 9, but by then the toxic mixture of dense fog and coal smoke had suffocated an estimated 12,000 Londoners.

This horrible event came about because of a toxic combination of weather and pollution. What formed over London was an *anticyclone*, a circulation of winds around a region of high atmospheric pressure, rotating clockwise in the Northern Hemisphere and counterclockwise in the Southern Hemisphere. The anticyclone caused a temperature inversion – a thin layer of the atmosphere where the normal decrease in temperature with height switches to the temperature increasing with height. Cold, stagnant air becomes trapped under an upper layer of warm air. On that tragic date in London, due to the lack of circulation, the lower, cooler air mixed with pollutants such as chimney smoke and vehicle exhausts, forming a thick fog.

During this time period, the coal being used was low grade, containing large amounts of sulfur. When the coal of this quality is burned, the sulfur is oxidized to form sulfur dioxide, which is released to the air. At this point, the sulfur dioxide can react with water and other substances in the atmosphere to form sulfuric acid. This acid, along with soot particulates in the fog, caused lung inflammation and ultimately suffocation of the exposed Londoners.

This calamity led to several improvements in environmental regulations, including the Clean Air Act of 1956. This Act introduced measures to reduce air pollution, including relocating power stations away from cities, making some chimneys taller, and shifting home heating toward the use of cleaner coals, electricity, and gas.

In the United States, a temperature inversion occurred on October 27, 1948, over Donora, Pennsylvania, trapping a thick smog that killed 20 people and made up to half of the town's population of 14,000 sick. The *New York Times* described the event as "one of the worst air pollution disasters in the nation's history."

Nail Polish or Nail Poison?

Continual exposure to certain chemicals found in nail polish has been found to have deleterious effects on the health of manicurists. In fact, manicurists, hairdressers, and makeup artists experience disproportionately high rates of Hodgkin's disease, low birth-weight babies, and multiple myeloma. What chemicals are related to these problems? And why are they in nail polishes? If they can cause health problems, why haven't they been banned?.

The three chemicals in nail polish that are of particular concern are toluene, dibutyl phthalate (DBP), and formaldehyde. Nail polish is a polymer resin dissolved in solvents, along with color pigments. After the solvent evaporates, a hard polymer film that sticks to the nail's surface remains. This is why nail polish has to be allowed to dry. Toluene, found in paints and paint thinners, is one of the solvents used in nail polish formulation. The Environmental Protection Agency claims that overexposure to toluene can cause dizziness and impaired kidney function (Figure 3.20).

In addition, DBP is added to some nail polishes as a plasticizer. Plasticizers are small molecules that embed themselves in the polymer matrix of the hardened nail polish. This inhibits the ability of the polymer molecules to stick to each other, making the nail polish softer, less brittle, and less likely to chip. Plasticizers are used in most plastic products, so exposure to plasticizers is of great concern. In California, DBP is listed as a suspected teratogen (any agent or factor that affects developing fetuses) and has been banned from children's toys in concentrations of 1000 ppm or greater. Lastly, nail polishes can contain

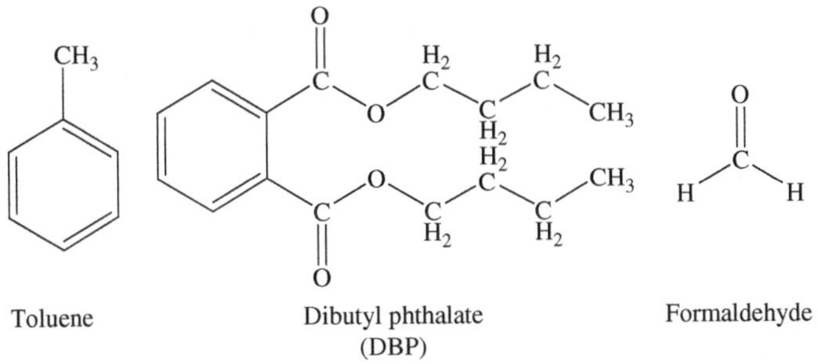

Figure 3.20 The toxic components of nail polish.

Figure 3.21 Nail polish or nail poison?

formaldehyde as a preservative to kill microorganisms. Best known for its use as embalming fluid, formaldehyde was identified as a human carcinogen in 2011 by the National Toxicology Program.

Why do so many cosmetics contain potentially harmful chemicals? The truth is that the regulatory power of the FDA is limited when it comes to cosmetic products. Many cosmetic products do not need FDA premarket approval, and cosmetic companies are not required to prove the safety of products or ingredients. In 2005, California lawmakers tried to ban DBP from cosmetic products; however, the bill failed after cosmetic companies and other stakeholders lobbied heavily against it.

Occasional limited exposure to the chemicals in nail polish is safe. However, with constant and heavy exposure, health problems may develop. In the face of these growing concerns, many nail polish companies have removed some chemicals from their products. These products are typically labeled "3-free" or "5-free," depending on the number of chemicals removed from the formulation (Figure 3.21).

Game Board Danger

Urushiol is the generic name for a class of oily organic allergens found in plants belonging to the family Anacardiaceae, which includes the notorious species

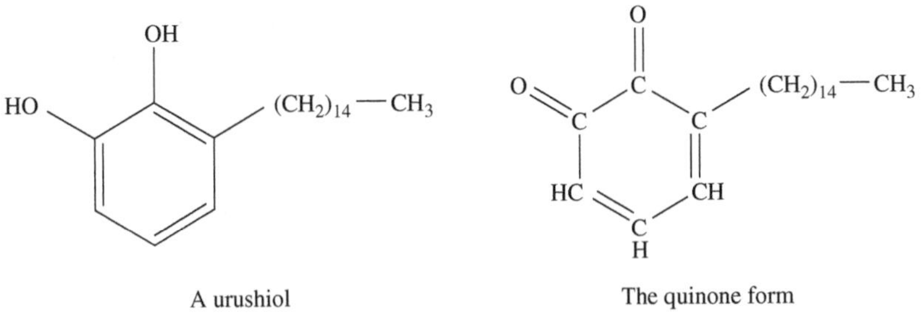

Figure 3.22 The structure of an urushiol and its quinone form.

poison oak, poison ivy, and poison sumac. The urushiol can be seen on these plants with the naked eye, appearing as a shiny substance on the surface of the leaves. Approximately 50–75% of the population will develop an allergic reaction when exposed to urushiol-containing plants. It can take as little as 50 μg of urushiol to cause urushiol-induced contact dermatitis, a condition characterized by a weeping, itchy rash. Up to 50 million Americans are affected every year, with the allergic rash causing more than 7.1 million doctor visits (Figure 3.22).

The problems start when urushiols are oxidized by air to a quinone form of the molecule, which chemically react with cells and proteins on the exposed skin. The resulting changes in the appearance of these proteins and cells fool our immune system into thinking that a foreign body is present. T-cells are constantly on the lookout for invaders such as germs. Once our skin cells are altered by urushiol, our bodies mount a T-cell-mediated immune response, attacking the cells by calling in an army of white blood cells. The white blood cells attack everything in the vicinity, leading to severe tissue damage, blisters, and the rash (Scheme 3.12).

The name "urushiol" is derived from the Japanese word for the Urushi lacquer tree, *Toxicodendron vernicifluum*. The composition of the raw Urushi sap from Japan is about 65% urushiol and has been used as a lacquer to create a durable coating material in Asian countries for thousands of years. In fact, "japanning" is an old name for applying several coats of this sap/varnish,

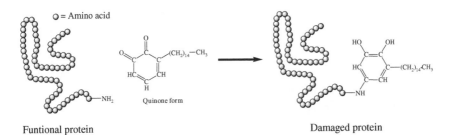

Scheme 3.12 The reaction of the quinone form of an urushiol with a protein.

each of which is subsequently dried and polished. After application of the lacquer, a copper-containing protein called *laccase* causes the oxidation and subsequent polymerization of the urushiol in the sap to form a hard lacquer. Because the main component of the lacquer is urushiol, contact with lacquer sap and lacquerware causes contact dermatitis in lacquer workers and others in the lacquerware industry. In addition, there have been reports of people developing contact dermatitis after being exposed to urushiol-lacquered items that were not completely cured, such as wooden vases, wooden pens, and game boards.

In addition to poison oak and poison ivy, other plants of the Anacardiaceae family contain detectable amounts of urushiol. The leaves of the mango tree and skin of the mango fruit both contain urushiol. Some people, and especially those who have had prior exposure to poison oak or ivy, can develop allergic contact dermatitis after peeling and eating mangoes. Urushiol is also present on the leaves of the cashew tree as well as in the raw cashew shells. The urushiol is found in the shell oil but not in the nuts. Handling cashew shells or eating a cashew nut with shell oil on it can cause an allergic reaction, which is why cashews are not sold raw. Even cashews labeled as "unroasted" are steamed to remove the urushiol.

As if global warming itself was not bad enough, the poison ivy vine grows faster and bigger as carbon dioxide levels in the air increase. When poison ivy was exposed to the concentration of carbon dioxide that has been predicted for 2050, it grew 150% faster, three times larger, and produced not only more urushiol, but a more allergenic form of this molecule. Furthermore, although both trees and poison ivy grew faster when exposed to higher concentrations of CO_2, the poison ivy grew seven times faster than a tree did. Poison ivy, like other woody vines, has adaptively evolved to take advantage of larger amounts of CO_2. Scientists predict that in time, woody vines will smother saplings to create the forests of the future.

What Molecule Killed "Weird Al" Yankovic's Parents?

On April 9, 2004, the elderly parents of the parody recording artist Al Yankovic were found dead in their Fallbrook, California home, apparently from carbon monoxide (CO) poisoning. They had lit a fire in their fireplace but had not opened the flue. In 2003, the stepdaughter of country singer Johnny Cash – Rozanna Nix-Adams – died from CO leaking from six propane space heaters that were being used in a poorly ventilated bus. In 1994, tennis player Vitas Gerulaitis died when an improperly installed pool heater caused CO to seep into his guesthouse. More recently, on March 9, 2007, Bradley Delp, lead vocalist for the rock band Boston, took his life by lighting two charcoal grills inside his sealed master bathroom (Figure 3.23).

Figure 3.23 The structure of carbon monoxide.

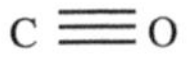

$$C \equiv O$$

Carbon monoxide

CO is an odorless gas formed by the incomplete combustion of fuels such as oil, propane, natural gas, coal, wood, kerosene, gasoline, diesel fuel, and charcoal. Most cases of accidental CO poisoning come from using gas heaters as an energy source or by barbecuing indoors using charcoal briquettes. It is not surprising that most CO poisonings occur between November and February, when people are trying to stay warm. Defective or poorly vented furnaces, fireplaces, wood stoves, gas water heaters, and generators can produce lethal amounts of CO. Other sources of CO include gasoline-powered tools, which lack catalytic converters and therefore produce toxic exhaust, and the burning of charcoal, which smolders and releases large amounts of CO.

According to the Centers for Disease Control and Prevention, CO is the leading cause of poisoning deaths in the United States. In 2007 alone, CO poisoning caused an estimated 2733 deaths and more than 20,000 emergency room visits. The deadliest characteristic of CO gas is that it is clear and odorless, leading to it being called "the invisible killer." For this reason, every building should have a CO detector that will sound when CO is near a dangerous level. Massachusetts and New York state laws require all homes to have these detectors. This legislation was passed after multiple cases where children died of CO poisoning.

The toxicity of CO is directly related to the mechanism by which our bodies remove oxygen from the air. In our lungs, the hemoglobin in red blood cells binds to oxygen and carries it throughout the body. In fact, the iron-containing protein hemoglobin makes up 97% of the dry weight of red blood cells. CO binds to hemoglobin about 250 times more strongly than oxygen. This means that the hemoglobin in red blood cells picks up CO faster than it picks up oxygen. If there is an even a slight amount of CO in the air, it will replace oxygen in the bloodstream. CO levels as low as 0.16% can be lethal, despite the fact that outdoor air is typically about 21% oxygen.

The initial symptoms of CO poisoning resemble those of the flu and include headaches, nausea, fatigue, dizziness, shortness of breath, and burning of the eyes. Prolonged exposure can lead to disorientation, convulsions, unconsciousness, and ultimately death.

The most insidious side of CO is its use in suicide and homicide. The CDC estimated that approximately two-thirds of the 2733 CO poisoning deaths in the United States in 2007 were intentional. Until the 1950s, fatal levels of CO were readily accessible in this country by simply turning on a gas oven. For this reason, CO inhalation was one of the most common means of suicide in the United States, England, and Western Europe. In the 1960s, the conversion to natural gas meant that the domestic gas supply no longer contained significant levels of CO, and the number of suicides by CO poisoning temporarily plummeted.

In the past, the inhalation of automobile exhaust was responsible for nearly all CO suicides in the United States. Because CO is a greenhouse gas, the United States set federal emission standards in 1986. These standards, coupled with the introduction of the catalytic converter in the 1970s, reduced the CO concentration in automobile exhaust by about 95%. Although still lethal, the exhaust from modern automobiles in this country does not kill as quickly as the exhaust from older vehicles did. The number of suicides from automobile exhaust inhalation has therefore dropped significantly.

CO has been implicated in several murders. In Ohio, Dr Mark Wangler was convicted for the 2006 murder of his wife, who mysteriously died in her home from CO poisoning. He claimed that his wife was killed by the gas that leaked from a faulty water heater. However, authorities tested the water heater and could not find a leak. The prosecution argued that the true exposure resulted from Wangler pumping automotive exhaust from an engine in the garage into his wife's bedroom while she slept.

Deadly Popcorn

In 2000, eight workers from a popcorn factory in Jasper, Missouri, were diagnosed with a rare lung disease called *Bronchiolitis obliterans*. Because of this disease, small airways in the lungs called *bronchioles* get plugged with scar tissue, resulting in shortness of breath, wheezing, and a dry cough. This devastating disease is irreversible and often requires a lung transplant. After a formal investigation by the National Institute of Occupational Safety and Health, it was found that nearly 200 workers at microwave popcorn and other flavoring plants had developed this lung disease, with at least three of them dying as a result. Because many of them worked in the popcorn industry, the disease has been called "popcorn lung."

The investigation linked the lung disease to exposure to the molecule diacetyl and other butter-flavoring agents. *Diacetyl*, found naturally in many foods, is artificially produced and used as a less expensive way to enhance flavor or impart the taste of butter in many foods, including microwave popcorn. Diacetyl is produced as part of many fermentations, so it occurs naturally in many wines and some beers. At low levels, diacetyl gives alcoholic beverages a slippery feeling in the mouth, and at higher levels, it makes the beverage taste like butter. Some winemakers deliberately promote the production of diacetyl for this reason. Many diacetyl-containing California chardonnays are known as "butter bombs."

Popcorn lung is suspected to be caused by inhalation of diacetyl vapor. This form of exposure is implicated owing to the safe use of diacetyl in many foods

Figure 3.24 The structure of diacetyl.

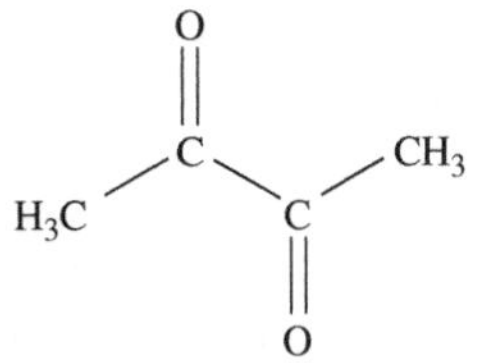

Diacetyl

as a flavoring. The production of diacetyl vapor through heating is the lethal component leading to popcorn lung (Figure 3.24).

More than 150 former popcorn plant workers have sued companies supplying or making the butter flavoring, alleging that it destroyed their lungs while the companies concealed their knowledge of the danger. To date, more than $100 million has been awarded in jury verdicts and paid in settlements, and many suits are still unresolved. In 2007, Wayne Watson, a dedicated consumer of microwave popcorn, was diagnosed with this disease. He reported having eaten two or three bags a day for more than 10 years, adding that he would always inhale the buttery aroma upon opening each bag. He filed suits against the producer of the popcorn and the grocery store where he had purchased it, for not warning consumers of the risks. In 2012, he won a $7.2 million verdict.

In the wake of these lawsuits, many food companies including Orville Redenbacher, Act II, Pop Secret, and Jolly Time microwave popcorn stopped using diacetyl. In addition, in 2007, H.R. 2693 was introduced in the House of Representatives, proposing stricter control of the use of diacetyl. The California Legislature is banning the use of diacetyl in certain commercial products.

One unexpected place where diacetyl has shown up is in electronic cigarette vaporizing liquids (e-liquids). E-liquids are a mixture of nicotine and flavoring agents dissolved in a solvent, such as propylene glycol and glycerin. Safety concerns were raised when some e-liquid flavors, such as butterscotch, caramel, coconut, and popcorn, were found to contain diacetyl to add a buttery flavor to the vapor. After many electronic cigarette users campaigned for disclosure, most major companies now have statements on their websites boasting diacetyl-free flavors. However, a 2014 study showed that more than 74% of the samples tested from seven countries still contained significant amounts of diacetyl.

Even Water Can Be Poisonous

A common misconception is that many things are completely safe. This idea leads to a dangerous sense of complacency and even death in many cases. Virtually, any substance, even water, can be lethal when consumed in sufficient doses. Water has a measured LD_{50} Oral Rat: >90 ml/kg. An LD_{50} is a measure of toxicity, defined as the dose that kills 50% of a given population. In this LD_{50} for water, the population is rats. Rats and other small mammals are used for testing of toxicity because humans cannot be used. The results in small mammals are then compared to humans. The LD_{50} of water for a 150-pound person is about 6.1 l or 1.5 gallons. One must remember, however, that it is not how much you drink, but how fast you drink it! Drinking that much water slowly over several hours is usually safe. Guzzling is not.

The study of toxicity is one of the most interesting aspects of chemistry. It is easy to say that something is toxic and can be lethal, but chemistry allows us to understand a defined chemical reason for the cessation of life. Our bodies are chemically balanced systems. The acidity, sugar levels, and salt concentrations in our blood are all maintained within a narrow range. Any major imbalance can cause illness and even death. The medical term for over-hydration is *hyponatremia* and is the result of low salt concentration in the blood. Drinking too much water causes the blood to become slightly diluted. It is important for our blood and our cells to have similar sodium ion concentrations to prevent osmosis. *Osmosis* is the movement of fluid through the cell wall from a solution with a low concentration to a solution with a higher concentration until the concentrations are equal.

The normal sodium concentration in the blood is 0.145 M (M stands for molarity, which is a measure of concentration). A patient is diagnosed with hyponatremia when the sodium concentration drops to 0.135 M. A mere 0.01 M drop in an element level puts a person in mortal danger, showing how fragile our systems are (Figure 3.25).

In a healthy human, the concentration of sodium in the cells and the blood is the same, so osmosis does not occur. However, if large amounts of water are consumed, the concentration of sodium in the blood is less than the concentration in the cells, so water moves from the blood into the cells. As the cells absorb water, they swell. This is a problem when it happens in the brain, where swelling can cause confusion and seizures. Eventually, the brain becomes too large for the skull, which produces pressure in the connection between the brain and spinal cord, leading to coma and death. Treatments for hyponatremia are the administration of a diuretic to increase urination and salt or saline to try and increase the sodium concentration in the blood.

Deaths from hyponatremia are frequently reported in the news. Many of these deaths were due to ignorance. In 2005, at California State University at Chico, a student was killed in a hazing incident when he was coerced to rapidly drink two gallons of water using a beer bong. In 2007, a radio station in Sacramento, California, held a contest called "Hold your WEE for a Wii." The contestant who drank the most water without going to the bathroom would win the gaming system. Unfortunately, one contestant died as a direct result of the contest. Her survivors were awarded more than $16 million in a wrongful death suit. Lastly, artist Andy Warhol died on February 22, 1987, from a sudden heart

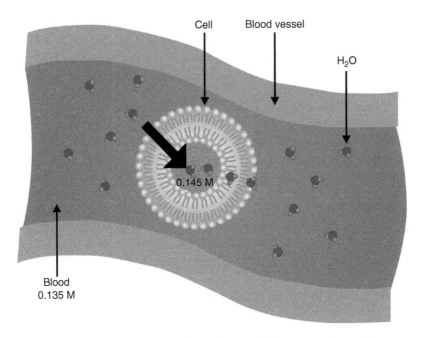

Figure 3.25 The movement of water during hyponatremia.

attack while recovering from routine gallbladder surgery. It turned out that the hospital staff had failed to monitor his condition and overloaded him with fluids, causing fatal water intoxication.

Further Reading

Why is Antifreeze Lethal?

Bronstein, AC, Spyker, DA, Cantilena, LR, Rumack, BH & Dart, RC 2012 '2011 annual report of the American Association of Poison Control Centers National Poison Data System (NPDS): 29th annual report', *Clinical Toxicology*, vol. 50, pp. 911–1164.

Chaytor, R 2012, *Kate Knight guilty of plot to kill husband with anti-freeze*, Mirror, Viewed 17 September 2016, http://www.mirror.co.uk/news/uk-news/kate-knight-guilty-of-plot-to-kill-290712

Cleveland 19 Digital Team 2014, *THIS JUST IN: Appeal denied for a woman in antifreeze poisoning death*, WDAM, Viewed 17 September 2016, http://www.wdam.com/story/27488134/this-just-in-appeal-denied-for-a-woman-in-antifreeze-poisoning-death

Conn, R 2008, *Husband guilty in antifreeze poisoning murder*, The MetroWest Daily News, Viewed 17 September 2016, http://www.metrowestdailynews.com/state/x2113790772/Husband-guilty-in-antifreeze-poisoning-murder

Conroy, S 2007, *Woman sentenced for antifreeze murder*, CBS News, Viewed 17 September 2016, http://www.cbsnews.com/2100-201_162-2780496.html

Fratz, DD 2010, *Your views: Number of deaths tied to antifreeze inflated in story*, Gainesville Times, http://www.gainesvilletimes.com/archives/29646/

Humane Society Legislative Fund 2012, *Antifreeze and engine coolant being bittered nationwide*, Humane Society Legislative Fund, Viewed 17 September 2016, http://www.hslf.org/news/press-releases/antifreeze-and-engine-coolant.html

Preventative Vet 2016, *Pet emergency statistics and veterinary costs*, Preventative Vet, Viewed 17 September 2016, http://www.preventivevet.com/pet-emergency-statistics

Rodrigues, A & Meyersohn, J 2010, *'Black Widow' to appeal guilty verdict*, ABC News, Viewed 17 September 2016, http://abcnews.go.com/2020/black-widow-convicted-antifreeze-murder-appeal/story?id=10751510

Aquadots: What a Difference a Carbon Makes!

Associated Press 2000, *Three men guilty of manslaughter in GHB poisoning case*, Daily Leader, Viewed 20 September 2016, http://www.dailyleaderextra.com/article_a40f10e5-863b-5657-9a64-1cd6af1140c9.html

Bay City News Service 2012, *Santa Rosa: Teen who died at sleepover last summer was poisoned by GHB drug*, East Bay Times, Viewed 20 September 2016, http://www.insidebayarea.com/california/ci_19869547

Benzer, T 2015, *Gamma-hydroxybutyrate toxicity*, Medscape, Viewed 20 September 2016, http://emedicine.medscape.com/article/820531-overview

Brady, J, Carroll, J, Dolan, L, O'Neill, J & Wiggins, L 2007, *Report: China halts export of bead toys tainted with toxic drug*, CNN, Viewed 20 September 2016, http://www.cnn.com/2007/WORLD/asiapcf/11/09/toy.recall/index.html

Brennan, R & Van Hout, M 2014, 'Gamma-hydroxybutyrate (GHB): A scoping review of pharmacology, toxicology, motives for use, and user groups', *Journal of Psychoactive Drugs*, vol. 46, pp. 243–251.

Consumer Product Safety Commission 2008, *Spin master recalls aqua dots - children became unconscious after swallowing beads*, Consumer Product Safety Commission.

Drug Enforcement Administration 2016, *GHB*, Drug Enforcement Administration, Viewed 20 September 2016, http://www.justice.gov/dea/druginfo/drug_data_sheets/GHB.pdf

Office of Diversion Control 2013, *Gamma hydroxbutyric acid*, Drug Enforcement Administration, Viewed 20 September 2016, http://www.deadiversion.usdoj .gov/drugs_concern/ghb/ghb.htm

How can Visine® Kill You?

Lev, R & Clark, RF 1995, 'Visine overdose: Case report of an adult with hemodynamic compromise', *Journal of Emergency Medicine*, vol. 13, pp. 649–652.

Mikkelson, D 2016, *FALSE: Visine causes diarrhea*, Snopes.com, Viewed 20 September 2016, http://www.snopes.com/medical/myths/visine.asp

MedlinePlus 2016, *Tetrahydrozoline poisoning*, U.S. National Library of Medicine, Viewed 20 September 2016, http://www.nlm.nih.gov/medlineplus/ency/article/ 002590.htm

Mohney, G 2013, *Man arrested after alleged visine poisoning*, ABC News, Viewed 20 September 2016, http://abcnews.go.com/US/man-arrested-allegedly- poisioning-girlfriend-visine/story?id=18693000

Taylor, K & Maslov, Y 2013, 'Accidental pediatric exposure to imidazoline derivatives', *Journal of Emergency Nursing*, vol. 39, pp. 59–60.

The Huffington Post 2011, *Luciana Reichel sentenced: 'Wedding Crashers'-inspired visine prankster to serve 90 days*, The Huffington Post, Viewed 20 September 2016, http://www.huffingtonpost.com/2011/09/19/luciana-reichel-sentenced- visine-prank_n_969992.html

Stahl, R 2015, *Vengeful bartender: 5 shots for that customer you hate (Pt. II)*, Examiner.com, Viewed 10 August 2015, http://www.examiner.com/article/ vengeful-bartender-5-shots-for-that-customer-you-hate-pt-ii

Synerholm, ME, Jules, LH & Sahyun, M 1956, U.S. Patent 2,731,471.

Visene 2016, *VISINE® original*, Visene.

Death by BENGAY®

Chin, RL, Olson, KR & Dempsey, D 2007, 'Salicylate toxicity from ingestion and continued dermal absorption', *The California Journal of Emergency Medicine*, vol. 8, pp. 23–25.

Crossland, AM 1948, 'Methyl Salicylate Poisoning', *Canadian Medical Association Journal*, vol. 58, pp. 75–77.

Drugs.com 2016, *Wintergreen*, Drugs.com, Viewed 20 September 2016, http:// www.drugs.com/npp/wintergreen.html

Mundell, EJ 2005, *Bengay death highlights OTC dangers*, ABC News, Viewed 20 September 2016, http://abcnews.go.com/Health/Healthday/story?id=4507580&page=1#.ULhHfYY8Vfw

Poirier, R & Corbet, R 1952, 'Salicylate poisoning in children', *Canadian Medical Association Journal*, vol. 67, pp. 117–120.

Sinsbury, SJ 1991, 'Fatal salicylate toxicity from bismuth subsalicylate', *Western Journal of Medicine*, vol. 155, pp. 637–639.

It Is in 93% of People in the United States

Biedermann S, Tschudin P, Grob K. 2010, 'Transfer of bisphenol A from thermal printer paper to the skin', *Analytical and Bioanalytical Chemistry*, vol. 398, pp. 571–576.

Calafat, AM, Ye, X, Wong, LY, Reidy, JA & Needham, LL 2008, 'Exposure of the U.S. population to bisphenol A and 4-tertiary-octylphenol: 2003–2004', *Environmental Health Perspectives*, vol. 116, pp. 39–44.

Feller, S 2016, *BPA alternative may not make plastic safer*, UPI, Viewed 17 September 2016, http://www.upi.com/Health_News/2016/02/02/BPA-alternative-may-not-make-plastic-safer/1041454339847/

Konkel, L 2013, *Thermal reaction: The spread of bisphenol S via paper products*, environmental health perspectives, Viewed 17 September 2016, http://ehp.niehs.nih.gov/121-a76/

Mittelstaedt, M 2010, *Canada first to declare bisphenol A toxic*, The Globe and Mail, Viewed 17 September 2016, http://www.theglobeandmail.com/technology/science/canada-first-to-declare-bisphenol-a-toxic/article1214889/

National Institute of Environmental Health Science 2016, *Bisphenol A (BPA)*, National Institute of Environmental Health Science, Viewed 17 September 2016, http://www.niehs.nih.gov/health/topics/agents/sya-bpa/

Nuwer, R 2011, *Check your receipt: It may be tainted*, New York Times, Viewed 17 September 2016, http://green.blogs.nytimes.com/2011/11/01/check-your-receipt-it-may-be-tainted/

Okada, H, Tokunaga, T, Liu, X, Takayanagi, S, Matsushima, A & Shimohigashi, Y 2008, 'Direct evidence revealing structural elements essential for the high binding ability of bisphenol A to human estrogen-related receptor-gamma', *Environmental Health Perspectives*, vol. 116, pp. 32–38.

Rubin, BS 2011, 'Bisphenol A: An endocrine disruptor with widespread exposure and multiple effects', *The Journal of Steroid Biochemistry and Molecular Biology* vol. 127, pp. 27–34.

Schreder, E 2010, *On the money BPA on dollar bills and receipts*, Washington Toxics Coalition.

U.S. Food and Drug Administration 2010, *Update on bisphenol A for use in food contact applications, U.S. Food and Drug Administration*, Viewed 17 September 2016, http://www.fda.gov/downloads/NewsEvents/PublicHealthFocus/UCM197778.pdf

Vandenberg, LN, Hauser, R, Marcus, M, Olea, N & Welshons, WV 2007, 'Human exposure to bisphenol A (BPA)', *Reproductive Toxicology*, vol. 24, pp. 139–177

The Dreaded...Apricot Pits?

Agency for Toxic Substances & Disease Registry 2006, *Cyanide*, Agency for Toxic Substances & Disease Registry, Viewed 17 September 2016, http://www.atsdr.cdc.gov/toxfaqs/tf.asp?id=71&tid=19

Agency for Toxic Substances & Disease Registry 2007, *Chemical and physical information*, Agency for Toxic Substances & Disease Registry, Viewed 17 September 2016, https://www.atsdr.cdc.gov/toxprofiles/tp8-c4.pdf

Bronstein, AC, Spyker, DA, Cantilena, LR, Rumack, BH & Dart, RC 2012, '2011 annual report of the American association of poison control centers national poison data system (NPDS): 29th annual report', *Clinical Toxicology*, vol. 50, pp. 911–1164

Jones, DA 1998, 'Why are so many food plants cyanogenic?', *Phytochemistry*, vol. 47, pp. 155–162.

Karp, D 2002, *What to use when you can't get the real thing*, LA Times, Viewed 17 September 2016, http://articles.latimes.com/2002/feb/20/food/fo-almondside20

Lutz, D 2010, *Beware the smell of bitter almonds why do many food plants contain cyanide?*, Washington University in St. Louis, Viewed 17 September 2016, http://news.wustl.edu/news/Pages/20916.aspx

Morse, DL, Harrington, JM & Heath, CW Jr, 1976, 'Laetrile, apricot pits, and cyanide poisoning', *New England Journal of Medicine*, vol. 295, p. 1264.

National Cancer Institute 2016, *Laetrile/amygdalin (PDQ ®)–patient version*, National Cancer Institute, Viewed 17 September 2016, http://www.cancer.gov/cancertopics/pdq/cam/laetrile/patient/allpages#Section_20

Shagg, TA, Albertson, TE & Fisher, CJ 1982, 'Cyanide poisoning after bitter almond ingestion', *Western Journal of Medicine*, vol. 136, pp. 65–69.

Staff 2014, *Almonds sold at reno whole foods recalled*, KOLO 8, Viewed 17 September 2016, http://www.kolotv.com/home/headlines/Almonds-Sold-at-Reno-Whole-Foods-Recalled-282167971.html

Vetter, J 2000, 'Plant cyanogenic glycosides', *Toxicon*, vol. 38, pp. 11–36.

Honey Intoxication

Gunduz, A, Turedi, S, Russell, RM & Ayaz, FA 2008,'Clinical review of grayanotoxin/mad honey poisoning past and present', *Clinical Toxicology*, vol. 46, pp. 437–442.

Jansen, SA, Kleerekooper, I, Hofman, ZLM, Kappen, IFPM, Stary-Weinzinger, A, & van der Heyden, MAG 2012, 'Grayanotoxin poisoning: 'Mad Honey Disease' and beyond', *Cardiovascular Toxicology*, vol. 12, pp. 208–215. http://www.ncbi.nlm.nih.gov/pmc/articles/PMC3404272

Koca, I & Koca, AF 2007, 'Poisoning by mad honey: a brief review', *Food and Chemical Toxicology*, vol. 45, pp. 1315–1318.

Robertsona, LM, Edlinb, JS & Edwards, JD 2010,'Investigating the importance of altitude and weather conditions for the production of toxic honey in New Zealand', *New Zealand Journal of Crop and Horticultural Science*, vol. 38, p. 87100.

The DMSO Patient

New York Times, 1994, *Hospital fumes that hurt 6 are tied to nerve gas*, New York Times, Viewed 17 September 2016, http://www.nytimes.com/1994/11/05/us/hospital-fumes-that-hurt-6-are-tied-to-nerve-gas.html

Rippey, JCR & Stallwood, MI 2005, 'Nine cases of accidental exposure to dimethyl sulphate – a potential chemical weapon', *Emergency Medicine Journal*, vol. 22, pp. 878–879.

Stone, R 1995, *Analysis of a toxic death*, Discover, Viewed 17 September 2016, http://discovermagazine.com/1995/apr/analysisofatoxic493

The National Institute for Occupational Safety and Health (NIOSH), 2016, *Dimethyl sulfate, Center for Disease Control and Prevention*, Viewed 17 September 2016, http://www.cdc.gov/niosh/npg/npgd0229.html

Deadly Helium Balloons

BBC News, 2010, *Tributes to 'helium death' teenager from Newtownabbey*, BBC News, Viewed 17 September 2016, http://www.bbc.co.uk/news/uk-northern-ireland-11795984

Fox News Health 2012, *Teen dies after inhaling helium at party*, Fox News Health, Viewed 17 September 2016, http://www.foxnews.com/health/2012/02/23/teen-dies-after-inhaling-helium-at-party/

Ramesh, R 2012, *Helium and barbiturates contribute to drug death statistics*, The Guardian, Viewed 17 September 2016, http://www.guardian.co.uk/society/2012/aug/29/helium-barbiturates-drug-death-statistics

The 2007 Pet Food Recall

The Associated Press 2007, *104 deaths reported in pet food recall*, The New York Times, Viewed 17 September 2016, http://www.nytimes.com/2007/03/28/science/28brfs-pet.html?ex=1176264000&en=8ee0fb91fd221e4b&ei=5070&_r=0

European Food Safety Authority 2008, *Statement of EFSA on risks for public health due to the presences of melamine in infant milk and other milk products in China*, EFSA Journal.

McDonald, M 2009, *Death sentences in China milk case, International Herald Tribune*, Viewed 17 September 2016, http://web.archive.org/web/20090122215601/http://www.iht.com/articles/2009/01/22/news/23MILK.php

Weise, E 2007, *Nestlé Purina, Hills join pet food recall*, USA Today, Viewed 17 September 2016, http://usatoday30.usatoday.com/news/nation/2007-03-30-pet-food-recall_N.htm

Mercury in Vaccines and Eye Drops?

Baker, JP 2008, 'Mercury, vaccines, and autism: One controversy, three histories', *American Journal of Public Health*, vol. 98, pp. 244–253. http://www.ncbi.nlm.nih.gov/pmc/articles/PMC2376879/

Bigham, M, Copes, R 2005, 'Thiomersal in vaccines: Balancing the risk of adverse effects with the risk of vaccine-preventable disease', *Drug Safety*, vol. 28, pp. 89–101.

Center for Biologics Evaluation and Research 2015, *Thimerosal in vaccines*, U.S. Food and Drug, Viewed 17 September 2016, http://www.fda.gov/BiologicsBloodVaccines/SafetyAvailability/VaccineSafety/UCM096228

Kharasch, M 1928, *Alkyl mercuric sulphur compound and process of producing it*, U.S. Patent 1,672,615, Viewed 17 September 2016, http://www.who.int/immunization/newsroom/thiomersal_information_sheet/en/index.html

The World's Deadliest Frog

Hantak, MM, Grant, T, Reinsch, S, Mcginnity, D, Loring, M, Toyooka, N, Saporito, RA 2013,'Dietary alkaloid sequestration in a poison frog: An experimental test of alkaloid uptake in Melanophryniscus stelzneri (Bufonidae)', *Journal of Chemical Ecology*, vol. 39, pp. 1400–1406.

National Geographic 2015, *Golden poison dart frog*, National Geographic, Viewed 19 September 2016, http://animals.nationalgeographic.com/animals/amphibians/golden-poison-dart-frog/

Wallace, S 2014, *Batrachotoxin*, ChemistryWorld, Viewed 19 September 2016, http://www.rsc.org/chemistryworld/2014/02/batrachotoxin-poison-dart-frog-podcast

Leaded Candy

Centers for Disease Control and Prevention 2002, *Childhood lead poisoning associated with tamarind candy and folk remedies – California*, 1999–2000, Centers for Disease Control and Prevention, Viewed 19 September 2016, http://www.cdc.gov/mmwr/preview/mmwrhtml/mm5131a3.htm
Food and Drug Branch 2016, *Lead in candy*, California Department of Public Health, Viewed 17 September 2016, http://www.cdph.ca.gov/programs/Pages/FDB%20Lead%20In%20Candy%20Program.aspx

Why not Drink "Real" Root Beer?

Liu, TY, Chen, CC, Chen, CL, Chi, CW 1999, 'Safrole-induced oxidative damage in the liver of Sprague–Dawley rats', *Food and Chemical Toxicology*, vol. 37, pp. 697–702.
National Toxicology Program, 2009, *Safrole*, US Department of Health and Human Services, Viewed 19 September 2016, http://ntp.niehs.nih.gov/ntp/roc/twelfth/profiles/Safrole.pdf
Scientific Committee on Food, 2001, *Opinion of the Scientific Committee on Food on the safety of the presence of safrole (1-allyl-3,4- methylene dioxy benzene) in flavourings and other food ingredients with flavouring properties*, European Commission.
Wiseman, RW, Fennell, TR, Miller, JA & Miller, EC 1985, 'Further characterization of the DMA adducts formed by electrophilic esters of the hepatocarcinogens 1'-hydroxysafrole and 1'-hydroxyestragole *in vitro* and in mouse liver *in vivo*, including new adducts at C-8 and N-7 of guanine residues', *Cancer Research*, vol. 45, pp. 3096–3105.

The Killer Fog

Bell, ML, Bell, ML, Davis, DL & Fletcher, T 2004, 'A retrospective assessment of mortality from the London smog episode of 1952: The role of influenza and pollution', *Environmental Health Perspectives*, vol. 112, pp. 6–8.
Nielsen, J 2002, *The killer fog of '52*, NPR, Viewed 19 September 2016, http://www.npr.org/templates/story/story.php?storyId=873954

Nail Polish or Nail Poison?

Allen, JE 2012, *Safer mani-pedis: Steps you can take*, ABC News, Viewed 9 October 2016, http://abcnews.go.com/Health/Health/nail-polish-safety/story?id=16108721

Boyle, M 2015, *More bad news about nail polish*, Environmental Working Group, Viewed 9 October 2016, http://www.ewg.org/enviroblog/2015/10/more-bad-news-about-nail-polish

Nir, SM 2015, *Perfect nails, poisoned workers*, The New York Times, Viewed 9 October 2016, http://www.nytimes.com/2015/05/11/nyregion/nail-salon-workers-in-nyc-face-hazardous-chemicals.html

Trotto, S 2015, *Dangerous beauty: Salon worker health is in the spotlight*, Safety and Health Magazine, Viewed 9 October 2016, http://www.safetyandhealthmagazine.com/articles/print/12930-nail-salon-workers-health

US Environmental Protection Agency 2014, *Toxicological review of toluene*, US Environmental Protection Agency, Viewed 9 October 2016, https://www.epa.gov/sites/production/files/2014-03/documents/toluene_toxicology_review_0118tr_3v.pdf

US Consumer Product Safety Commission 2015, *Phthalates*, US Consumer Product Safety Commission, Viewed 9 October 2016, https://www.cpsc.gov/Business--Manufacturing/Business-Education/Business-Guidance/Phthalates-Information

Game Board Danger

Aguilar-Ortigoza, CJ, Sosa, V, Aguilar-Ortigoz, M 2003, 'Toxic phenols in various Anacardiaceae species', *Economic Botany*, vol. 57, pp. 354–364

Associated Press 2006, *Another global warming gift: itchier poison ivy*, NBC News, Viewed 19 September 2016, http://www.nbcnews.com/id/13046200/ns/us_news-environment/t/another-global-warming-gift-itchier-poison-ivy/#.Vm4kL1LbxX8

Hapten Sciences, Inc. 2015, *Hapten sciences to begin clinical trials for novel poison ivy vaccine*, PR Newswire, Viewed 19 September 2016, http://www.prnewswire.com/news-releases/hapten-sciences-to-begin-clinical-trials-for-novel-poison-ivy-vaccine-300181513.html

Hashida, K, Tabata, M, Kuroda, K, Otsuka, Y, Kubo, S, Makino, R, Kubojima, Y, Tonosaki, M & Ohara, S 2014, 'Phenolic extractives in the trunk of Toxicodendron vernicifluum: Chemical characteristics, contents and radial distribution', *Journal of Wood Science*, vol. 60, pp. 160–168.

Hershko, K, Weinberg, I, Ingber, A 2005, 'Exploring the mango – poison ivy connection: The riddle of discriminative plant dermatitis', *Contact Dermatitis*, vol. 52, pp. 3–5.

Kumanotani, J 1995, 'Urushi (oriental lacquer) – a natural aesthetic durable and future-promising coating', *Progress in Organic Coatings*, vol. 26, pp. 163–195.

Lu, R, Yoshida, T, Miyakoshi, T 2013, 'Oriental lacquer: A natural polymer', *Polymer Reviews*, vol. 53, pp. 153–191.

RongLu, X & Miyakoshi, T. 2012, 'Recent advances in research on lacquer allergy', *Allergology International*, vol. 61, pp. 45–50.

Tsujimoto, T, Ando, N, Oyabu, H, Uyama, H & Kobayashi, S 2007,'Laccase-catalyzed curing of natural phenolic lipids and product properties', *Journal of Macromolecular Science, Part A: Pure and Applied Chemistry*, vol. 44, pp. 1055–1060

Ziska, LH, Sicher, RC, George, K & Mohan, JE 2007, 'Rising atmospheric carbon dioxide and potential impacts on the growth and toxicity of poison ivy (Toxicodendron Radicans)', *Weed Science*, vol. 55, pp. 288–292.

What Molecule Killed "Weird Al" Yankovic's Parents?

Blasser, K, Tatschner, T & Bohnert, M 2014, 'Suicidal carbon monoxide poisoning using a gas-powered generator', *Forensic Science International*, vol. 236, pp. e19–e21.

Burkhart, C 2014, *Keeping your family safe from carbon monoxide poisoning*, ABC News.

CBS News 2011, *Ohio killer's son, wife continue innocence crusade*, CBS News, Viewed 20 September 2016, http://www.cbsnews.com/news/ohio-killers-son-wife-continue-innocence-crusade/

Chen, Y, Bennewith, O, Hawton, K, Simkin, S, Coopers, J, Kapur, N & Gunnell, D 2012 'Suicide by burning barbecue charcoal in England', *Journal of Public Health*, vol. 35, pp. 223–227.

Goldstein, M 2008, 'Carbon monoxide poisoning', *Journal of Emergency Nursing*, vol. 34, pp. 539–542.

Hampson, NM, Bodwin, D. 2013, 'Toxic co-ingestions in intentional carbon monoxide poisoning', *The Journal of Emergency Medicine*, vol. 44, pp. 625–630.

Hnatov, MV 2009, *Non-fire carbon monoxide deaths associated with the use of consumer products 2009 annual estimates*, U.S. Consumer Product Safety Commission, Viewed 20 September 2016, http://www.cpsc.gov/PageFiles/136146/co12.pdf

Hari, A 2014, *Colder temperatures, higher chance of carbon monoxide poisoning*, WBNG News.

Moreland, J 2004, *Parody star's parents die in Fallbrook*, North County Times.

Routley, V 2007, 'Motor vehicle exhaust gas suicide', *Crisis*, vol. 28, pp. 28–35.

Schmitt, MW, William, TL, Woodard, KR & Harruff, RC 2011, 'Trends in suicide by carbon monoxide inhalation in King County, Washington: 1996–2009', *Journal of Forensic Sciences*, vol. 56, pp. 652–655.

Truesdell, J 2014, *Kelli Stapleton sentenced to 10–22 years for attempt on daughter with Autism's life*, People Magazine, Viewed 20 September 2016, http://www.people.com/article/kelli-stapleton-sentenced-prison?xid=rss-topheadlines

Weed, RI, Reed, CF & Berg, G 1963, 'Is hemoglobin and essential structural component of human erythrocyte membranes?', *Journal of Clinical Investigation*, vol. 42, pp. 581–588.

Deadly Popcorn

Castellano, A 2012, *Popcorn lung' lawsuit nets $7.2M award*, ABC News, Viewed 19 September 2016, http://abcnews.go.com/blogs/headlines/2012/09/popcorn-lung-lawsuit-nets-7-2m-award/

Department of Health and Human Services, 2003, *Preventing lung disease in workers who use or make flavorings*, Centers for Disease Control or Prevention, Viewed 19 September 2016, http://www.cdc.gov/niosh/docs/2004-110/

Farsalinos, KE, Kistler, KA, Gilman, G & Voudris, V 2014, 'Evaluation of electronic cigarette liquids and aerosol for the presence of selected inhalation toxins', *Nicotine & Tobacco Research*, vol. 17, pp. 168–174.

Fujioka, K, Shibamoto, T 2006, 'Determination of toxic carbonyl compounds in cigarette smoke', *Environmental Toxicology*, vol. 21, pp. 47–54

Hamill, J 2014, *The health claims of E-cigarettes are going up in smoke, forbs*, Viewed 19 September 2016, http://www.forbes.com/sites/jasperhamill/2014/08/31/the-health-claims-of-e-cigarettes-are-going-up-in-smoke/

Schneider, A 2006, *Disease is swift, response is slow Government lets flavoring industry police itself, despite damage to workers' lungs*, Baltimore Sun, Viewed 19 September 2016, http://defendingscience.org/sites/default/files/upload/Baltimore_Sun_Schneider_2006.pdf

The National Institute for Occupational Safety and Health (NIOSH) 2016, *Flavorings-related lung disease*, Viewed 19 September 2016, http://www.cdc.gov/niosh/topics/flavorings/

Even Water can be Poisonous

Associated Press 2007, *Woman dies after water-drinking contest*, NBC News, Viewed 9 October 2016, http://www.nbcnews.com/id/16614865/ns/us_news-life/t/woman-dies-after-water-drinking-contest/

Farooq, S 2009, *"Hold Your Wee for a Wii" costs radio station dearly*, NBC Bayarea, Viewed 9 October 2016, http://www.nbcbayarea.com/news/local/Hold-Your-Wee-for-a-Wii-Costs-Radio-Station-Dearly-67345657.html

Korry, E 2005, *A fraternity hazing gone wrong*, National Public Radio, Viewed 9 October 2016, http://www.npr.org/templates/story/story.php?storyId=5012154

Moritz, ML & Ayus, JC 2003, 'The pathophysiology and treatment of hyponatraemic encephalopathy: An update', *Nephrology, Dialysis, Transplantation*, vol. 18, pp. 2486–2491.

Sciencelab.com 2013, *Water MSDS*, Sciencelab.com, Viewed 9 October 2016, http://www.sciencelab.com/msds.php?msdsId=9927321

Simon, EE 2016, *Hyponatremia treatment & management*, Medscape, Viewed 9 October 2016, http://emedicine.medscape.com/article/242166-treatment

Sullivan, R 1991, *Care faulted in the death of Warhol*, New York Times, Viewed 9 October 2016, http://www.nytimes.com/1991/12/05/nyregion/care-faulted-in-the-death-of-warhol.html

4

Why Old Books Smell Good and Other Mysteries of Everyday Objects

As a college professor, I am often asked why I chose chemistry as a career. My answer is always the same: I wanted to study a science that allows me to make things I can hold in my hand. From pharmaceuticals, to polymers, to nanoparticles, everything I have ever made as a chemist was real and tactile. Just think, any commercial item or product that involves a chemical has a chemist to thank for its production. Elements and chemicals are used to build virtually everything in our world. Because of this, a knowledge of chemistry gives me a unique understanding into items we see every day.

The Smell of Old Books and the Hidden Vanilla Extract Underworld

I cannot be the only person who inhales appreciatively when I enter a library or used bookstore to catch a whiff of that familiar, sweet mustiness. I have even been known to stick my nose between the pages of an old tome to enjoy more of this pleasant scent. Everyone should take time to stop and smell the old books though, I have often wondered where the smell comes from.

The sweet part of the "old book smell" comes from the molecule lignin, which is present in all wood-based paper. Books printed after about 1850 use paper made from soft or hard woods. Prior to this, most books were made from cotton or linen. The switch was made simply to save money; trees are much less expensive to harvest in great quantities than cotton. Being wood based, most modern paper is composed primarily of cellulose and also significantly (25–33% by weight) of lignin molecules. A molecule of *lignin* – one of the most abundant organic polymers on the Earth – has a complex structure, with a molecular weight in excess of 10,000 g/mol (grams per mole). By comparison, most organic molecules weigh only a few 100 g/mol. Portions of the molecule react to environmental stressors, resulting in lignin degrading as the years pass due to the effects of ultraviolet (UV) light, humidity, oxygen, and even acids in the

Strange Chemistry: The Stories Your Chemistry Teacher Wouldn't Tell You, First Edition. Steven Farmer.
© 2017 John Wiley & Sons, Inc. Published 2017 by John Wiley & Sons, Inc.

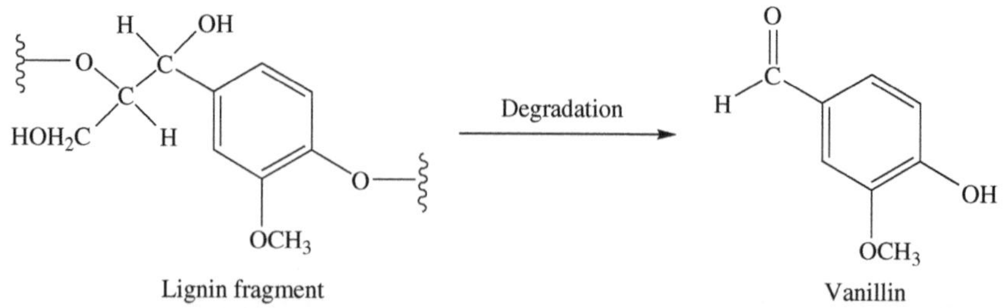

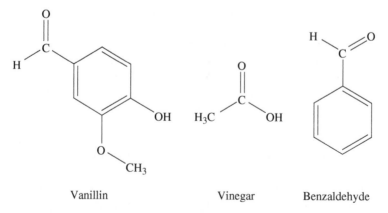

Scheme 4.1 The synthesis of vanillin from lignin.

paper. By "degrade," I mean that chemical bonds break, resulting in smaller compounds splitting away from the larger lignin molecule. These smaller pieces are released as a gas having the odor we associate with old books (Scheme 4.1).

Certain parts of the molecular structure of lignin (see diagram) are very similar to vanillin, the primary molecule found in vanilla bean extract. As it breaks down, lignin releases vanillin, giving old books that faint, sweet scent of vanilla. The degradation of lignin also imparts a vanilla flavor to wines, spirits, and other products stored in wooden barrels. Similarly, cooking over wooden fires can give a vanilla edge to the flavor of food. Besides lignin, other compounds in paper break down and release hundreds of other volatile molecules, making the smell of an old book quite complex, much like a fine wine. A couple of other molecules contributing to the old book smell are acetic acid (found in vinegar), which lends a slightly acidic scent, and benzaldehyde, which smells like almonds (Figure 4.1).

The degradation of lignin also explains why modern paper eventually turns yellow. A yellow, nonvolatile compound is produced during the breakdown of lignin, and this compound remains in the paper. As you would expect,

Figure 4.1 The structure of some of the molecules that produce the "Old Book" smell.

yellowing occurs only in books made since about 1850, when wood-based paper containing lignin started being used. Books made before the nineteenth century used cotton-based paper, and the absence of lignin made them much more resistant to yellowing.

If you are an e-book reader who misses the smell of an actual book, you are in luck. "Book smell" has been captured in an air freshener, aerosol spray, and even a perfume, all of which can be purchased online. Personally, I prefer the old fashion method of roaming through a library.

One interesting side note is that the imitation vanillin used to produce vanilla extract was once synthesized from a lignin-containing by-product produced by the wood pulp industry. In fact, by 1981, a single Canadian paper company was providing 60% of the then current world market. Artificial sources of vanillin were sought out because of the expensive processing required to produce natural vanilla extract. In particular, the vanilla orchids need to be pollinated and harvested by hand to produce the coveted vanilla beans. This causes natural vanilla extract to cost up to 200 times more than the man-made substitute. This high demand has caused manufacturers to seek out even cheaper ways of making vanillin. Although more than 12,000 tons of vanillin is produced each year, less than 1% of this actually comes from the vanilla bean. Currently, most artificial vanillin is made from the molecule guaiacol, which is obtained from petroleum (Scheme 4.2).

The high price of natural vanilla extract caused many unscrupulous venders to try and pass off the artificial product as being the real thing. The quest to detect the artificial product has created a cat and mouse game between the scientists and the fakers. Initially, it was easy because the natural vanilla extract contains molecules other than vanillin, such as vanillic acid, *p*-hydroxybenzaldehyde, and *p*-hydroxybenzoic acid. These molecules are easy to test for, but the fakers simply started adding these molecules to their

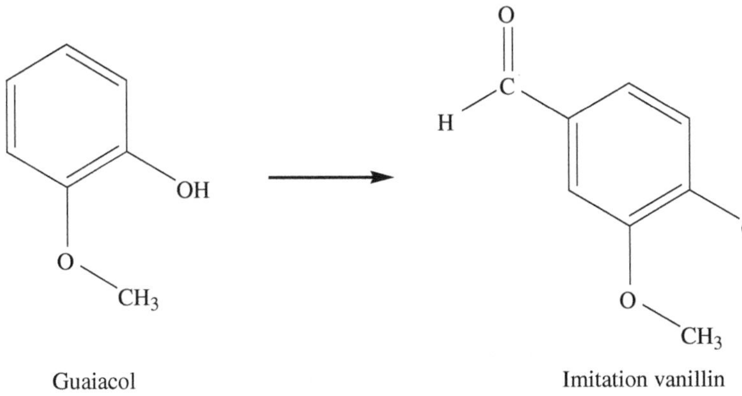

Guaiacol   Imitation vanillin

Scheme 4.2 The formation of "imitation" vanillin from guaiacol.

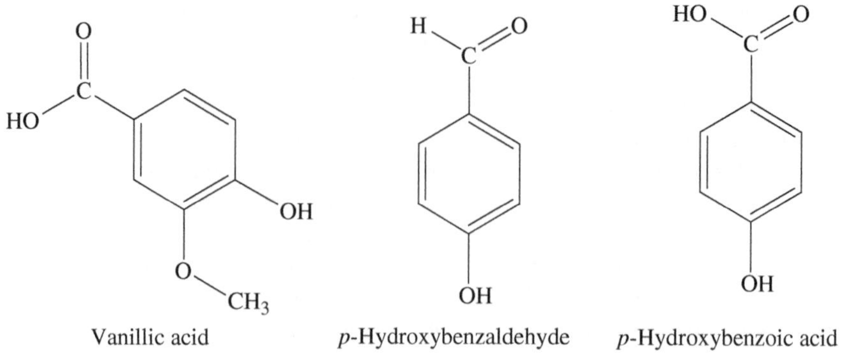

Vanillic acid   *p*-Hydroxybenzaldehyde   *p*-Hydroxybenzoic acid

Figure 4.2 The structure of some of the molecules contained in vanilla extract.

imitation product. Scientists then started to test suspicious vanilla extract samples for the content of the radioactive carbon-14 isotope. Most carbon atoms are carbon-12; however, a miniscule amount is the radioactive isotope carbon-14. Because carbon-14 is continually being formed in the Earth's atmosphere, virtually every live thing on the planet absorbs and contains a certain level of carbon-14. If the vanilla extract is derived from natural plants source, it should also contain a certain level of carbon-14

Figure 4.3 The "Old Book" smell.

that can be detected. However, any vanilla extract sample that has been artificially made from molecules obtained from crude oil, such as guaiacol, should contain virtually no carbon-14 atoms. The plants that form crude oil died and stopped absorbing carbon-14 millions of years ago. Because the half-life of carbon-14 is 5730 years, it should have decayed away during the time oil was trapped underground. Subsequently, the fakers started spiking their imitation vanilla product with carbon-14 atoms and scientists are now in the process developing tests to stop them. I eagerly look forward to watching how this game plays out (Figures 4.2 and 4.3).

That Smell Is You!

Have you ever decided to cash in your jars of loose change and noticed that your hands had a metallic smell to them when you finished counting the coins. Perhaps you have noticed a similar scent from other times you have had handled metals, especially copper. The smell must have come from some volatile molecule, but metals are not volatile and do not evaporate (fortunately!). The odor molecules associated with metals must come from a different source.

It was not until fairly recently that scientists learned that the characteristic metallic smell comes from volatile compounds produced by a chemical reaction between oils on the skin and the metal. The volatile molecule produced by the reaction creates a sensory illusion that the metal is producing the smell, even though metals have no natural odor because they are nonvolatile. For example, when iron comes into contact with our skin, perspiration quickly corrodes some of the iron producing Fe^{2+} ions. Eventually these Fe^{2+} ions, along with oxygen from air, break down the compounds in skin oils to form new molecules. These molecules, called aldehydes and ketones, are volatile and produce the smell we associate with the metal. The key odorant, responsible for about one-third of the odor concentration, is 1-octen-3-one, which some people describe as smelling like mushrooms or metal. Once again, this is an illusion because the molecule does not smell like metal; the reacting metal smells like this molecule. We have smelled 1-octen-3-one coming from metals for so long that we mistake the smell of this molecule as the smell of metal itself (Figure 4.4).

Figure 4.4 The structure of 1-octen-3-one.

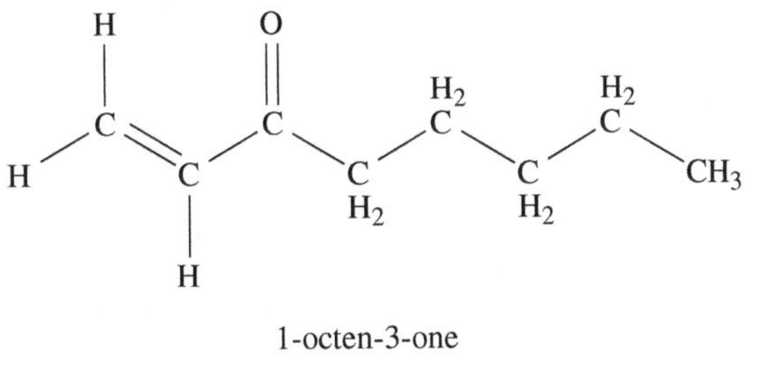

1-octen-3-one

Because the formation of these compounds involves an electrochemical process (movement of electrons), it makes sense that different metals would have varying abilities to produce these odor molecules. In the example, iron (Fe) loses two electrons to become Fe^{2+} and initiates the chemical reactions. The ability to give up electrons is distinctive to each metal, with some giving electrons up more easily than others. Copper and brass produce compounds with odors similar to those from iron. Zinc and stainless steel only produce weak metallic smells, while gold, magnesium, nickel, and aluminum create no smell at all. On a side note, most modern US silver-colored coins are made from a copper/nickel alloy, so they will indeed produce the characteristic metallic smell.

These kinds of chemical reactions also explain why blood produces a metallic odor when it comes into contact with skin. Our blood contains hemoglobin for the transport of oxygen, with the iron in hemoglobin existing as either Fe^{2+} or Fe^{3+}. Since iron ions are present in blood, the same chemical reaction produces odor molecules when blood touches skin. Though it may seem like we smell the iron in our blood, it is actually the volatile compounds from this reaction. In fact, humans can detect the key odorant, 1-octen-3-one, in concentrations as low as 0.005 ppb, which makes the molecule one of the most sensitive to our noses. This is much more sensitive than the sulfur containing compounds used to give natural gas an odor. Actually, for this one molecule, human noses are as sensitive as a dog's. An easy way to test this is to try to get the smell out of a hand full of change.[1] Our unusually keen sensitivity to this odor is thought to be linked to an evolutionary advantage developed by humans over time. The heightened detection of the scent of blood's reaction with skin oils possibly gave early humans a better ability to track wounded prey.

Electric Blue

We have all seen a lightning bolt and its distinctive bluish purple color. However, have you ever noticed that the spark you get from a door knob is of the same color? The color of sparks and that of lightning both come from the ionization of the nitrogen gas molecules in air, resulting in ionized air glow or electric blue.

The process by which a spark produces color is very similar to the way a neon sign gives off its distinctive red glow. When a sufficient voltage of electricity passes through a gas, the electrons in the gas molecules go from a *ground state* into what is called an *excited state*. Because the excited state is unstable, the gas tends to release this energy in the form of light (photons) when its

---

1 I once tried to prove to some students that it was possible to remove the smell from change but it was not easy. Multiple washings were not successful and it eventually took baking the change in an oven to successfully remove the volatile compounds. I was truly impressed at how I could smell the smallest amount and I can say that clean coins are in fact scentless.

electrons return to the ground state. How much energy is being released, and consequently, the color of light being emitted is dependent on the type of gas molecule. In other words, the electrons in a gas are caused to emit light by the application of electricity and the color of light emitted is specific to the type of gas being excited.

In the case of sparks and lightning, electricity is traveling through air, composed mostly of nitrogen. When this gaseous nitrogen is excited, it emits a purple color. On other planets, such as Jupiter, where the atmosphere is primarily made up of hydrogen, lightning bolts would probably appear orange in color.

As you can imagine, the process of exciting a gas requires enormous amounts of voltage. A typical number for sparks is about 30 kV per centimeter of distance traveled in air. This means that if you are shocked by a spark that has traveled 1 cm through the air, you have received roughly 30,000 V of electricity. For the sake of comparison, if you were struck by lightning, your exposure would come in at about 100 million volts!

Neon signs work on basically the same principle. In neon signs, the two main gases used are neon, which radiates red when excited; and argon, which radiates blue. Different colors are created by varying the color of the glass tubes and changing the relative amounts of neon and argon. The constant flow of electricity within neon signs causes the gas molecules to be in a continually excited state, resulting in their ongoing emission of light. In order to work, neon signs need large amounts of voltage, requiring that they be run on a transformer that makes that familiar (or annoying) humming noise.

Another way the air can become excited is by interaction with strongly radioactive materials, such as radium or polonium. Radioactive decay produces ionizing radiation, such as alpha or beta particles, which can cause the oxygen and nitrogen molecules in air to become excited. Correspondingly, samples of these elements are said to have an eerie blue glow. The blue glow from ionizing radiation was reported to be seen at the Trinity nuclear test explosion and at the Chernobyl nuclear power plant disaster. Human curiosity over this blue glow has produced devastating results. Marie Curie used to carry test tubes containing radioactive isotopes in her pocket remarking on their blue glow. She later died due to her exposure to radiation. On 13 September 1987, in Goiânia, Brazil scavengers broke open a radiation therapy machine in an abandoned medical clinic. The rice-sized caesium-137 pellets they found inside gave off a blue glow in the dark. Interest in this blue glow was directly attributed to the radioactive caesium being distributed to members of several families. In the end, over 200 persons were contaminated and 4 died.

Ionization radiation also explains the Aurora Borealis. The sun also produces ionizing radiation, called solar wind, which impacts the Earth's upper atmosphere causing a green glow. The green color is produced due to the increased oxygen content of the Earth's upper atmosphere.

The World's Most Abundant Organic Compound

The world's most common organic compound is *cellulose*, the structural component in green plants. Cellulose from wood pulp is a polymer made up of between 300 and 1700 glucose monosaccharide (sugar) units and its primary function is to give form and strength to cell walls. Some people mistakenly think of all polymers as being man-made, but actually, there are many types of polymers found in nature, including cellulose, starch, proteins, latex, and even deoxyribonucleic acid (DNA). Plants use the cellulose polymer to form structures in the same way as artificial polymers – for example, polyethylene – provide the structure of many consumer products. Cellulose makes up approximately 33% of all plant matter, 40–50% of dry wood weight, and 90% of cotton. Accounting for about 1.5×10^{12} tons of the total annual biomass production on the Earth, it is considered an almost inexhaustible source of raw material. It is hard to believe that the forests and plants of the world are being held together by a sugar.

Because of its abundance, cellulose has many commercial applications. Have you ever eaten wood? Cellulose derived mainly from wood is a popular food additive. Although some animals, such as cows, can digest cellulose, humans cannot. This is why cellulose is considered an indigestible carbohydrate, sometimes called dietary fiber. Cellulose is often added to diet foods because it has no calories but still creates a sensation of fullness in the stomach. Some Weight Watchers® and Skinny Cow® products contain cellulose.

Starch, which is found in potatoes, rice, and wheat, is also a polymer of glucose units but is digestible by humans and other animals. The difference is in how the glucose units are connected. Starch has what is called an α acetal linkage between glucose units, and cellulose has a β acetal linkage. This small difference results in a major difference in digestibility in humans. Humans lack the enzymes needed to break the β acetal linkage so cellulose cannot be digested. Animals such as cows, horses, sheep, goats, and termites can digest cellulose because they have symbiotic bacteria in the intestinal tract that possess the necessary enzymes to break the β acetal linkage (Figure 4.5).

At a cost of only about $3 per pound, cellulose is an inexpensive means of thickening, stabilizing, or increasing the volume of various foods. The US Food and Drug Administration (FDA) considers cellulose safe for human consumption and has set no maximum limit on the amount that can be used in food products. However, the Department of Agriculture has set a maximum of 3.5% for the use of cellulose in meats. Consumer awareness of the issue was raised in 2011, when a class action lawsuit was filed against Taco Bell Corporation for its claim of using "seasoned beef" in its products. The plaintiffs claimed that the products contained so many binders and extenders that it was only 35% meat and therefore did not meet the US Department of Agriculture's minimum requirement to be labeled as beef. The lawsuit was dismissed when

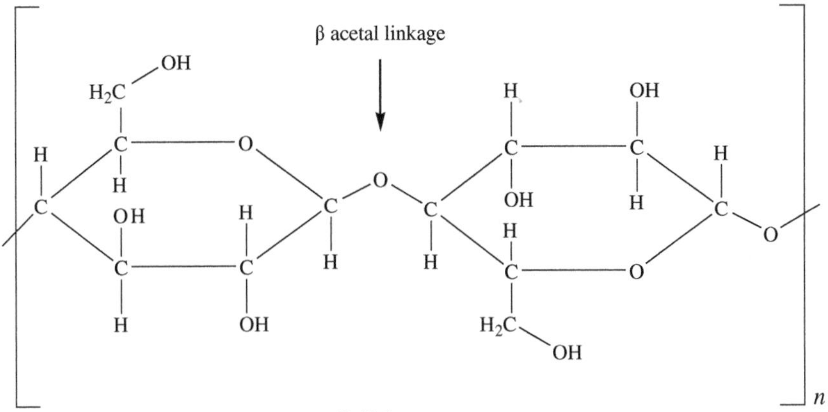

Cellulose

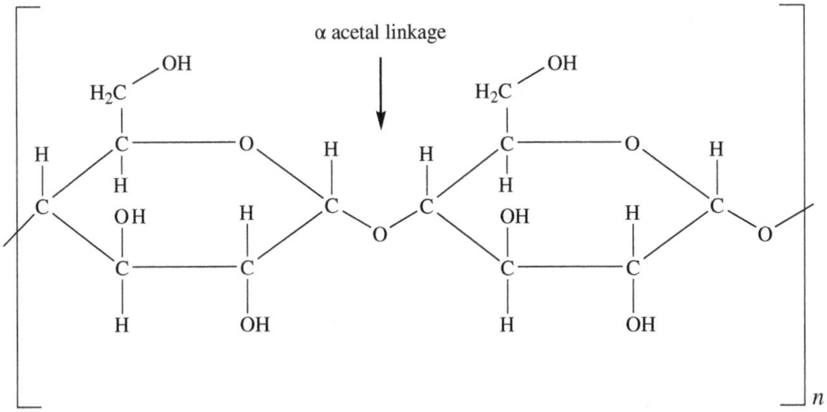

An amylose molecue from starch

Figure 4.5 A comparison of the structures of cellulose and starch.

Taco Bell proved that its seasoned beef was 88% meat; the other 12% was cellulose and other ingredients. Some other well-known products that contain added cellulose are Aunt Jemima's® Original Syrup, MorningStar Farms® Chik'n Nuggets, Eggo® Blueberry Waffles, many Betty Crocker® whipped frostings, many Duncan Hines® cake mixes, McDonald's® Fish Filet Patty, Taco Bell's® corn tortillas, Kraft® Foods Macaroni & Cheese Thick 'n Creamy, and many Nestle® Hot Cocoa Mixes. Cellulose is also added to ice cream as a thickening agent and because it adds creaminess. Many fast-food ice creams contain added cellulose, including McDonald's McFlurries® and Triple Thick

Milkshakes®, Jack-in-the-Box® ice cream shakes, Wendy's Frosties™, and Sonic® ice cream. It is a common practice for food companies to use a small amount of cellulose to keep grated parmesan from clumping, with 2–4% being considered acceptable in the industry.

Another commercial use of cellulose is its chemical reconstitution into cellophane. The polymeric properties of cellulose allow it to be turned into a thin film. Since the mid-1930s, cellophane has been used to wrap and pack foods. Cellophane is also the base of Scotch™ tape and the packaging material for cigars. The reconstitution process starts with dissolving cellulose, usually from wood, in alkali and carbon disulfide to make a solution called viscose. This solution is then extruded into a bath of sulfuric acid and sodium sulfate, where it is reconverted to cellulose. Although the sales of cellophane have declined in the past, it has recently returned as a popular food wrapping because it is 100% biodegradable.

Have you ever worn a piece of wood? Rayon is an artificial textile material composed of regenerated cellulose, which is typically derived from wood. Rayon was the first man-made fiber and was originally developed as a substitute for silk. First produced commercially in the United States in 1910, rayon is commonly used for clothing, medical surgery products, and even the filling in Zippo® lighters. Because of their abundance and ease of cultivation, pine, spruce, and hemlock are major species of tree used to provide the wood fibers needed in the manufacturing of rayon. Although rayon is derived from natural sources, the product is still considered artificial due to the extensive processing required in its creation. The processing starts with turning the wood into a pulp. Cellulose is then removed from the pulp and purified. The purified cellulose is converted to cellulose xanthate through a reaction with carbon disulfide. The cellulose xanthate is dissolved in sodium hydroxide, forced through spinnerettes, and then introduced in a bath of acids and salt. The acid bath causes the loss of carbon disulfide that converts cellulose xanthate back to cellulose, causing it to coagulate and form a solid filament. The filaments can be manipulated during the manufacturing process to control luster, strength, and size. Rayon is used because it is more economical than other fiber alternatives, such as cotton. In 2010, cotton prices surged 91%, which has caused rayon to make a return in popularity. The price of rayon is relatively cheap at $2.35 per yard compared to $4 per yard for cotton.

Chalk Used to Be Alive

Chalk is made from the fossilized remains of marine microorganisms. Formed in deep ocean conditions, chalk is a sedimentary rock that arises from the accumulation of shells shed from organisms such as plankton, algae, sponges, brachiopods, echinoderms, and mollusks. The shells of these microorganisms are made of a form of calcium carbonate ($CaCO_3$). As they die, their shells are deposited in layers on the bottom of the ocean. Over millions of years, the weight of the sediment transforms these shells into a type of rock. After learning this fact, I have never been able to look at a piece of chalk quite the same. Every time I use a chalkboard I feel like I am using a bone to write with.

Chalk is quite common and comprises roughly 4% of the Earth's crust. An example can be seen in the England's White Cliffs of Dover, which are made of chalk. Chalk is just one form of calcium carbonate. Other forms include limestone, which is a more compact form of calcium carbonate, and marble, which is a crystalline form of calcium carbonate, created when this molecule is recrystallized under high temperature and pressure.

In addition to its use on chalkboards, calcium carbonate is commercially applied as a dietary calcium supplement, antacid, and filler material for pills. Sometimes, it is added to soy and almond milk products to enhance their calcium content. In toothpaste, it functions as a mild abrasive, and in wine, it tempers acidity. Calcium carbonate used for industrial purposes is typically mined or quarried. If intended for human consumption, as in food or pharmaceuticals, it is extracted from a pure source, such as marble.

Decaffeinated? Try Deflavored!

If you are an avid coffee drinker, you may have often wondered about the odd and rather disappointing taste of decaffeinated coffee. It is well known that caffeine, which is bitter and provides some flavor, is removed from decaffeinated blends, but does that alone explain the drastic change in flavor?

The first thing to note is that a cup of coffee contains more than just caffeine. Some of the molecules in coffee are responsible for its aroma and taste. In fact, more than 950 compounds have been identified in roasted coffee. A good cup of coffee is a complex but delicate balance of all these molecules. Coffee manufacturers actually hire taste testers to make sure that the combination is correct for their product. Before you start filling out a job application, you should know that taste testers do not actually drink the coffee but end up spitting out the samples.

To understand the problem with decaffeinated coffee, it is necessary to be familiar with the process by which caffeine is removed. When coffee is brewed, the hot water separates the caffeine and flavor-producing molecules from the coffee grounds. Similarly, during the decaffeination process, caffeine is taken out of the coffee, but unfortunately, many of the flavor molecules are also removed. Manufacturers attempt to compensate for the loss by capturing the removed flavor molecules and readding them to the coffee. However, in my opinion, the flavor is still not quite the same. The extracted caffeine is used for medicinal and other purposes described later (Figure 4.6).

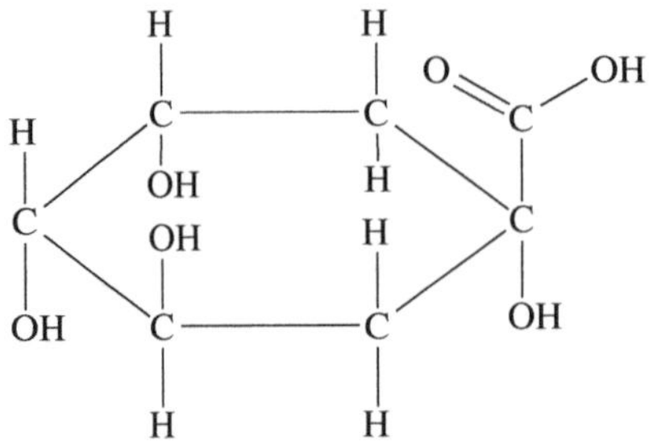

Quinic acid (a sour flavor) 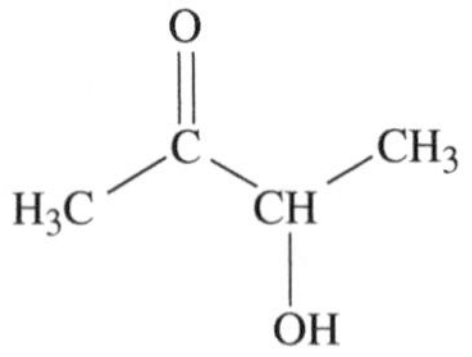   Acetoin (a buttery flavor)

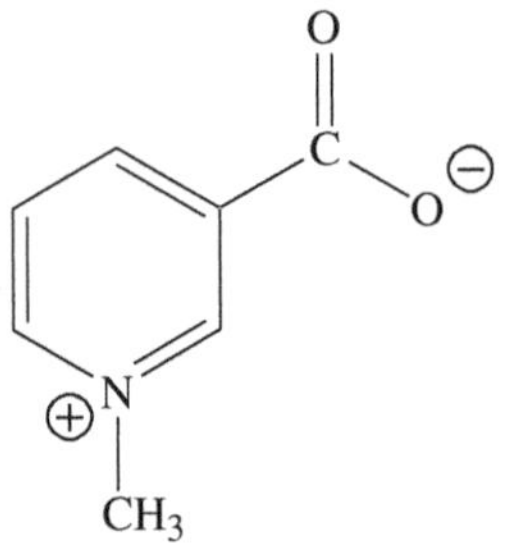

Trigonelline (an earthy flavor)

Figure 4.6 Some of the molecules responsible for the flavor of coffee.

Ludwig Roselius, the leader of a group of German researchers who originally developed decaffeinated coffee, registered the first patent for decaffeination in 1905. The European or traditional method of decaffeinating coffee, devised by Roselius and his scientists, made use of organic solvents while steaming and then soaking the beans. The caffeine is dissolved by the organic solvents and thereby removed. The beans are then dried and roasted like other green coffee beans. The solvents used are usually methylene chloride, now known to be a carcinogen, or ethyl acetate, another toxin. Although ethyl acetate is called a "natural solvent" because it can be found in nature, the ethyl acetate used in decaffeination is made artificially. Despite these solvents being removed as part of the process, their use worries some coffee lovers.

In the 1970s, a decaffeination process that uses carbon dioxide to remove the caffeine was patented by Kurt Zosel. His technique involves soaking presteamed beans in carbon dioxide in the supercritical phase, a state brought about by increasing temperature and pressure conditions until normally gaseous carbon dioxide takes on properties midway between a gas and a liquid. Carbon dioxide becomes supercritical at temperatures above 30.1 °C coupled with pressures above 72.8 atm. Supercritical fluids make excellent extraction solvents because they separate certain substances from a mixture, while leaving others unchanged. In addition, no harmful organic solvents are used. Zosel observed that caffeine is one of the substances that dissolve well in supercritical carbon dioxide and applied this property to coffee decaffeination. High-pressure vessels (operating at 250–300 times atmospheric pressure) are utilized to move the supercritical carbon dioxide through a bed of premoistened, green coffee beans. The carbon dioxide process is becoming one of the more popular methods of decaffeination. I cannot help but notice that the supercritical CO_2 extraction is also a popular method for extracting THC oil from marijuana.

Today, decaffeinated coffee accounts for approximately 12% of worldwide coffee consumption or nearly 1 billion pounds per year. The United States is the world's biggest consumer of decaf. Technically, decaffeinated coffee still contains traces of caffeine. The FDA requires that 97% of the caffeine be removed from coffee for it to be labeled "decaffeinated."

Now, what about decaffeinated soda? Quite simply, no caffeine is added. The term "caffeine free" is incorrect; a more accurate description would be "uncaffeinated." With the exception of colas (cola beans may contain a slight amount of caffeine), most other sodas that contain caffeine have it added by the manufacturers. Of course, most of the added caffeine is obtained from what is extracted from coffee beans during the decaffeination process!

Bad Blood

In 1901, Karl Landsteiner discovered the ABO groups while experimenting with blood transfusions. At that time, blood transfusions were risky because they often caused fatal clots in the recipient's body. In his research, Landsteiner collected blood samples from many people and separated the serum from the red blood cells. He mixed each donor's blood serum with a sample of someone else's red blood cells and then noted which combinations resulted in clotting and which did not. He was thus able to identify the first three blood types, that made it so that the recipients of blood transfusions could receive blood from donors who shared their blood type. This made blood transfusions much safer, and they are now a routine medical procedure that saves countless lives. Landsteiner received the Nobel Prize in Physiology and Medicine in 1930, partly for the discovery of these blood groups. Later on, a fourth blood type was identified.

The four blood groups are A (~43% of the population), B (~12%), AB (~5%), and O (~43%). The differences among the blood groups are caused by sugar molecules attached to the surfaces of red blood cells. The sugar molecules act as biochemical labels and must be in a certain configuration, or a person runs the risk of their body generating antibodies that will attack the red blood cells. Antibodies are the specialized cells of the immune system that recognize and destroy organisms such as bacteria, viruses, and fungi, which invade the body. A healthy person's blood does not attack itself, but if someone receives blood from a donor with a different blood type, there can be life-threatening consequences. A person with type A blood will produce antibodies against the red blood cells in type B blood and vice versa. Type AB blood produces no antibodies; people who have it are called *universal recipients* because they can receive blood from anyone. A person with type O blood produces both A and B types of antibodies and is called a *universal donor* because his or her blood can be given to anyone who needs a transfusion. An interesting example in popular media occurs in the 2015 movie, *Mad Max: Fury Road*, where the main character, Max Rockatansky, is captured by a gang called the War Boys. After determining that Max is type O, he is used as a "blood bag" to supply blood to a sick War Boy named Nux (Figures 4.7–4.9).

When someone receives a transfusion from a donor with an incompatible blood type, an *acute hemolytic transfusion reaction* (AHTR) can occur. The recipient's antibodies attack the donor's red blood cells, causing them to hemolyze (break open) and release harmful substances into the bloodstream.

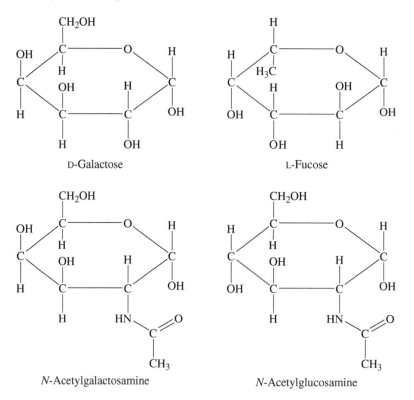

Figure 4.7 The structure of sugars found on the surface of red blood cells.

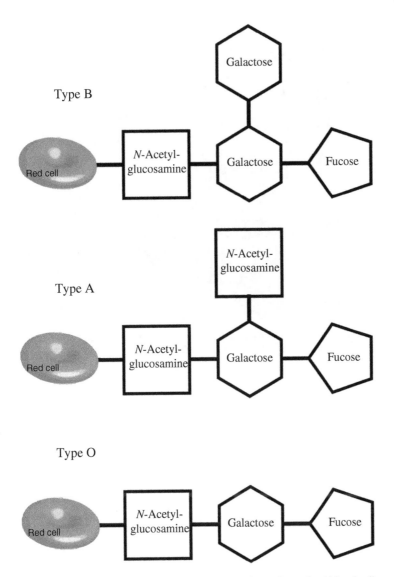

Figure 4.8 The orientation of different sugars on the surface of red blood cells.

The onset of AHTR symptoms can result from receiving as little as 5–20 ml of blood and usually occur within 24 hours of the transfusion. These symptoms include chills, fever, back pain, nausea, and death. The kidneys may become damaged enough to require dialysis. Fortunately, AHTRs are rare, occurring in approximately 1 of every 77,000 transfusions and causing about 20 deaths per year in the United States (Table 4.1).

He's a universal donor.

Figure 4.9 Another definition of a "universal donor."

Table 4.1 The compatibility of different blood groups.

Blood type	Can donate blood to	Can receive blood from
O	AB, A, B, O	O
AB	AB	O, AB, A, B
A	AB, A	O, A
B	AB, B	O, B

The Problem with Dry Cleaning

The process of dry cleaning involves the use of an organic solvent instead of water to remove dirt and stains from clothes. As most of us have experienced, many greasy stains cannot be washed away with water. The OH bonds in water allow it to be able to interact with other molecules and itself with a relatively strong intermolecular force called hydrogen bonding. Other molecules with an OH bonds, such as ethanol or sugar, tend to be able to dissolve in water because they can share the hydrogen-bonding intermolecular force (Figure 4.10).

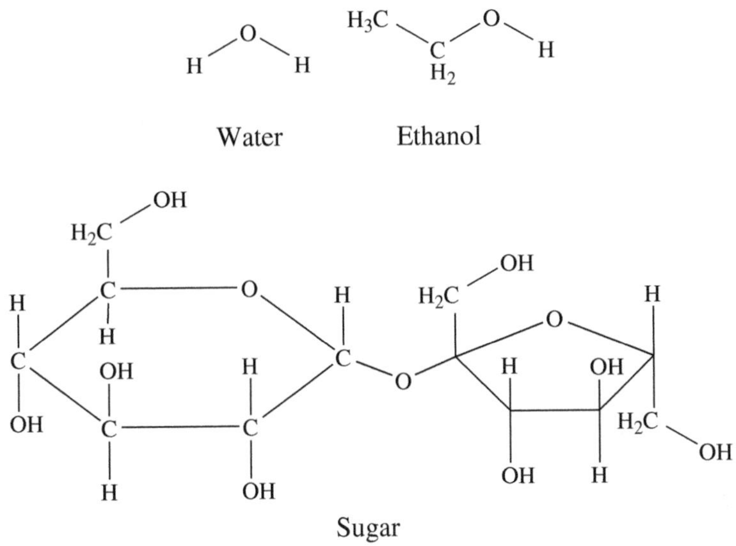

Figure 4.10 A figure showing that water, ethanol, and sugar all contain –OH groups.

Grease molecules interact with other molecules with a much weaker inter-molecular force call an instantaneous dipole. In order for grease to dissolve in water, it would have to overcome the hydrogen-bond interaction in water, which is not energetically favorable. Water would rather interact with itself so the grease molecules are literally pushed out of the water. This effect is clearly seen when you see drops of oil floating on water. Traditionally, detergent is added during washing to allow the grease to dissolve. However, some fabrics are adversely affected by contact with water. The strands of some fabric, such as cotton, are held together by the hydrogen-bonding intermolecular force. Because water can also form hydrogen bonds, it can affect how these fabric stands interact causing shrinkage, wrinkles, or damage, so an alternative to water is desirable. Many organic solvents lack hydrogen-bonding groups so they can easily dissolve the stains on clothing without causing damage to the fabric (Figure 4.11).

Jean Baptiste Jolly invented dry cleaning in the mid-nineteenth century after he noticed that his tablecloth was cleaner after his maid accidently spilled kerosene on it. Early dry cleaners used petroleum-based solvents, such as gasoline and kerosene, to clean clothes. These flammable solvents predictably caused many accidental explosions and fires. After World War I, dry cleaners began using less flammable and more effective chlorinated organic solvents. By the mid-1930s, the dry cleaning industry had started using *tetrachloroethylene* (PERC) as the solvent because it was stable, fairly nonflammable, had excellent cleaning power, and is gentle to most fabrics. PERC removes fats, greases, and

Cellulose

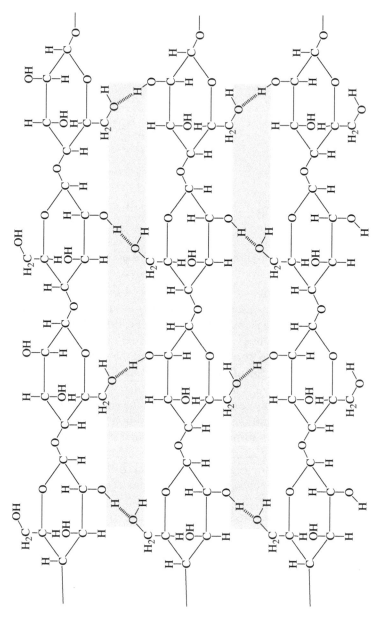

Figure 4.11 How cellulose fibers are held together by hydrogen bonds. The hydrogen bonds are dashed and highlighted in gray.

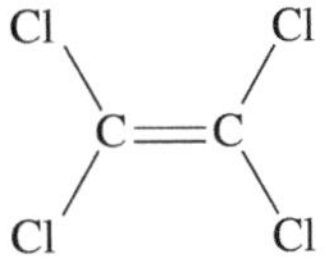

Figure 4.12 The structure of PERC.

$$\begin{array}{c} \overset{\text{Cl}}{\diagdown}\overset{\text{Cl}}{\diagup} \\ \text{C} = \text{C} \\ \overset{\diagup}{\text{Cl}}\overset{\diagdown}{\text{Cl}} \end{array}$$

Figure 4.13 The structure of carbon dioxide.

$$O = C = O$$

oils without harming natural or manmade fibers. In addition, PERC lacks an –OH group, which means that it cannot form hydrogen bonds or affect the hydrogen bonds between the cellulose strands. It is PERC you smell when you go into a dry cleaning establishment and catch a whiff of that sharp, sweet smell. In 1992, the US annual production capacity of PERC was estimated at 400 million pounds (Figure 4.12).

The issue with the use of PERC in dry cleaning is that this solvent, a contaminant, is released into the environment. Methods of recycling PERC produce contaminated sludge that, when disposed of in a landfill, leaches PERC into the soil and eventually into the water supply. Data from 1992 reveals that an estimated 12.3 million pounds of PERC were released to the atmosphere, and 10,000 pounds were leached to surface water in the United States alone. Studies have confirmed that much of the water supply in the United States has some PERC contamination, and the Center for Disease Control estimates that 74% of Americans have measurable amounts of PERC in their blood.

On 25 January 2007, California became the first state to ban PERC and other chlorinated solvents for dry cleaning. By 2023, no dry cleaning machines that use PERC will be permitted in the state. In 2011, the US Environmental Protection Agency listed PERC as a carcinogen to humans. Drycleaners will have to adopt greener cleaning solvents such as supercritical carbon dioxide, created when this gas is subjected to high temperatures and pressures (see previous story discussing decaffeinating coffee). A supercritical fluid has properties of both a liquid and a gas. The liquid property of supercritical carbon dioxide allows it to act as a solvent that removes stains from fabric, and because carbon dioxide is already present in the atmosphere, there are no real detrimental environmental effects if some escapes during the cleaning process. Carbon dioxide, like PERC, does not contain a hydrogen bonding –OH group and therefore does not damage clothes like water (Figure 4.13).

The Smell of Dead Fish

In order to survive, sea creatures must find a way to counteract *osmosis*, water's tendency to pass through membranes, such as a cell, to equalize the concentrations of materials on either side. If the fluids on either side of a membrane do not have equal concentrations of material, osmosis will cause water to pass through the membrane to the more concentrated side until the concentrations are equal. Osmosis is a problem for marine species because the high salt concentration of seawater will dehydrate their cells causing them harm. Trimethylamine oxide (TMAO) is an *osmolyte*, an organic compound found in the tissues of marine organisms to fight osmosis. TMAO and other osmolytes allow marine organisms to prevent osmosis and maintain a constant water content.

After a fish dies, bacteria begin to degrade TMAO to trimethylamine (TMA), which is partially responsible for the "fishy" odor of dead fish. In fact, the presence of TMA is used as a spoilage indicator for commercial fish. A truly fresh fish should not have an odor because bacteria have not had a chance to degrade the TMAO.

The structure of the TMAO molecule contains both positive and negative charges. The ionic nature of TMAO causes very strong intermolecular interaction between the molecules, making it relatively nonvolatile. Its melting point is between 428 and 432 °F, and its boiling point has not been established. Because smells are caused by evaporated molecules, TMAO has virtually no detectable odor. However, after a fish dies, bacteria begin a decomposition process in which the ionic portions of TMAO are removed, leaving TMA. Held together with weaker, nonionic dipole intermolecular forces, TMA is much more volatile than TMAO and readily releases molecules to the atmosphere (Scheme 4.3).

The volatility of TMA is reflected in its low boiling point of only 37 °F. Since room temperature is typically 68 °F, any TMA generated by bacterial decomposition rapidly evaporates. That is when the dead fish starts to stink.

How do we get rid of the fishy smell and taste in foods? A fishy smell can typically be removed from foods by adding a weakly acidic solution such as vinegar or lemon juice, which is why these are often served as condiments with seafood dishes. TMA is a base, so the addition of an acid starts a chemical reaction. The nitrogen in TMA can accept the hydrogen from the acidic condiment to form what is called an ammonium salt. In particular, after reacting with vinegar, the TMA becomes trimethyl ammonium acetate (TMAA), with a boiling point of

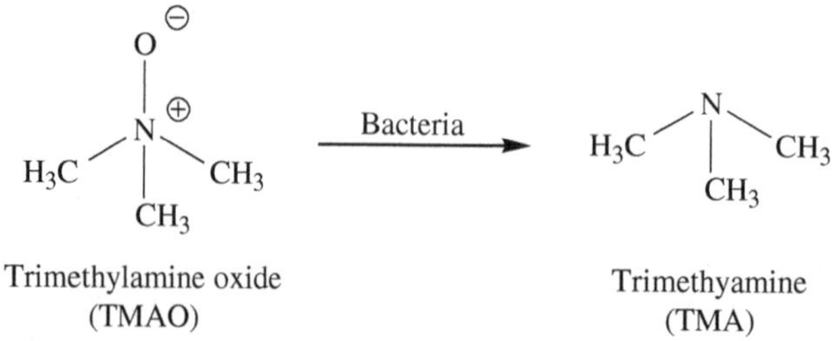

Scheme 4.3 The synthesis of TMA from TMAO.

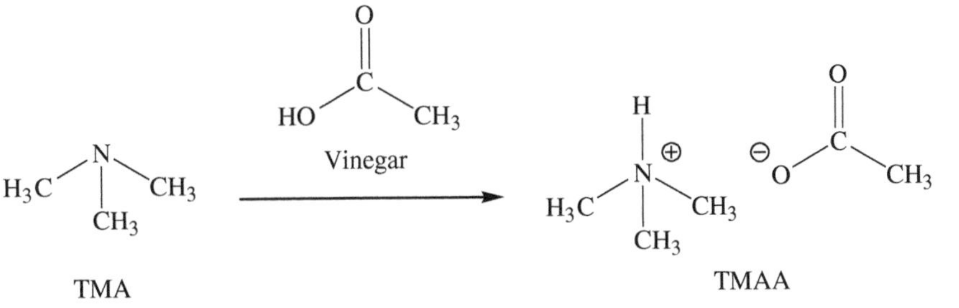

Scheme 4.4 The synthesis of TMAA from TMA.

328 °F. As with TMAO, an ammonium salt has ionic characteristics that make it much less volatile and therefore virtually odorless and tasteless (Scheme 4.4).

How to Make a Spark

Lighters use a combination of ferrocerium and steel to create the sparks that ignite flames. Invented in 1903, *ferrocerium* is an alloy containing iron (~19%), the rare-earth element cerium (~38%), and lanthanum (~22%), along with small amounts of neodymium, praseodymium, and magnesium. Because an alloy of rare-earth elements is too soft to be used for the purpose in mind, iron oxide and magnesium oxide are added to form a harder, more brittle material. Sparks are generated when the ridged steel wheel in a lighter shaves off a bit of ferrocerium. Because the cerium in the alloy has a low ignition temperature – between 302 and 356 °F – the shavings are ignited by the heat produced by friction. This property of ferrocerium rods makes them frequently used in the strikers for gas welding, cigarette lighters, and in fire starters in emergency survival kits with about 860 tons of ferrocerium being produced in the United States in 2012.

Ferrocerium is often mistaken for flint. Flint and steel produce sparks in a similar manner. However, with flint, which is harder than steel, striking the two causes small steel shavings to be chipped off and ignited. This same phenomenon also gives rise to sparks under other conditions. A prime example is a machinist using a grinder on a piece of metal. In this case, the hard grinding stone produces enough friction to ignite the small pieces of metal that are being ejected.

The "New Car Smell"

Polymers, paints, grease, and adhesives in a car's interior emit volatile organic compounds that produce the "new car smell." Unfortunately, there is no

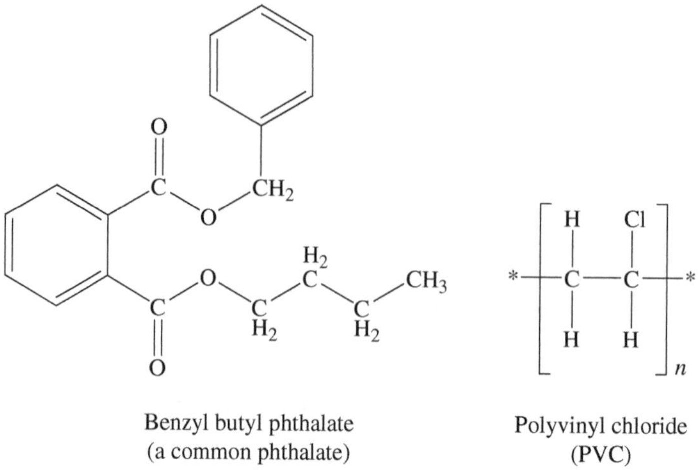

Benzyl butyl phthalate
(a common phthalate)

Polyvinyl chloride
(PVC)

Figure 4.14 The structures of benzyl butyl phthalate and PVC.

mandatory testing or regulation of chemicals inside vehicles sold in the United States, so it is basically up to car buyers to be aware of the hazardous materials to which they may be exposed. Testing has shown that there are more than 100 organic compounds that can be found in the air space of a new car. Many parts within a car's interior are made of the plastic, polyvinyl chloride (PVC). PVC by itself is an extremely hard plastic, and to make it more flexible, plasticizers are added. Plasticizers are small organic molecules that embed themselves in between the PVC polymer chains so that they cannot interact with each other as well. This reduced interaction makes the polymer more pliable. In PVC, the typical plasticizers are in a class of molecules called phthalates (Figure 4.14).

Because phthalates are slightly volatile, they can leach out and evaporate, producing part of the new car smell. The gradual removal of phthalates causes the PVC plastic to revert to its more ridged original form, which cracks. If you ever doubt that phthalates are volatile, leave a new car in the sun on a hot day. The phthalates will produce a white haze on the windshield!

Unfortunately, phthalates are carcinogens and have the ability to affect the endocrine system of mammals. Although there has been little investigation into the effects on humans, exposure to phthalates has been reported to result in increased incidence of developmental abnormalities such as cleft palate, skeletal malformations, and increased fetal death in experimental animal studies. The most sensitive system is the immature male reproductive tract, with phthalate exposure resulting in increased incidence of undescended testes.

Car companies have made a great effort to show that the phthalates are at safe levels, but their calculations have not taken into account hot days or the car

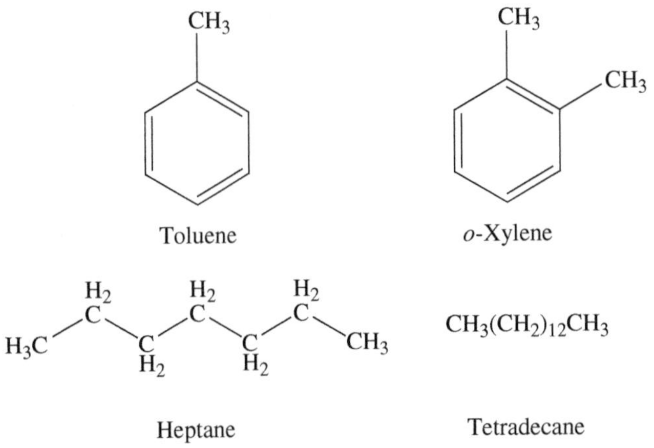

Figure 4.15 The structure of some of the molecules found in the new car smell.

being stored for long periods of time. A study in Japan found that the volatile organic chemicals in a new minivan were more than 35 times the safe level on the day after its delivery.

In fact, most plastic items – water bottles, for instance – contain phthalates. There are no legal limits for phthalates in bottled water because this particular industry blocked the FDA's proposal to set a legal limit. On 15 October 2007, California was the first state to ban the use of the plasticizers in toys designed for children under the age of 3. The ban was passed after consumers learned that phthalates had been added to baby bottles and teething rings in order to make them softer.

In addition to the phthalates in plastic, adhesives and paints used in a car's interior contain volatile organic compounds. Some of the major molecules found were toluene, xylene, tetradecane, and heptane. Unfortunately, these molecules are also intoxicants that produce a "high" when inhaled, leading some toxicology experts to compare "new car smell" to "huffing." To mitigate the effect, these experts suggest leaving the windows of a new car slightly open to air it out, keeping the car in a cool garage, or if parking outside, keeping it out of direct sunlight and putting up a windshield shade. If you like the "new car smell," however, there are commercially available aerosol sprays and air fresheners that attempt to reproduce it (Figure 4.15).

A Gecko Cannot Stick to It!

Teflon is the DuPont® trade name for the polymer polytetrafluoroethylene (PTFE). The chemical structure of Teflon is similar to the structure of

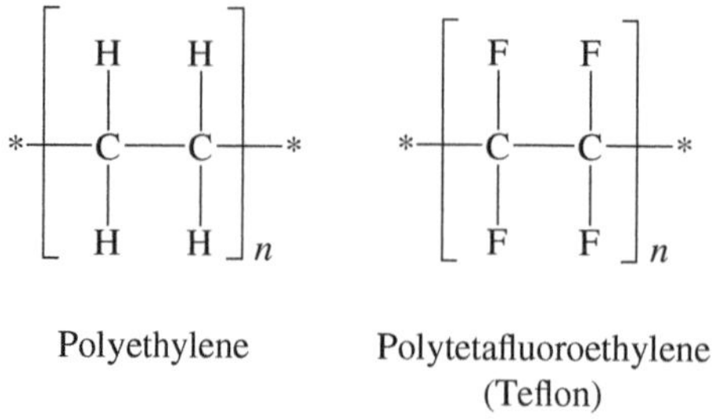

Figure 4.16 A comparison of the structures of polyethylene and Teflon.

Polyethylene

Polytetafluoroethylene
(Teflon)

polyethylene, the polymer used to make plastic bottles. The difference is that PTFE has fluorine atoms instead of hydrogen atoms (Figure 4.16).

In order for something to stick to something else – like a stamp on a letter, for example – there needs to be some kind of interaction on a molecular level. The molecules on the stamp and the molecules on the letter have to be attracted to each other by an intermolecular force. For many polymers, such as Teflon, the major intermolecular force is called instantaneous dipoles. A molecule is made up of atoms, all of which have positively charged protons in the nucleus, surrounded by negatively charged electrons. The electrons travel around the nuclei at close to the speed of light. Because of their high mobility, electrons can sometimes crowd – for a brief instant – one side of the molecule, causing a separation of charge (a dipole). The result is that one part of the molecule has a positive charge (from the protons) and another part has a negative charge (from the crowded electrons). The intermolecular force comes about when the positive charge on one molecule is attracted to the negative charge on a different molecule.

Teflon is distinguished by the presence of fluorine atoms. Fluorine is the most electronegative element on the periodic table, electronegativity being defined as the measure of an element's ability to draw electrons to itself in a chemical bond. Because fluorine is so electronegative, the electrons in Teflon are strongly pulled toward the fluorine, which limits the ability of Teflon to create the charge separation needed to create instantaneous dipoles. What this means is that Teflon has only a very limited ability to create intermolecular forces with other objects. This is the reason why it appears slippery.

Another aspect of Teflon's slipperiness is its inertness to chemical reactions. The large fluorine atoms shield the interior carbon atoms, making them difficult to react with. In addition, the carbon–fluorine bond is the strongest bond in organic chemistry so it is very difficult to break. In fact, Teflon is so slippery, it is the only known substance to which even a gecko cannot stick! (Figure 4.17).

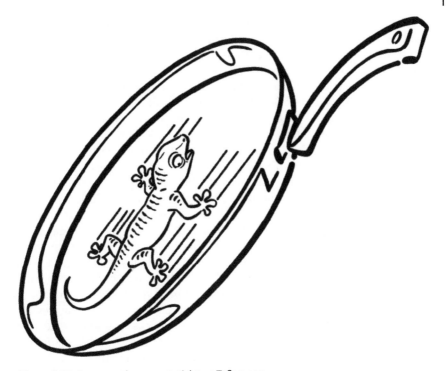

Figure 4.17 Even a gecko cannot stick to a Teflon pan.

Why Are Day Glow Colors and Highlighter Pens So Bright?

The color of some day glow inks and highlighters pens can be so electrically intense, it is almost painful to look at them. If you have ever wondered how this effect is achieved, the answer lies in optical brighteners that are added to these colored compounds. *Optical brighteners* are organic molecules that can *fluoresce*, meaning that their molecules absorb light of one type and emit light of a different, lower energy type.

As you are probably aware, there are many kinds of light, each with its own corresponding energy. These range from dangerous, very high-energy gamma rays, X-rays, and UV to lower energy and relatively safe visible light that we can see, as well as microwaves and radio waves.

Bringing the focus to just a couple of these, UV light can be compared to visible light. UV light is not part of the visible spectrum, is relatively of high energy, and can actually burn the skin. Although UV rays are associated with the sun, small amounts are present in artificial lighting, such as fluorescent lighting. By contrast, visible light can be seen by the human eye and is made up of the colors of the rainbow: red, orange, yellow, green, blue, and violet. Visible light is lower in energy than UV, and it cannot damage the skin. Most kinds of artificial light are primarily visible light, along with minimal quantities of invisible UV light.

All objects that appear colored to our eyes absorb and reemit visible light. However, objects that contain optical brighteners have an additional fluorescence process at work. During fluorescence, the optical brighteners absorb the UV light, which is invisible to our eye, and emit it as colored visible light, which we can see with our eyes. This converted UV light combines with the light normally reflected to create the unusually brilliant day glow effect. In other words, objects that contain optical brighteners emit more visible light than they absorb, thus appearing unusually bright. The process is depicted in Figure 4.18.

The fluorescence process has numerous commercial applications due to the eye-catching color brightness it produces. Day glow colors are used for advertising and warning signs. Highlighter pens are used to mark writing, and even tennis balls contain fluorescent molecules to help players' eyes track them. In fact, just about anything that appears unusually bright probably contains optical brighteners. Many objects containing the type of fluorescent molecules discussed here glow when placed under a black light. In case you were wondering, a black light is just a common fluorescent bulb that has a special coating that blocks all the generated light except the UV and a small amount of purple light. Under a black light, objects that contain fluorescent compounds will glow in a wide variety of colors. All other objects will have a dull purple hue.

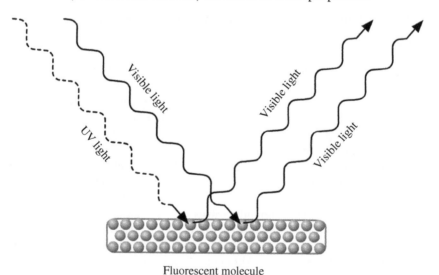

Fluorescent molecule

Figure 4.18 A representation of the fluorescence process in optical brighteners.

In fact, some tropical birds, such as parakeets and parrots, have fluorescent molecules in their plumage. This adaptation is beneficial because the males of many species use their brightly colored feathers to attract females. The fluorescent molecules actually make their feathers look even brighter to potential mates. Scientists have even tested the effect of fluorescent plumage by masking it in study groups. Sure enough, they found that the male birds with less day glow plumage were not as likely to attract female mates. It can be assumed that the male birds involved were not impressed by these scientific results.

This process is also used in counterfeit protection. The idea being that it would be almost impossible for counterfeiters to incorporate fluorescent inks into their product. If only a small amount of dye is used, the amount of UV light in ambient light is not sufficient to cause the color to appear. However, under a black light, they glow. Currently, $20 dollar bills have a strip that fluoresces green. You may see tellers put your bills under a light to check for this strip. Many credit and ATM card have a fluorescent emblem on them, typically a "V" or an eagle. Probably, the most impressive use is in driver's licenses. Modern California driver's licenses have both your date of birth and your portrait in fluorescent ink. In fact, it is possible to get a tattoo with fluorescent ink that only shows up under a black light.[2]

Another important commercial use of this process is seen in light sticks. Light sticks contain fluorescent molecules that can easily be seen by placing them under a black light. They will glow regardless of their being activated or not. In fact, activating the light sticks causes a chemical reaction that excites the fluorescent molecules in a similar manner as UV light. When you crack a light stick, what you are hearing is the breaking of a glass tube that allows two sets of chemicals to mix. A reaction subsequently occurs and colored light is produced.

Why Your White Clothes Are not Really White?

In the case of clothes, the cotton used to make fabric for clothing – even though it is chemically bleached – normally contains natural yellow/brown tones. *Whitening agents*, which are organic molecules that have the property of fluorescence, are added to white clothes to remove these tones. As discussed in the previous story, fluorescence is a process by which molecules absorb higher energy light and emit lower energy, visible light. During this fluorescence, cotton fabrics containing whitening agents convert the small amount of UV light in ambient lighting into blue light. The emitted blue light combines with the yellow/brown tones being emitted by the clothes to produce white light.

---

2 I made a promise to get one of these tattoos after the publication of this book. If you ever meet me, ask to see it.

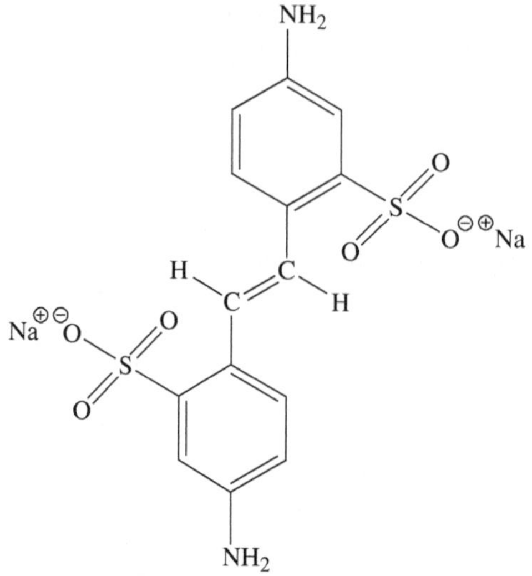

Figure 4.19 The structure of an optical brightener.

Disodium diaminostilbene disulfonate

This hides the yellow/brown tones and makes the treated fabric appear whiter. In the past, people would remove the dingy natural brown tones by dying white clothes with a bluing agent (ask your grandmother about this one!). Currently, whitening agents are added to most white cloth sold commercially.[3] In addition, whitening agents are added to many laundry detergents to replace the whitening agents removed from clothes during repeated washing. Looking at the website for the laundry detergent Tide®, you can see that they add disodium diaminostilbene disulfonate to some of their formulations to make clothes appear brighter and whiter (Figure 4.19).

This same process is used in white paper. Even though the paper pulp is bleached, a slight yellow color remains. Whitening agents are added to remove these yellow colors and make the paper appear white. Another interesting commercial application is the addition of fluorescent molecules to cheap teeth whiteners as an inexpensive way to get rid of the yellow stains on teeth.

The fact that white clothing, white paper, and laundry detergents contain fluorescent whitening agents is evidenced when these items are held under a black light. If you perform this experiment, you will note that they all glow a similar bluish color.

---

3 I once spent the better part of an hour at a fabric store shining a handheld black light on white cloth samples trying to find one that did not already have whitening agents added. There was only one in the entire store!

How Can a Spray-on Sunscreen Be Dangerous?

On 19 October 2012, the makers of Banana Boat® sun care products voluntarily recalled 23 varieties of UltraMist® sunscreen. Five people had caught fire after applying the product and then coming into contact with open flame. One man stood too close to a barbeque and caught fire after using the sunscreen causing severe burns on his chest, ear, and back. A woman wearing the sunscreen caught fire while welding and sustained second- and third-degree burns. The makers of UltraMist® – Energizer Holdings, Inc. – blamed the mishaps on a faulty spray valve that dispensed too much of the product, claiming that this caused the lotion to take longer to dry and increased its flammability.

Many aerosol hairsprays, deodorants, and sunscreens contain flammable ingredients such as grain alcohol. In fact, most spray-on sunscreens contain as much as 75–90% alcohol. Used in this way, the alcohol is a volatile solvent. After the application of the sunscreen, the solvent quickly evaporates, and only the product remains. However, since the solvent is volatile, it is also flammable. Some Banana Boat® products contain other flammable substances such as propane and isobutene. This is why UltraMist® products now carry the warning: "Keep away from sources of ignition – no smoking"! In a sad side note, it is a problem for severe alcoholics to puncture cans of aerosol products, including disinfectant spray and hair spray, to drain out and drink the contents to get at the alcohol contained within.

There Is Ink in That Paper

Most modern receipts, like the ones that come from automated teller machines (ATMs), are generated with thermal printing. *Thermal printing*, introduced by National Cash Register Company in 1968, is a cheap, clean printing method for cash registers, ATM receipts, fax machines, and medical instruments such as an electrocardiograph. This type of printing has several commercial advantages, the most significant of which is the elimination of ink supply and maintenance required for transfer printing.

Thermal printing utilizes a special type of paper impregnated with chemicals that produce a color change when exposed to heat. During the printing process, the paper passes over a thermal printer head that heats parts of the paper, producing a black image. The thermal paper contains a molecule called a *leuco dye*, which is either colorless or black, depending on the presence or absence of an acid. In addition to the dye, the paper contains an acid such as bisphenol A. The acid and leuco dye are both solids and therefore do not react with each other. However, when the paper is heated, the two molecules melt and mix, causing a reaction. During the reaction, the colorless leuco dye is converted to a form which is black in color. This form remains when the reaction matrix solidifies, so the black color is retained. Thermal paper is actually quite sensitive to heat, which is why receipts turn black if left too long in the sun. Typically, this reaction occurs at 109 °F, so it can easily be demonstrated by putting a receipt under a hairdryer; the receipt will turn black. Moreover, you may have seen someone "magically" write on a receipt using their fingernail. In this case, the friction from their fingernail rubbing on the paper provides enough heat to cause this reaction to occur (Figure 4.20; Scheme 4.5).

The chemistry involved in the leuco dye/acid reaction makes a great science fair project or home experiment. First, get some thermal paper. The paper used in fax machines is often thermal paper; if you do not have access to a fax machine, thermal fax paper is available in office supply stores. Then, place the paper in a cup with enough rubbing alcohol – preferably 100% isopropyl alcohol – to completely submerge the paper. The alcohol acts as a solvent and removes the leuco dye from the paper. Remove the paper and you will see that the solution is still clear. Next, add a small amount of white vinegar, which is a weak acid. This initiates the reaction, and the solution will turn dark. It gets even more fun because the reaction is reversible. Adding ammonia, a base, will neutralize the acid and turn the solution pale yellow.

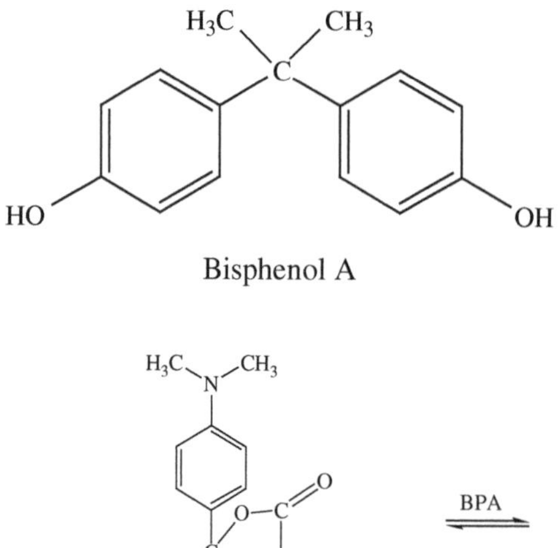

Figure 4.20 The structure of bisphenol A.

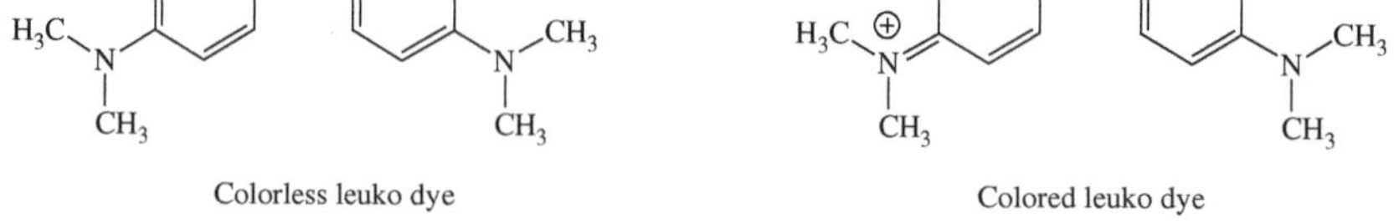

Bisphenol A

Colorless leuko dye

Colored leuko dye

Scheme 4.5 The formation of the colored form of a leuko dye.

Vomit and Sunless Tanners

The FDA has approved only one type of sunless tanning product, consisting of formulations containing *dihydroxyacetone* (DHA). DHA is a type of sugar typically isolated from sugar cane or derived by the fermentation of glycerin (Figure 4.21).

In the 1950s, Eva Wittgenstein, a scientist at the University of Cincinnati, studied DHA as a treatment for children with glycogen storage disease. The treatment required the children to receive large oral doses of DHA, sometimes as much as 1 g/kg of their body weight. Because of the large doses, sometimes the children spat out, spilled, or vomited the substance onto their skin and that of the health workers. It was noticed that the affected skin then turned brown after this would happen. This discovery was developed into the first sunless tanner, which Coppertone® introduced to the marketplace in the 1960s. This shows that even getting vomited on can be the mother of invention.

DHA produces a sunless tan by chemically reacting with the proteins found on the dead top layer of skin. Amino acids in the proteins react with DHA and produce molecules called *melanoidins*, which range in color from yellow to brown. Melanoidins are similar in color to melanin, the natural substance found in our deeper skin layers that provide a traditional tan color. Because melanoidins are chemically bonded to the proteins, the skin becomes stained. However, as only the outermost part of the skin is involved, the coloring sloughs off in 7–10 days. In case you are wondering, melanoidins provided by a sunless tan are only moderately effective for protecting against ultraviolet radiation, with a protection factor of 2–5 SPF.

The reaction used by sunless tanning formulations to form melanoidins is similar to the *Maillard reaction*, which describes the chemistry behind how bread and meat turn brown when heated. During the Maillard reaction, sugars and proteins in the meat or bread react to produce brown-colored compounds that represent the cooked product. In breads, multiple proteins called gluten, and sugar in the form of starch, are the precursors for this reaction. The compounds made by the Maillard reaction provide some of the flavor and smells of roasted meats and baked breads. Those of you who have ever tried to toast "gluten-free" products have noticed that it is virtually impossible. This is because the lack of gluten prevents the Maillard reaction from occurring.

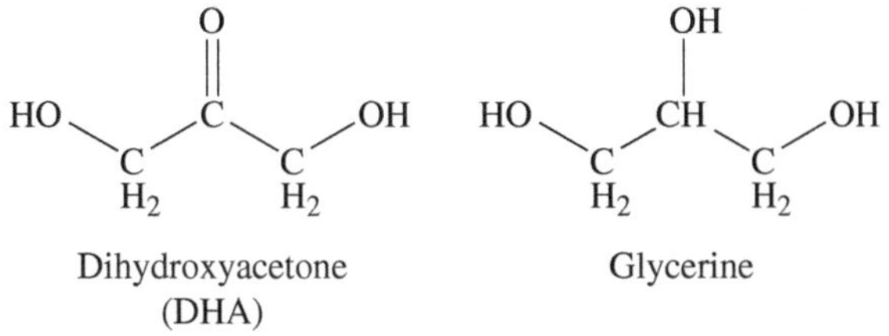

Figure 4.21 A comparison of the structures of DHA and glycerine.

Dihydroxyacetone
(DHA)

Glycerine

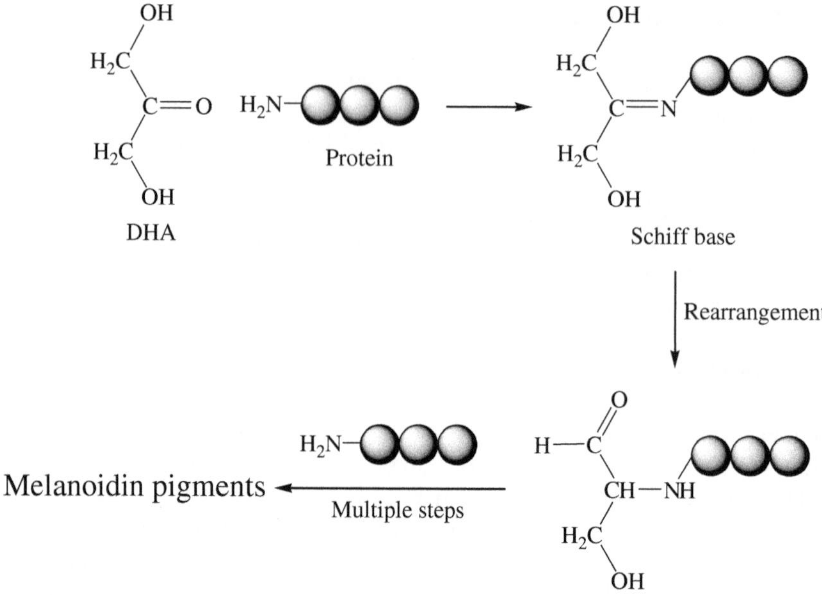

Scheme 4.6 The formation of melanoidin pigments from DHA and proteins.

Moreover, some of you may have noticed that tanning salons sometimes have a bread-like smell. I feel that this is probably due to the Maillard reaction being used. Finally, the Maillard reaction is also largely responsible for the browning of sugar-containing foods such as fruits.

The melanoidin pigmentation that gives a sunless tan develops over a period of hours after the application of DHA. This process is initiated by condensation of the DHA with an amino group, typically on a protein. Dehydration leads to the formation of an intermediate compound called a Schiff base. Subsequently, a rearrangement reaction follows leading to a compound called Heyns product. Because the C=O is reformed, this product continues to react with other amine-containing compounds in a similar series of reactions to produce complex, colored molecules (Scheme 4.6).

Formaldehyde: Funerals, Flooring, and Outer Space

Embalming is the introduction of a chemical agent into a corpse to preserve it long enough for transport and viewing. Initially, arsenic was used for embalming but was later replaced by *formaldehyde*, which, according to undertakers, was almost as good. Formaldehyde kills the cellular proteins in bacteria, temporarily preventing the decomposition of bodily organs. In addition, it fixes tissue in the body by irreversibly connecting amine groups in protein molecules via a $-CH_2-$ linkage, which results in the tissue firming up and appearing more "lifelike." This cross-linking stiffens and toughens the tissue so that it is less likely to be broken down by any remaining bacteria. Studies indicate that formaldehyde tends to react with collagen proteins, in particular the amino side chain of lysine and the nitrogen atom at the end of the protein chain. This reaction is very similar to the Maillard reaction discussed in the previous story (Figures 4.22 and 4.23; Scheme 4.7).

Today's embalming fluid is a mixture of formaldehyde, other less toxic chemicals, and water. Embalming fluid has a concentration of formaldehyde between 5% and 50%. An average of three gallons of embalming solution is used to prepare a corpse. Given that the typical funeral home embalms about 150 bodies

Figure 4.22 The structure of formaldehyde.

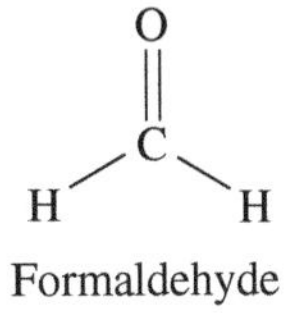

Formaldehyde

Figure 4.23 The structure of lysine.

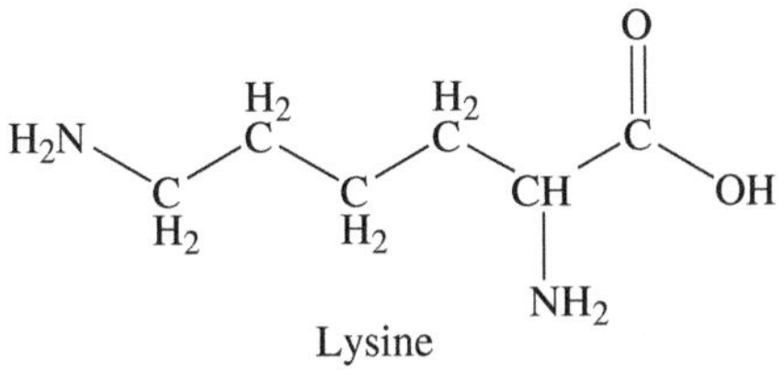

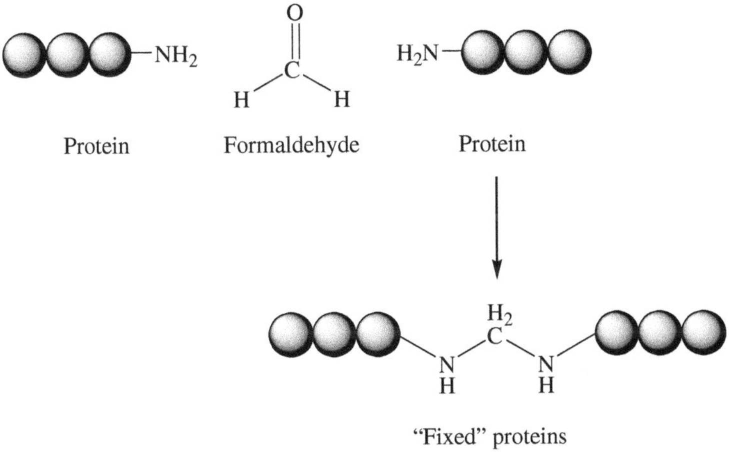

Lysine

"Fixed" proteins

Scheme 4.7 The reaction of proteins and formaldehyde.

per year, it is clear that a significant amount of formaldehyde is used in the United States.

After an embalmed body is buried and decomposes, the embalming fluid eventually seeps into the soil and then into the water ecosystem. Formaldehyde exposure has been shown to cause cancer in laboratory animals. In addition, several studies have found that embalmers and medical professionals who use formaldehyde have an increased risk of developing leukemia. Formaldehyde is carcinogenic in humans and animals due to the same cross-linking reaction that stiffens tissue as part of the embalming process. Because DNA also has amine groups, it can react with formaldehyde and thereby become dysfunctional. The damaged DNA is impaired in halting its own replication, and the unwanted replication of cells can lead to cancer (Scheme 4.8).

Formaldehyde is classified as a strong sensitizer substance that can cause an allergic reaction through recurring or prolonged contact. Formaldehyde release is such a problem that it is featured on the US Environmental Protection Agency's list of top 10% most environmentally hazardous chemicals. The growing awareness of the negative effects of embalming fluid on the environment has caused some companies to market "green burials" using

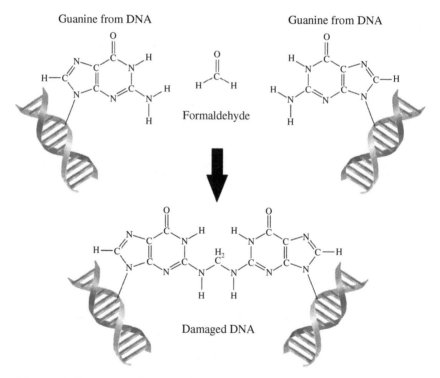

Scheme 4.8 The reaction of DNA and formaldehyde to form damaged DNA.

less toxic chemicals such as phenol for embalming. However, phenols are sanitizing agents, not preservatives, and are commonly used in mouthwash formulations. Furthermore, these chemicals do not work as well, cost nearly three times as much, and lack the cross-linking ability that gives corpses an everlasting effect.

Unfortunately, cremation does not solve the problem either. Formaldehyde released from the cremation process enters the atmosphere and remains suspended for up to 10 days. Because formaldehyde is soluble in water, it eventually bonds with the moisture in the atmosphere and enters precipitation. Human exposure to formaldehyde produces symptoms including excessive tearing, sore throat, runny nose, sneezing, asthma, skin rashes, blisters, headaches, and nausea.

Besides its use in funeral homes, formaldehyde is a common chemical in thousands of products such as wallpaper and paints. In particular, formaldehyde is in the glues used to bind wood particles to make plywood, particleboard, and laminate flooring. Low levels are normally present in these products, and the construction industry is monitored by the US government. Anyone doing home improvements who is interested in limiting exposure to formaldehyde should steer clear of pressed-wood products or install wood products labeled as ULEF (ultra-low-emitting formaldehyde), NAF (no added formaldehyde), or CARB (California Air Resources Board) Phase 1 or Phase 2 compliant.

The presence of formaldehyde in these products came to light in 2015 when Lumber Liquidators, an American retailer of hardwood flooring, was found selling a product that contained unsafe amounts. At the time, Lumber Liquidators obtained much of its laminate flooring from China, where there is less regulation of chemicals than in the United States. On average, the flooring contained formaldehyde levels 6–7 times the state standard, with some samples measured at 20 times as high.

During the investigation, employees openly admitted that they used core boards with higher levels of formaldehyde, thereby saving the company 10–15% in material costs. In addition, employees admitted falsely labeling the company's laminate flooring as CARB 2. The US Center for Disease Control estimated that the formaldehyde produced by the flooring at that time would cause 6–30 cancer cases per 100,000 people exposed. Considering 100 million square feet of the company's cheaper laminate flooring is installed in American homes every year, the potential for exposure is quite high. Investigators speculated that the noncompliant wood was used to save money and increase profits. In fact, based on these profits, Lumber Liquidators' stock price soared from $13 per share in 2011 to $119 in 2013. Since the investigation, however, Lumber Liquidators' stock price has plummeted to $11 per share.

Finally, some smokers dip their tobacco or marijuana cigarettes into embalming fluid, dry and then smoke them. The formaldehyde supposedly produces a hallucinogenic effect and also causes the cigarette to burn more slowly. Each

Figure 4.24 Formaldehyde: interstellar traveler.

formaldehyde-treated tobacco cigarette is sold for about \$20 and is known on the street as "wet," "fry," or "illy."

After reading this story, it may seem like formaldehyde is everywhere. It is fairly ubiquitous, present in wood, kerosene, oil, natural gas, gasoline, and cigarettes. It can even be generated by personal computers, laser printers, and photocopiers. A little-known fact is that formaldehyde was the first polyatomic organic molecule detected in interstellar space! (Figure 4.24).

Further Reading

The Smell of Old Books and the Hidden Vanilla extract Underworld

Bensaid, FF, Wietzerbin, K & Martin, GJ 2002, 'Authentication of natural vanilla flavorings: Isotopic characterization using degradation of vanillin into guaiacol', *Journal of Agricultural and Food Chemistry*, vol. 50, pp. 6271–6275.

Heitner, C 1993, *Photochemistry of Lignocellulosic Materials*, American Chemical Society, Washington, DC, Chapter 1, pp. 2–25 ACS Symposium Series, Volume 531.

Kamlet, J & Mathieson, O 1953, *Manufacture of vanillin and its homologues*, U.S. Patent 2,640,083. U.S. Patent Office.

Mosca Conte, A, Pulci, O, Knapik, A, Bagniuk, J, Del Sole, R, Loiewska, J & Missori, M 2012, 'Role of cellulose oxidation in the yellowing of ancient paper', *Physical Reviews Letters*, vol. 108, p. 158301.

Pearl, IA 1942, 'Vanillin from lignin materials', *Journal of the American Chemical Society*, vol. 64, pp. 1429–1431.

Strlic, M, Thomas, J, Trafela, T, Csefalvayova, L, Kralj Cigic, I, Kolar, J & Cassar, M 2009, 'Material degradomics: On the smell of old books', *Analytical Chemistry*, vol. 81, pp. 8617–8622.

That Smell is You!

Ball, P 2006, *A 'metallic' smell is just body odour*, Nature News, Viewed 4 October 2016, http://www.nature.com/news/2006/061023/full/news061023-7.html

Glindemann D, Dietrich, A, Staerk, HJ, Kuschk, P 2006, 'The two odors of iron when touched or pickled: (Skin) Carbonyl compounds and organophosphines', *Angewandte Chemie International Edition*, vol. 45, pp. 7006–7009.

Electric Blue

Uman, MA 1984, *All About Lightning*, Courier Dover Publications, New York.

The World's Most Abundant Organic Compound

Davies, GJ, Dodson, GG, Hubbard, RE, Tolley, SP, Dauter, Z, Wilson, KS, Hjort, C, Mikkelsen, JM, Rasmussen, G & Schulein, M 1993, 'Structure and function of endoglucanase V', *Nature*, vol. 365, pp. 362–364.

Delmer, DP, Haigler, CH 2002, 'The regulation of metabolic flux to cellulose, a major sink for carbon in plants', *Metabolic Engineering*, vol. 4, pp. 22–27.

Fink, H, Ganster, J & Lehmann, A 2013, 'Progress in cellulose shaping: 20 years industrial case studies at Fraunhofer IAP', *Cellulose*, vol. 21, pp. 31–51.

Klemn, D, Heublein, B, Fink, HP & Bohn, A, 2005, 'Cellulose: Fascinating biopolymer and sustainable raw material', *Angewandte Chemie International Edition*, Vol. 44, pp. 3358–3393.

Nassauer, S 2011, *Why wood pulp makes ice cream creamier*, The Wall Street Journal, Viewed 4 October 2016, http://online.wsj.com/article/SB10001424052748703834804576300991196803916.html

Pettersen, RC 1984, *The chemical composition of wood*, U.S. Department of Agriculture, Viewed 4 October 2016, http://www.fpl.fs.fed.us/documnts/pdf1984/pette84a.pdf

Reimer, M 2011, *15 food companies that serve you 'Wood'*, TheStreet, Viewed 4 October 2016, http://www.thestreet.com/story/11012915/2/cellulose-wood-pulp-never-tasted-so-good.html

TheStreet 2010, *Who's putting wood in your food?*, MSN, Viewed 4 September 2014, http://money.msn.com/shopping-deals/who-is-putting-wood-in-your-food-thestreet.aspx?cp-documentid=6792745

Chalk Used to be Alive

Goreau TF 1963, 'Calcium carbonate deposition by coralline algae and corals in relation to their roles as reef builders', *Annals of the New York Academy of Sciences*, vol. 109, pp. 127–167.

Industrial Minerals Association North America 2009, *What is calcium carbonate?*, Industrial Minerals Association North America.

Office of Dietary Supplements 2016, Calcium, *National Institutes of Health*, Viewed 4 October 2016, http://ods.od.nih.gov/factsheets/Calcium-QuickFacts/

Decaffeinated? Try Deflavored!

Blackstock, C 2004, *Scientists discover decaf coffee bean*, The Guardian, Viewed 4 October 2016, http://www.guardian.co.uk/science/2004/jun/24/food.research

Meyer, JF, Roselius, L & Wimmer, KH 1908, *Preparation of coffee*, US patent 897840.

Scientific American 2016, *How is caffeine removed to produce decaffeinated coffee?*, Scientific American, Viewed 4 October 2016, http://www.scientificamerican.com/article.cfm?id=how-is-caffeine-removed-t

Mazzafera, P 2012, 'Which is the by-product: Caffeine or decaf coffee', *Food and Energy Security*, vol. 1, pp. 70–75.

Bad Blood

American Cancer Society 2016, *Blood transfusion and donation*, Viewed 4 October 2016, http://www.cancer.org/treatment/treatmentsandsideeffects/treatmenttypes/bloodproductdonationandtransfusion/index

Katz, EA 2009, 'Blood transfusion friend or foe', *AACN Advanced Critical Care*, vol. 20, pp. 155–163

Beaumont Hospital Kidney Center 2016, *Living donor transplantation*, Beaumont Hospital Kidney Center, Viewed 20 January 2017, http://www.beaumont.ie/kidneycentre-forpatients-aguide-living

Sarode, R 2016, *Complications of transfusion*, Viewed 4 October 2016, http://www .merckmanuals.com/professional/hematology_and_oncology/transfusion_ medicine/complications_of_transfusion.html?qt=&sc=&alt=

Timberlake, KC 2006, *An introduction to General, Organic, and Biological Chemistry*, 7th ed, Benjamin/Cummings Publishing, Menlo Park, CA.

The Problem with Dry Cleaning

Betts, KS 1999, 'Technology Update: CO_2 gets taken to the cleaners', *Environmental Science & Technology*, vol. 33, pp. 170A.

Janssen, S 2010, *Congress must protect people from toxic chemicals known to cause harm: Tricholoroethylene (TCE)*, Natural Resources Defense Council, Viewed 4 October 2016, http://www.nrdc.org/health/files/tricholoroethylene.pdf

US Environmental Protection Agency 1994, *Chemical summary for perchloroethylene*, Viewed 4 October 2016, http://www.epa.gov/chemfact/s_ perchl.txt

US Environmental Protection Agency 2011, *EPA releases final health assessment for TCE*, Viewed 4 October 2016, http://yosemite.epa.gov/opa/admpress.nsf/0/ B8D0E4D8489AD991852579190058D6C3

The Smell of Dead Fish

Hebard, CE, Flick, GJ & Martin, RE 1982, *Occurrence and significance of trimethylamine oxide and its derivatives in fish and shellfish*, In *Chemistry and Biochemistry of Marine Food Products*, AVI Publishing Company, Westport, CT.

Seibel, BA & Walsh, PJ 2002, 'Trimethylamine oxide accumulation in marine animals: relationship to acylglycerol storage', *Journal of Experimental Biology*, vol. 205, pp. 297–306.

Sotelo, CG & Rehbein, H 2000, *Seafood Enzymes*, Marcel Dekker, New York, NY.

Yancey, PH, Clarke, ME, Hand, SC, Bowlus, RD & Somero, GN 1982, 'Living with water stress: Evolution of osmolyte systems', *Science*, vol. 217, pp. 1214–1222.

How to Make a Spark

Erean 2015, *REE in cigarette lighters flint: Ferrocerium*, Erean, Viewed 6 October 2016, http://erean.eu/wordpress/ree-in-cigarette-lighters-flint-ferrocerium/

Shanghai Jiangxi Metal Co. LTD 1995, *FerroCerium+Mischmetal(RE)*, Shanghai Jiangxi Metal Co. LTD, Viewed 6 October 2016, http://www.jxmetals.com/sdp/ 316680/4/cp-1271724/0.html

Treibacher Insustrie AG 2009, *Material safety data sheet*, Zippo.com.

U.S. Geological Survey 2015, *Rare earth*, U.S. Department of the Interior, Viewed 6 October 2016, http://minerals.usgs.gov/minerals/pubs/commodity/rare_earths/mcs-2015-raree.pdf

The "New Car Smell"

Chien, Y 2007, 'Variations in amounts and potential sources of volatile organic chemicals in new cars', *Science of the Total Environment*, vol. 382, pp. 228–239.
Faber, J, Brodizik, K, Golda-Kopek, A & Lomankiewicz, D 2013, 'Air pollution in new vehicles as a result of VOC emissions from interior material', *Polish Journal of Environmental Studies*, vol. 22, pp. 1701–1709.
US Department of Health and Human Services 1997, *Toxicological profile for di-n-octyphthalate*, Agency for Toxic Substances and Disease Registry, Viewed 4 October 2016, https://www.atsdr.cdc.gov/toxprofiles/tp95.pdf
US Environmental Protection Agency 2007, *Phthalates TEACH chemical summary*, US Environmental Protection Agency.
Verriele, M, Plaisance, H, Vandenbilcke, V, Locoge, N, Jaubert, J & Meunier, G 2012, 'Odor evaluation and discrimination of car cabin and its components: Application of the "Field of Odors" approach in a sensory descriptive analysis', *Journal of Sensory Studies*, vol. 27, pp. 102–110.

A Gecko Can't Stick to it!

Autumn, K & Peattie AM 2003, 'Mechanisms of adhesion in geckos', *Integrative and Comparative Biology*, vol. 42, pp. 1081–1090.
Autumn, K, Sitti, M, Liang, YA, Peattie, AM, Hansen, WR, Sponberg, S, Kenny, T, Fearing, R, Israelachvili, JN & Full RJ 2002, 'Evidence for van der Waals attachment by gecko foot-hairs inspires design of synthetic adhesive', *PNAS*, vol. 99, pp. 12252–12256.
O'Hagan, D. 2008, 'Understanding organofluorine chemistry. An introduction to the C–F bond', *Chemical Society Reviews*, vol. 37, pp. 308–319.
The Chemours Company 2016, *Fluoroplastic comparison - typical properties*, The Chemours Company, Viewed 5 October 2016, http://www2.dupont.com/Teflon_Industrial/en_US/tech_info/techinfo_compare.html

Why Are Day Glow Colors and Highlighter Pens So Bright? Why Your White Clothes are not Really White?

Blacklightworld 2016, *Black light uses the many uses for black lights*, Blacklightworld, Viewed 8 October 2016, http://www.blacklightworld.com/blacklight_uses.htm

Kuntzleman, TS, Rohrer, K, Emeric, S 2012, 'The chemistry of lightsticks: Demonstrations to illustrate chemical processes', *Journal of Chemical Education*, vol. 89, pp. 910–916.

Pearn, SM, Bennett, ATD & Cuthill, IC 2001, 'Ultraviolet vision, fluorescence and mate choice in a parrot, the budgerigar Melopsittacus undulates', *Proceedings of the Royal Society B-Biological Sciences*, vol. 268, pp. 2273–2279.

Pearn, SM, Bennett, ATD & Cuthill, IC 2003, 'The role of ultraviolet-A reflectance and ultraviolet-A-induced fluorescence in budgerigar mate choice', *Ethology*, vol. 109, pp. 961–970.

Procter & Gamble 2016, *Tide liquid*, Procter & Gamble, Viewed 4 October 2016, http://www.pgproductsafety.com/productsafety/ingredients/household_care/ laundary_fabric_care/Tide/Tide_Liquid_Original.pdf

Yilmaz, MD 2016, 'Layer-by-layer hyaluronic acid/chitosan polyelectrolyte coated mesoporous silica nanoparticles as pH-responsive nanocontainers for optical bleaching of cellulose fabrics', *Carbohydrate Polymers*, vol. 146, pp. 174–180.

What's in it? 2016, Diaminostilbene disulfonate disodium salt, consumer product information database, Viewed 20 December 2016, https://www .whatsinproducts.com/chemicals/view/1/4433

Zhang, G, Zheng, H, Guo, M, Du, L, Guojun Liu, G & Wang, P 2016, 'Synthesis of polymeric fluorescent brightener based on coumarin and its performances on paper as light stabilizer, fluorescent brightener and surface sizing agent', *Applied Surface Science*, vol. 367, pp. 167–173.

How Can Spray-On Sunscreen be Dangerous?

Jaslow, R 2012, *Man catches fire after applying spray-on sunscreen before grilling*, CBS News, Viewed 5 October 2016, http://www.cbsnews.com/8301-504763_ 162-57447309-10391704/man-catches-fire-after-applying-spray-on- sunscreen-before-grilling/

O'Connor, A 2012, *Is sunscreen flammable?*, The New Your Times, Viewed 5 October 2016, http://well.blogs.nytimes.com/2012/06/06/is-sunscreen- flammable/

Perrone, M 2012, *Banana boat recalls sunscreen due to fire risk*, The Huffington Post.

Playtex 2015, *Playtex products*, Playtex, Viewed 5 October 2016, http://www .playtexproducts.com/MSDS_BB.aspx

US Food and Drug Administration 2012, *Energizer Holdings Inc., announces the voluntary nationwide market withdrawal of several banana boat sun care products*, US Department of Health & Human Services, Viewed 5 October 2016, http://www.fda.gov/Safety/Recalls/ucm324824.htm

There is Ink in that Paper

Jasuja, OM & Singh, G 2009, 'Development of latent fingermarks on thermal paper: Preliminary investigation into use of iodine fuming', *Forensic Science International*, vol. 192, pp. e11–e16.

Hall, CS & Kurt, R 2012, *Tissue temperature indicating element for ultrasound therapy*, US patent, US 8323221 B2

Seeboth, A, Loetzsch, D, Potechius, E & Vetter, R 2006, 'Thermochromic effects of leuco dyes studied in polypropylene', *Chinese Journal of Polymer Science*, vol. 24, p. 363.

Bourque, AN & White, M 2014, 'Control of thermochromic behaviour in crystal violet lactone (CVL)/alkyl gallate/alcohol ternary mixtures', *Canadian Journal of Chemistry*, vol. 93, pp. 22–31.

Vomit and Sunless Tanners

Kennedy, P 2012, *Who made that spray tan?*, The New York Times, Viewed 7 October 2016, http://www.nytimes.com/2012/09/09/magazine/who-made-that-spray-tan.html?_r=0

Nguye, B & Kochevar, IE 2003, 'Infuence of hydration on dihydroxyacetone-induced pigmentation of stratum corneum', *Journal of Investigative Dermatology*, vol. 120, pp. 655–661.

Quispe, CAG, Coronado, CJR, Carvalho, JA 2013, 'Glycerol: Production, consumption, prices, characterization and new trends in combustion', *Renewable and Sustainable Energy Reviews*, vol. 27, pp. 475–493.

US Food and Drug Administration 2000, *Sunscreen labelling: Docket number 78N0038*, US Food and Drug Administration, Viewed 7 October 2016, http://www.fda.gov/ohrms/dockets/dailys/00/sep00/090700/c000578.pdf

US Food and Drug Administration 2015, *Sunless tanners & bronzers*, US Food and Drug Administration, Viewed 7 October 2016, http://www.fda.gov/Cosmetics/ProductsIngredients/Products/ucm134064.htm

Wittgenstein, E & Berry, HK 1960, 'Staining of skin with dihydroxyacetone', *Science*, vol. 132, pp. 894–895.

Wittgenstein, E & Berry, HK 1961, 'Reaction of dihydroxyacetone (DHA) with human skin callus and amino compounds', *Journal of Investigative Dermatology*, vol. 36, pp. 283–286.

Formaldehyde: Funerals, Flooring, and Outer Space

American Cancer Society 2014, *Formaldehyde what is formaldehyde?*, American Cancer Society, Viewed 5 October 2016, http://www.cancer.org/cancer/cancercauses/othercarcinogens/intheworkplace/formaldehyde

Bernard, TA 2010, *When wrinkle-free clothing also means formaldehyde fumes*, The New York Times, Viewed 5 October 2016, http://www.nytimes.com/2010/12/11/your-money/11wrinkle.html?_r=0

Chen, NH, Djoko, KY, Veyrier, FJ & McEwan, AG 2016, 'Formaldehyde stress responses in bacterial pathogens', *Frontiers in Microbiology*, vol. 7, pp. 1–17.

Seeboth, A, Loetzsch, D, Potechius, E & Vetter, R 2006, 'Thermochromic effects of leuco dyes studied in polypropylene', *Chinese Journal of Polymer Science*, vol. 24, p. 363.

Cooper, A 2015, *Lumber Liquidators linked to health and safety violations*, CBS News, http://www.cbsnews.com/news/lumber-liquidators-linked-to-health-and-safety-violations/

National Cancer Institute 2015, *Formaldehyde and cancer risk*, National Cancer Institute, Viewed 5 October 2016, http://www.cancer.gov/about-cancer/causes-prevention/risk/substances/formaldehyde/formaldehyde-fact-sheet

National Industrial Chemicals Notification and Assessment Scheme 2016, *Formaldehyde in embalming safety FactSheet*, Australian Government Department of Health.

US Drug Enforcement Administration 2015, *Marijuana and embalming fluid*, US Drug Enforcement Administration, Viewed 5 October 2016, http://www.dea.gov/pubs/states/newsrel/newark_intel_bulletin_embalming.html

Zuckerman, B, Buhl, D, Palmer, P & Snyder, LE 1970, 'Observation of interstellar formaldehyde', *Astrophysical Journal*, vol. 160, pp. 485–506.

5

Bath Salts and Other Drugs of Abuse

It seems odd that illicit drugs are not discussed more in chemistry classes. Illegal drugs were never discussed in the lectures I attended in my studies to become a chemistry professor, and they were mentioned only in passing in textbooks. During my chemistry career, their existence was only brought up when an ordered chemical was found to be on the FDA's (US Food and Drug Administration's) controlled substance list because it was a known precursor in the synthesis of an illicit drug. Peoples' interest in this area is demonstrated by the general reaction to the television show "Breaking Bad," the popularity of which is evidence that chemistry is fascinating if presented in a way with which people can connect. I have capitalized on this in my courses by engaging students in frank, scientific discussions about illicit drugs. Considering that the illegal manufacturing of drugs uses the same techniques taught in organic chemistry laboratories, the topic would appear quite relevant. Furthermore, because drug use is so prevalent in society and popular culture, it is one of the most relatable connections people understand between organic chemistry and everyday life.

What Are the Dangers of Bath Salts?

A few years back, there was a rash of news stories about people using bath salts to get high. At the time, it seemed likely that this was just another urban legend. In fact, these so-called *bath salts* are purposely mislabeled and have no actual use as bath additives. In fact, they contain synthetic variations of cathinone, an amphetamine-like stimulant found in the Khat plant (*Catha edulis*).

Khat is a flowering evergreen that grows wild in the Horn of Africa and in the southwest Arabian Peninsula. For centuries, the chewing of fresh Khat leaves for their gratifying stimulant effects has been a tradition in those regions. There were an estimated 10 million Khat users worldwide in 2006. The molecule cathinone was identified as the main psychoactive compound found in Khat leaves. Several structurally similar compounds that replicate the effect of this natural

Strange Chemistry: The Stories Your Chemistry Teacher Wouldn't Tell You, First Edition. Steven Farmer.
© 2017 John Wiley & Sons, Inc. Published 2017 by John Wiley & Sons, Inc.

amphetamine-like compound have been artificially created and they are used as recreational drugs. These cathinone derivatives first appeared in the European recreational drug scene in the early 2000s, and have since spread to the US market.

The most commonplace cathinone variations are *mephedrone, methylone,* and *methylenedioxypyrovalerone* (MDPV). Unfortunately, these cathinone derivatives were being legally sold in the United States without FDA approval because they were packaged as bath salts and labeled "not for human consumption." The FDA only intervenes when a molecule is designated for human consumption or is known to be abused, so for years "bath salts" could be legally bought almost anywhere, especially at head shops and online. In fact, the popularity of bath salts in the United States is attributed to a decrease in the availability of the more common drugs of abuse such as ecstasy, methamphetamine, and cocaine (Figure 5.1).

Commercial bath salts have been sold under brand names designed to entice consumers such as Ivory Wave, Bliss, White Lightning, and Hurricane Charlie. All of these contain MDPV, a molecule that can be snorted or injected to produce euphoria, hypervigilance, and increased sociability. The similarity in the effects of synthetic cathinones, amphetamine, and ecstasy is attributed to the resemblance of their chemical structures (Figure 5.2).

Although clinical evidence indicates that recreational doses of bath salts elevate the mood and increase alertness, high doses or chronic use can lead to rapid heart rate and increased body temperature, which is why users often

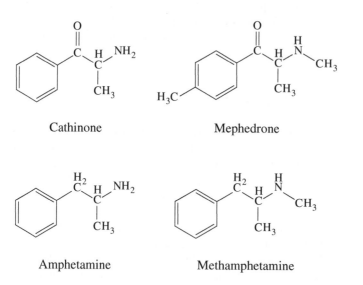

Cathinone

Mephedrone

Amphetamine

Methamphetamine

Figure 5.1 A comparison of the structures of cathinones and amphetamines.

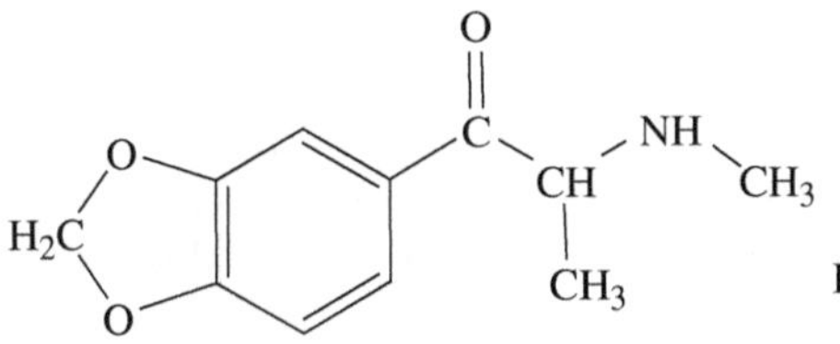

Methylone
(a bath salt)

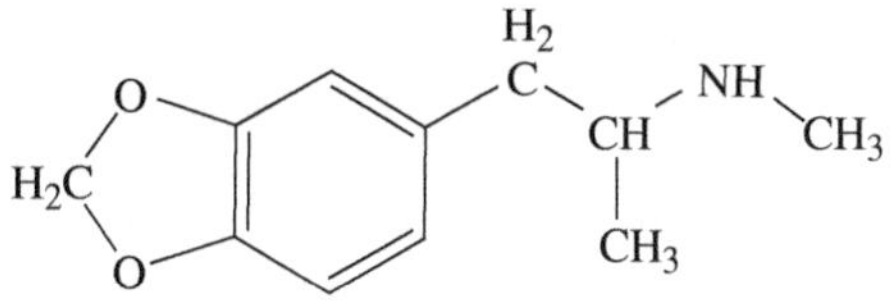

MDMA
(ecstasy)

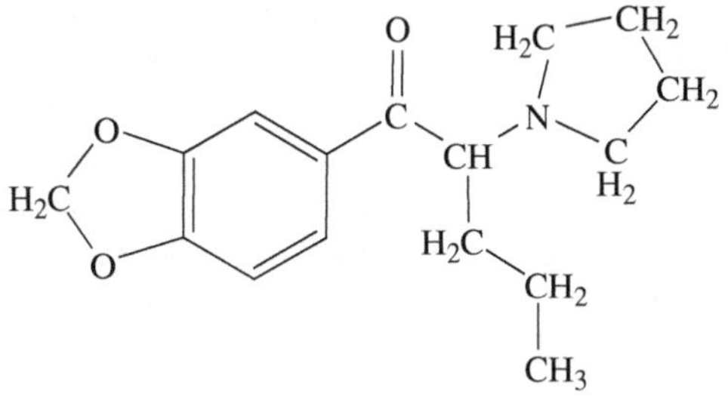

MDPV
(a bath salt)

Figure 5.2 Comparison of the structures of ecstasy and various bath salts.

remove their clothes. Chronic users of bath salts experience hallucinations, disassociate from reality, and can become paranoid or psychotic.

The negative side effects of bath salt use have made for some rather intense news stories. One user from Mississippi carved up his face with a knife. After police put him in handcuffs, he scraped his teeth across the hood of their patrol car, scratching the paint down to the metal. After being transported to a hospital, he repeatedly asked to be killed. A West Virginia man who told police he was high on bath salts was found in women's underwear and standing over a goat's carcass. By far, the most egregious example was the infamous Miami cannibal attack of 2012. On May 26, police in Miami, Florida, shot and killed a nude homeless man who was allegedly chewing off the face of another man on a busy highway. Before being killed by four police bullets, according to authorities, he had bitten off most of his victim's face, including his left eye, leaving him blind. In 2013, an attendee of the Ultra Music Festival in Miami, Florida, died after overdosing on a bath salt called alpha-PVP. In September of that same year, two high school students in Westland, Michigan, were hospitalized after ingesting a bath salt called Cloud 9.

Illicit drug manufacturers have resorted to cutting ecstasy tablets with synthetic cathinone derivatives to dilute their purity and thereby increase profits. There has been a documented drop of MDMA content in ecstasy tablets, from greater than 90% MDMA content before 2009, to currently just below 50%. Some tablets have no MDMA at all. It has been hypothesized that the substitution of bath salts in ecstasy tablets has been responsible for the rash of overdoses and deaths.

According to the American Association of Poison Control Centers, the number of reported human poisonings with synthetic cathinones increased almost 20-fold, from 306 during the year of 2010 to 6137 during 2011. Correspondingly, the Drug Abuse Warning Network (DAWN) reported 22,904 emergency department visits because of bath salt exposure. The 2011 rise in the abuse of bath salts was labeled "America's new drug problem."

In October, 2011, the US government responded to these alarming reports by classifying mephedrone, MDPV, and methylone as *Schedule I drugs*, which means they have no current therapeutic use, a high susceptibility for abuse, and may lead to psychological or physical dependence. This step seems to have curtailed the abuse of bath salts; in 2013, there were only 995 calls to poison control centers regarding bath salts.

Unfortunately, unscrupulous chemists have circumvented drug scheduling by synthesizing an endless array of new replacement cathinones. Only specific molecular structures can be banned by the government. Changing one atom makes the molecule legal, even if the substance still has the same effects. By constantly modifying the chemical structure of the molecules sold as bath salts, dishonest manufacturers stay one step ahead of the law. These new, technically legal, molecules created to circumvent laws are commonly referred to as "designer drugs." In 2012, 31 new synthetic cathinones were identified, up from just four in 2009 (Figures 5.3 and 5.4).

This strategy of switching cathinone structure violates the ethical development of drugs. In order for a drug to obtain FDA approval, drug development follows a distinct set of guidelines and steps including animal testing, pharmacokinetic studies, safety and efficacy studies, and human trials. These strict guidelines were set in place to protect the public from unexpected harmful effects from drugs. The opposite is occurring with bath salts. Illicit manufacturers make small variations to a chemical structure and the molecule is then packaged, distributed, and consumed by the public without FDA approval or any form of safety testing. These new cathinones are most likely causing the extreme reactions seen in some users of bath salts.

Another unforeseen problem comes with the ability of these drugs to be detected in the body. For a drug to be found in someone's system, it has to be specifically tested for. A change in even one carbon atom can deceive a

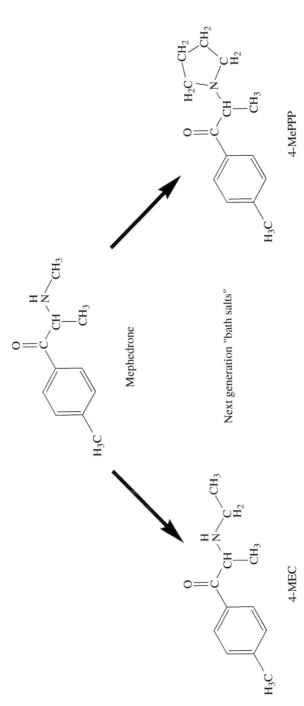

Figure 5.3 The next generation of mephedrone bath salts.

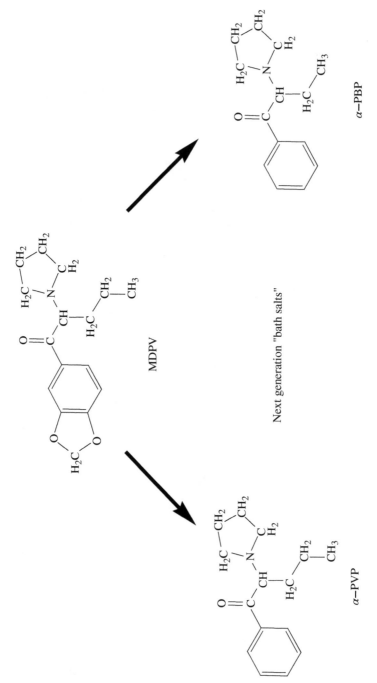

Figure 5.4 The next generation of MDPV bath salts.

toxicology test. One of the allures of bath salts is that most drug tests will miss them, so users subject to regular drug testing can use these new cathinones without fear of being detected. There are hundreds of bath salt compounds, but toxicologists can only test for 40. Adding insult to injury, recent bath salt fatalities have led to wasteful expenditure of taxpayer dollars on testing. During the autopsy of an overdose death, identifying the drug compounds responsible can be difficult if they have never been identified by authorities. It has been estimated that testing for each synthetic drug-related death costs from $2000 to more than $5000.

What to Do If You Want Your Skin to Turn Blue

A condition called *argyria* is caused by overexposure to silver due to continued ingestion of silver-containing products, resulting in a gradual accumulation of silver particles in the skin. These accumulated silver particles react with sunlight to form ionic silver cations (Ag^+). Unfortunately, argyria will not make you glisten like the vampires in Twilight movies. In fact, the silver ions react with biological sources of sulfur to form silver sulfide (Ag_2S), which appears as a dark blue pigment in the skin. Particles of silver sulfide have been found in the skin of argyria sufferers during autopsies. Most proteins can act as a source of sulfur for this reaction due to sulfur-containing amino acids such as cysteine (Figure 5.5).

In addition, the silver particles may stimulate melanin synthesis, darkening the skin even more. This is demonstrated by the fact that the effects of argyria are most pronounced in sun-exposed areas such as the face, neck, arms, and hands. Unfortunately, argyria is believed to be irreversible, with avoiding sunlight being the only known treatment, which is only good news if you are a vampire. In virtually all cases, the patients retain the blue coloring for the rest of their lives.

Silver exposure can come from a variety of sources. In the past, suspensions of colloidal silver were used to treat certain diseases, but this was discontinued due to concerns over the side effects (including argyria). In 1999, the FDA declared that over-the-counter silver-containing preparations were not recognized as safe, due to the lack of substantial scientific evidence supporting

Figure 5.5 The structure of cysteine.

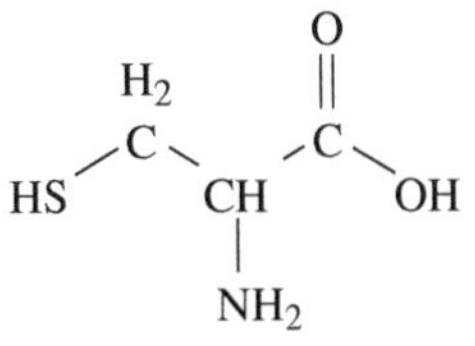

Cysteine

their use to treat diseases. Nonetheless, many health-food manufacturers have promoted colloidal-silver-based products as home treatments of ulcers, cancer, AIDS, diabetes mellitus, and herpetic infections, despite there being no evidence that they are effective. Because they are listed as a dietary supplement, these silver-containing formulations avoid the FDA guidelines for human consumption.

The amount of absorbed silver required to cause generalized argyria pigmentation is unknown; it usually requires 8 months to 5 years of regular exposure. In the most famous recent case, a man gave himself argyria by consuming a home-made silver concoction to treat acid reflux and rubbing a silver-containing salve on his skin to treat dermatitis. He made these silver solutions using a simple battery-operated chamber designed to leach silver from pure silver wire. After ingesting approximately 16 ounces (450 ml) of 450 ppm colloidal silver three times a day for 10 months, he developed argyria.

In addition, people who are regularly exposed to silver as part of their occupation, for example, silversmiths and workers in the silver-plating industry, can show signs of argyria. In fact, localized argyria can appear at sites of acupuncture needles and silver earrings. Exposure to silver may have other toxic effects such as liver and kidney damage and irritation of the eyes, skin, respiratory, and intestinal tracts. In addition, people who have had excessive exposure to silver exhibit *melanodacryorrhea*, or black tears.

A similar discoloration of skin, *chrysiasis*, can result from the ingestion of gold salts. In the early 1900s, gold was used to treat tuberculosis and rheumatoid arthritis. However, chrysiasis has rarely been seen since then.

With the discovery of silver *nanoparticles*, there has been a resurgence in the use of silver in commercial products. Some modern uses of silver capitalize on its ability as an antibacterial and antifungal agent. The use of silver nanoparticles, ranging in size from 1 to 100 nm, prevents argyria because the nanoparticles are too small to be absorbed by the human body. In addition, silver nanoparticles are used because the ionization of metallic silver is proportional to the surface area of particles. The release of Ag^+ from nanoparticles of 20 nm is more than 100 times higher than that from silver foil.

Silver produces antimicrobial effects by reacting with sulfur-containing proteins. These proteins are rendered inactive when the silver ions bind the sulfur groups of proteins to form *hemisilver sulfides*. This causes the precipitation of the protein, making it useless for the organism. By binding to the protein residues on cell membranes of sensitive bacteria, fungi, and protozoa, the subsequent inactivation of proteins and essential enzymes causes the metabolism of the organism to shut down, and the organism dies (Scheme 5.1).

Silver nanoparticles can be found in workout clothing to prevent the smell of sweat, in wound dressings, on the surfaces of household appliances, and in some brands of dental floss.

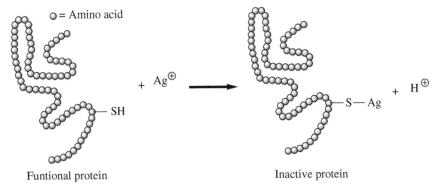

O = Amino acid

Funtional protein

Inactive protein

Scheme 5.1 The reaction of an Ag⁺ and a protein.

Figure 5.6 The structure of silver sulfadiazine.

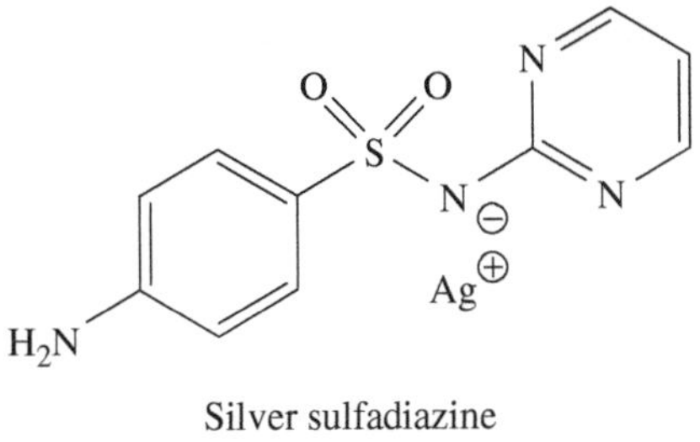

Silver sulfadiazine

Silver is used therapeutically in surgical instruments, dental prostheses, wound dressings, and catheters. Metallic silver and most inorganic silver compounds ionize in moisture, body fluids, and secretions to release biologically active Ag^+. Coating or impregnating medical devices and implants with silver compounds provides clinicians with an effective means of preventing infections. A prime example is silver sulfadiazine (1%), which was formulated in 1967 and is now one of the most frequently used topical agents for burn treatment (Figure 5.6).

The Flesh-Rotting Street Drug

Desomorphine is an opioid that was first synthesized in 1932 in the United States. It has no accepted medical use and has been a controlled substance since 1936. Because it is 8–10 times stronger than morphine and acts faster, it is highly addictive. Desomorphine is easily made from codeine, so it costs roughly 20 times less than heroin but produces a similar high (Figure 5.7).

The simple and cheap process for making homemade desomorphine involves boiling codeine with gasoline, hydrochloric acid, iodine, and red phosphorus (scraped from the striking surfaces on matchboxes). The created desomorphine

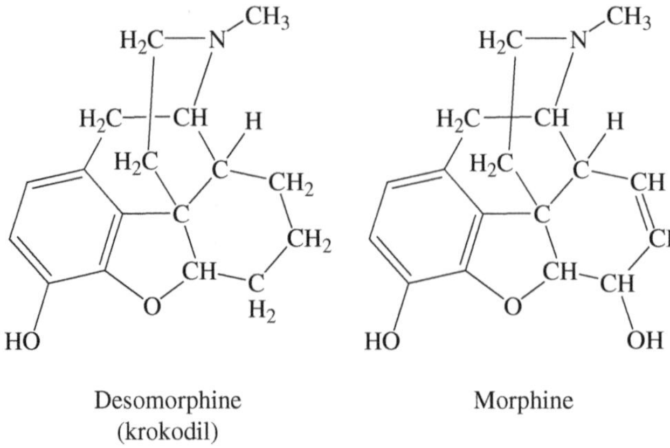

Desomorphine
(krokodil)

Morphine

Figure 5.7 A comparison of the structures of krokodil and morphine.

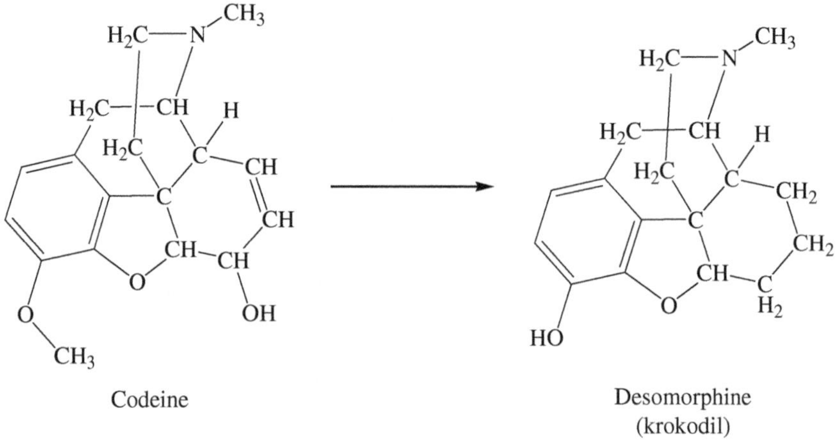

Codeine

Desomorphine
(krokodil)

Scheme 5.2 The conversion of codeine to krokodil.

and its toxic by-products are injected. Once injected, in addition to producing the high, homemade desomorphine causes damage to blood vessels and tissue, which causes flesh to rot from the inside out. In fact, the street name in Russia for homemade desomorphine is *krokodil* (Russian for crocodile) because the skin in long-term abusers may present as greenish and scaly, similar to a crocodile's scaly skin. Other reactions are rotting teeth, blood poisoning, brain damage, and death (Scheme 5.2).

The gasoline and hydrochloric acid in the injected liquid solution are blamed for the horrific side effects of using this drug. People using this drug have an estimated mean survival time of 2 years from first use of the drug, and many die even sooner. Because desomorphine has a shorter half-life than morphine, desomorphine users experience euphoria for an hour-and-a-half only, whereas heroin users experience the same effects for 4–8 hours. Considering that the home preparation of desomorphine takes roughly an hour to achieve, it traps its addicts in a 24-hour daily cycle of "cooking" and injecting to avoid withdrawal. In a sense, desomorphine abuse is a full-time job.

In 2003, desomorphine abuse made headlines in Russia, where codeine is available without a prescription. Today, it has been estimated that around 100,000 people in Russia and 20,000 in the Ukraine are using the drug. There are also reports of its use in Germany, the Czech Republic, France, Belgium, Sweden, and Norway. Because codeine is a controlled substance in the United States, it was thought that desomorphine would not spread here. However, in September, 2013, the Banner Poison Control Center in Phoenix, Arizona, reported the first instances of use of this drug in the United States. In October, 2013, seven krokodil-related hospitalizations were reported in Illinois, Utah, Arizona, and Oklahoma. The Oklahoma Bureau of Narcotics announced that autopsies revealed that two men had died from krokodil abuse.

How Does a Breathalyzer Detect a Blood Alcohol Level?

In 2013, 10,076 people died in drunk driving accidents in the United States, one death every 52 minutes. On average, two out of three people will be involved in at least one drunken driving accident. In 2012, more than 29 million people admitted to driving under the influence (DUI) of alcohol. In 2011, more than 1.2 million drivers were arrested for DUI of alcohol or narcotics.

With these statistics, it is clear to see why law enforcement needed an efficient way of determining alcohol intoxication in the field. *Alcohol intoxication* is legally defined by the *blood alcohol concentration* (BAC) level, determination of which requires drawing a blood sample for direct testing. It is impractical for police officers to take blood samples from drivers, and obtaining a urine sample is not much better. What was needed was a way to easily measure something equivalent to the BAC.

The first breath alcohol testing devices were developed for use by police in the 1940s. In 1954, Dr Robert Borkenstein of the Indiana State Police invented the breathalyzer™. Since then, breathalyzer has become the generic name for a piece of equipment that detects alcohol when a person breathes into a mouthpiece.

Alcohol remains in the bloodstream long after it has been drunk. As blood goes through the lungs, some of the alcohol moves across the membranes of the alveoli and evaporates in the breath, since it is a volatile compound. Subsequently, it is present in exhaled air, where it can be detected by a breathalyzer. Because the concentration of alcohol in exhaled air is related to the

$$CH_3CH_2OH_{(g)} + H_2O_{(l)} \rightarrow CH_3CO_2H_{(l)} + 4\,H^+{}_{(aq)} + 4e^-$$

Scheme 5.3 The oxidation ethanol to acetic acid.

$$O_{2(g)} + 4H^+{}_{(aq)} + 4e^- \rightarrow 2\,H_2O_{(l)}$$

Scheme 5.4 The reduction of oxygen to water.

$$CH_3CH_2OH_{(l)} + O_{2(g)} \rightarrow CH_3COOH_{(l)} + H_2O_{(l)}$$

Scheme 5.5 The overall reaction used to measure BAC by a breathalyzer.

concentration of alcohol in blood, a BAC can be calculated by measuring the alcohol content in breath. Blood has an alcohol content 2100 times more concentrated than breath, so the calculation is simple.

So how does a breathalyzer actually work to determine the concentration of alcohol in breath? It utilizes a *fuel cell*, which consists of two platinum electrodes with an acid electrolyte material between them. As the user exhales, any ethanol present is oxidized to acetic acid at one of the electrodes – the anode – and produces electrons (Scheme 5.3). At the other electrode – the cathode – atmospheric oxygen is reduced to water (Scheme 5.4). The overall reaction is the oxidation of ethanol to acetic acid and water, with an electrical current being produced (Scheme 5.5).

In a breathalyzer, the electric current produced by this reaction is measured, and a microprocessor calculates the approximate the blood alcohol content. The more alcohol in the breath, the greater the electrical current produced and the higher the calculated BAC.

The consequences of a conviction of a DUI have caused the demand for personal and professional breathalyzers to skyrocket. In 2011, the breathalyzer market was estimated at \$284.6 million, and it is expected to reach \$3.2 billion in 2018. A company called BACtrack® has created Vio, a smartphone breathalyzer for alcohol testing. The Vio is small enough to fit on a keychain and can transmit a user's BAC via bluetooth to his or her smartphone.

How to Become a Brewery

Auto-brewery syndrome, also called gut fermentation syndrome or endogenous ethanol fermentation, is a documented condition wherein patients become intoxicated without actually ingesting alcohol, which may or may not be the best thing in the world, you can decide. In one case, a patient who developed this syndrome had taken antibiotics following surgery for a broken foot. The antibiotics killed enough of the normal bacterial population in his stomach that he was vulnerable to an infection by *Saccharomyces cerevisiae*, also known as brewer's yeast. Exposure to brewer's yeast is common, since it's found in many foods, including breads, wine, and beer. When the patient ate or drank starch-containing foods, such as bagels, pasta, or even a soda, the yeast fermented the sugars into ethanol, and he would get drunk. Now that gives a whole new meaning to the term Beer Belly! To test for this, doctors monitored the blood alcohol level of this patient as he consumed carbohydrate-rich foods. They found that it rose to as high as 0.12%, which is 150% the typical legal limit for driving. To treat this condition, the patient received an oral course of fluconazole, an antifungal, followed by acidophilus tablets to recolonize the gut with bacteria which is normally present.

There have been other documented cases of auto-brewery syndrome, usually due to a person becoming infected after taking antibiotics, or suffering an illness that suppressed their immune system. One might wonder how someone can become infected with bacteria when the stomach is so acidic (either for scientific curiosity or for a more cost-effective way to become inebriated). Indeed, some bacteria can take root in the intestines, if they can reach them. Stomach acids kill most bacteria, but the presence of food can temporarily decrease the acidity, allowing bacteria to pass through. This is how many food-borne pathogens, such as *Escherichia coli* or salmonella, infect humans.

Another way "auto-brewery syndrome" can be brought about is through endogenous (produced by the body) ethanol production. In the human body, ethanol is constantly being formed from acetaldehyde as part of various metabolic processes. In fact, the spontaneous auto-production of ethanol constantly occurs in healthy subjects. Most of the generated ethanol is neutralized in the liver, however, small amounts of the endogenous ethanol does manage to pass through and is diluted in the total body mass. The author was surprised to find out that our bodies constantly maintain minute levels of ethanol. Yes, this means we are all just a little bit drunk! In normal humans, the values of blood ethanol are so low that they are almost undetectable by standard analytical methods. However, there are some exceptions, particularly for people with metabolic disorders. For example, endogenous ethanol is present in small amounts in urine of diabetic patients, even without the consumption of alcohol-containing drinks.

The main exceptions are ethnic groups from the Far East such as the Japanese. Typically, when endogenous ethanol is produced, it is rapidly removed from our systems through the activity of alcohol dehydrogenase (ADH) and acetaldehyde dehydrogenase (ALDH) enzymes found in the liver. Roughly, half of the Japanese population lacks the enzyme ALDH, which converts acetaldehyde into acetic acid. This condition is commonly referred to as Asian flush syndrome, due to the disproportionate number of Asians

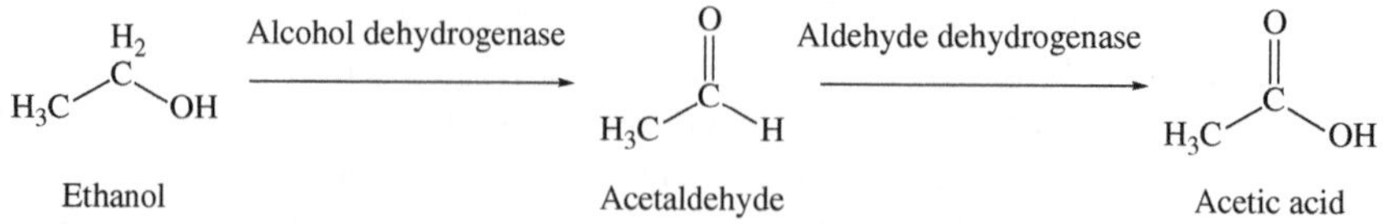

H₂C / Alcohol dehydrogenase / H₃C — C(=O) — H

Ethanol

Acetaldehyde

Aldehyde dehydrogenase

Acetic acid

Scheme 5.6 The conversion of ethanol to acetic acid.

with one or more of these mutations. These mutations, together with a diet consisting of about 80% carbohydrates, may be an important factor explaining why almost all cases of auto-brewery syndrome have been reported in Japan. In fact, auto-brewery syndrome was first described in 1952 in a Japanese patient. This syndrome is so common in Japan that it has been given the name "Meitei-sho" (Scheme 5.6).

On a side note, auto-brewery syndrome has been used as a defense against drunk driving charges, as justification that a person became unintentionally intoxicated. However, most states have passed "per se" laws against driving with a blood alcohol content of 0.08% or more. The legal term illegal *per se* means that the act is inherently illegal. This means that having the alcohol in your body is a crime, regardless of whether you have knowledge of its presence. Neither intent, nor even knowledge is required. The crime is simply having the alcohol in your body, even if you have had nothing to drink (Figure 5.8).

Figure 5.8 Making the best out of auto-brewery syndrome.

How Was a Painkiller Used to Free Hostages?

The Moscow theater hostage crisis occurred on 23 October 2002, when the Dubrovka Theater was stormed by approximately 40 terrorists who claimed allegiance to the militant separatist movement in Chechnya. The terrorists held 850 people inside the theater, demanding withdrawal of Russian forces from Chechnya and an end to the Second Chechen War. The standoff lasted for two-and-a-half days. On the third day, Russian authorities pumped an unknown gaseous chemical sedating agent into the building's ventilation systems and then raided it. During the raid, all the terrorists and about 130 hostages were killed. All but two of the hostages died from reactions to the sedating gas, by what was believed to be suffocation. The handling of the situation was considered a disaster.

Although the Russian government still has not disclosed what was used to incapacitate the terrorists, it is possible that the military made use of an aerosol form of powerful synthetic opioids consisting mainly of *carfentanil*. Researchers found traces of carfentanil using mass spectrometry analysis of clothing extracts from two survivors and urine from a third. Many critics blame the disaster on the failure of the Russian authorities to inform medical teams, who were consequently unprepared to treat survivors under the effects of an opioid. Naltrexone hydrochloride rapidly reverses the effects of carfentanil and could have been administered to save hundreds.

Carfentanil and another compound called *fentanyl* are examples of *synthetic opioids*, meaning they are not derived directly from a product found in nature. The structures of fentanyl and carfentanil are similar to other opioid painkillers, such as oxycodone. Fentanyl is the most widely used synthetic opioid in clinical practice, with 6.63 million fentanyl prescriptions dispensed in the United States in 2014. This amounts to 1.7 tons of fentanyl. First synthesized by Paul Janssen in 1960, fentanyl is estimated to be 80 times more potent than heroin and hundreds of times more potent than morphine. It is used as an anesthetic and treatment for chronic pain. Fentanyl is available in multiple formulations, including solutions for injection, transdermal patches, lozenges, and nasal sprays (Figure 5.9).

The potency of fentanyl makes it a great candidate for abuse. It was hoped, through the development of transdermal patches, that dosage could be better controlled and thus reduce the potential for abuse and/or overdose. However, deaths have still been reported. One man died after chewing a fentanyl patch. Two others died after they smoked 7 fentanyl patches, and one woman died after putting 11 fentanyl patches on her skin.

Carfentanil, first synthesized in 1974, is a derivative of fentanyl and is known as the most potent opioid used commercially. Carfentanil is 10,000 times more potent than morphine and 100 times more potent than fentanyl. As little as 20 µg (the tiniest drop) can be lethal for humans. Under the trade name Wildnil,

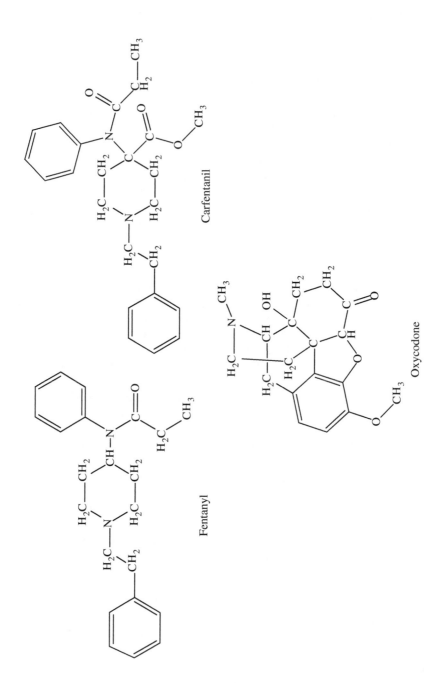

Figure 5.9 Comparison of the structures of fentanyl, carfentanil, and oxycodone.

carfentanil is used as a general anesthetic for large animals. A 10-mg dose is used to sedate an adult African elephant. Some humans have even been affected by the drug after eating animals that had been sedated by it.

The Secret Ingredient in Coca-Cola®

Coca-Cola® was the fifth most valuable brand in the world in 2013, with an estimated company worth of more than $78 billion. At that time, Coke® products could be found in more than 200 countries, with more than 1.8 billion servings being consumed each day. Originally, Coca-Cola® was a patent medicine that contained cocaine. Patent medicines were sold at the turn of the twentieth century and marketed with wild claims that their therapeutic value came from exotic ingredients such as swamp root and snake oil. The actual effects came from more common drugs such as cocaine, opium, or grain alcohol. Another fairly famous patent medicine is laudanum, which contained 40% alcohol with opium.

The original Coca-Cola® formula, concocted in 1886, was born directly out of the rising American coca culture of the late nineteenth century. Originally, Coca-Cola® was "Peruvian Wine Cola," an imitation of Vin Mariani, a Bordeaux coca tonic popular in Europe during the 1860s. Eventually, the wine in Coca-Cola® was replaced with caffeine from African kola nuts and ample sugar.

In 1910, cocaine was also removed from the formula after it was found to be addictive. However, not commonly known is that Coca-Cola® is still flavored with a non-narcotic extract from coca, the plant from which cocaine is derived. This top-secret "decocainized" leaf extract, called Merchandise No. 5, is the fifth of seven mystery ingredients found in coke syrup. The composition of Merchandise No. 5 came to light in a Federal show trial of Coca-Cola® in 1911–1912, which centered on charges of fraud. It was claimed that Coke® willfully duped consumers by still using the word "Coca" in its name, when it no longer contained cocaine. Coca-Cola® won the case by admitting Merchandise No. 5 was a decocainized extract of Peruvian coca leaf.

In 1914, Congress passed the Harrison Narcotics Act, banning the non-medical use of cocaine, as well as many other drugs, like marijuana. However, Section 6 of the Harrison Act explicitly exempted Coca-Cola®'s Merchandise No. 5 from the revenue controls of the law. In fact, the United States allows for coca leaves to be imported for processing into medicinal cocaine. The remaining leaves are sold to the Coca-Cola® Company, where flavoring agents are removed from them for adding to the Coca-Cola® beverages. It has been shown that there are still minute traces of cocaine in Coca-Cola® but the amount is so small that they cannot cause any harm.

The manufacture of medicinal cocaine is strictly controlled by the United States Drug Enforcement Administration (DEA) under importation regulations specified in law. A Stepan Company laboratory in Maywood, New Jersey, is the only legal commercial importer of coca leaves in the nation. Coca leaves are obtained mainly from Peru and, to a lesser extent, from Bolivia. The Stepan laboratory buys about 100 metric tons of dried coca leaves each year and extracts cocaine from them. Each shipment carries its own import permit, issued by the DEA. Once extracted, it is shipped to Mallinckrodt Incorporated, a pharmaceutical manufacturer that is the only company in the United States licensed to purify cocaine for medicinal use.

Coke® worked with the Federal Bureau of Narcotics (FBN) to secure exclusive access to legal coca imports into the United States after the federal government criminalized coca trafficking. The FBN approved special exemptions in federal legislation, allowing Coca-Cola® to purchase decocainized coca leaf extract, or coca leaf fluid leftover after cocaine processing. In addition, the FBN aggressively denied other applicants seeking legal coca imports into the United States, including many rival soft drink companies, which helped to create a monopsony (one buyer only) for Coca-Cola®, thereby keeping prices for this exotic ingredient down.

Cocaine is a *Schedule II controlled substance*, which means that it has a high potential for abuse but still has medicinal uses. Topical solutions of cocaine can be prescribed by a doctor to be used as a local anesthetic in certain cases of eye, ear, or throat surgeries. Cocaine produces a numbing effect by blocking conduction of nerve impulses. Although cocaine works very well to provide local aesthetic effects, the risk for abuse prevents its widespread clinical use.

Why Is Crack Cocaine So Addicting?

Cocaine is a powerfully addictive stimulant drug extracted from the leaves of the *Erythroxylon coca* plant, native to South America. After the 1990s, Colombia produced about 90% of the cocaine powder illicitly reaching the United States. The powdered form of cocaine is usually either inhaled through the nose (snorted), where it is absorbed through the nasal tissue, or is dissolved in water and directly injected into the bloodstream. *Crack* is a form of cocaine that has been processed in such a way that it can be smoked. Crack is heated to produce vapors that are absorbed into the bloodstream through the lungs. In fact, the street name "crack" comes from the crackling sound made by the mixture when it is smoked.

So how much of a problem is cocaine in the United States? By 2008, almost 15% of Americans had tried cocaine, with 6% having tried it by their senior year of high school. In addition, 1.9 million people had used cocaine in the past month. Of these, approximately 359,000 were current crack users. Cocaine was involved in 482,188 of the nearly 2 million visits to emergency departments for drug misuse or abuse in 2008. This translates to almost one in four of all drug misuse or abuse emergency department visits. In 2007, cocaine accounted for about 13% of all admissions to drug abuse treatment programs, 72% of whom were seeking treatment for crack abuse.

Powder cocaine is normally in the form of a *hydrochloride salt*. The cocaine molecule contains a nitrogen atom, which is basic. This means that when cocaine is mixed with an acid such as hydrochloric acid (HCl), the amine gains the proton from the acid to form an *ammonium salt*. The nitrogen thus becomes positively charged, and interacts with the negatively charged chloride to form an ionic bond. The ionic nature of the salt increases the intermolecular forces holding together the cocaine molecules, which also changes certain properties of the compound such as raising its melting point and boiling point. This is significant because during the smoking process, the drug is vaporized. Because the hydrochloride form of cocaine has a melting point of 190 °C (374 °F), it decomposes when heated and cannot be smoked.

In order for cocaine to be smoked, the nitrogen needs to have the hydrogen removed or deprotonated. Because the mixture cocaine–HCl is now acidic, the hydrogen attached to the nitrogen can be removed using the easily obtainable base, baking soda (NaHCO$_3$). When reacted with baking soda, the hydrogen on cocaine's nitrogen is removed, thereby freeing up the basic nitrogen. This is where the term "freebase" comes from. Once the ionic portion of the amine is removed, which is the case with crack cocaine, the intermolecular forces holding the molecules together greatly decrease. Crack cocaine actually vaporizes at 90 °C (194 °F), which means that the cocaine can now be smoked without it being decomposed (Figure 5.10; Scheme 5.7).

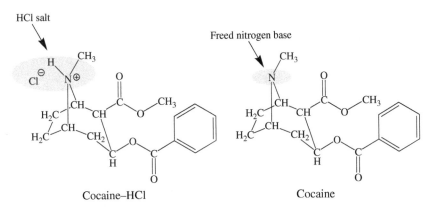

Figure 5.10 Comparison of cocaine–HCl and cocaine.

$$\text{Cocaine--HCl} \quad + \quad \text{NaHCO}_3 \quad \longrightarrow \quad \text{Cocaine} \quad + \quad \text{H}_2\text{CO}_3 \quad + \quad \text{NaCl}$$

Sodium bicarbonate   Free base

(baking soda)   (crack)

Scheme 5.7 The synthesis of crack using baking soda.

Smoking cocaine is much more addictive because of the faster onset time of the effects. *Addiction* is a very complex subject, and there are many factors that affect what makes something addicting. One characteristic of an addictive drug is how quickly the drug takes effect, or the so-called rush from using it. The sooner the effects are felt, the more addicting the drug. Cocaine is absorbed quickly when inhaled. This produces the "lift-off" rush described by crack users. Because crack cocaine can be smoked, it is much more addictive than the powder form, which is snorted.

Although smoking cocaine provides quicker and stronger pleasurable effects, the high is also shorter. The high from snorting powder cocaine may last 15–30 minutes, whereas the high from smoking crack is only one-third of this time – between 5 and 10 minutes. This requires more frequent administration to maintain the high with crack. Another odd difference between crack and powder cocaine is the laws against them. The sale of 500 g of powder cocain–HCl is punishable by a 5-year mandatory prison sentence, while only 5 g of crack is required for the same penalty.

Cocaine is a strong central nervous system stimulant and produces a high by increasing the levels of the neurotransmitter *dopamine*, particularly in the *nucleus accumbens*, one of the brain's key areas involved in reward. Increasing dopamine levels in the nucleus accumbens increases neural activity, resulting in a euphoric feeling.

In the normal neurotransmission process, dopamine is released by a presynaptic neuron into the synapse, the small gap between two neurons, where it binds to dopamine receptors on neighboring postsynaptic neurons, causing them to become active. Dopamine is then recycled back into the transmitting (presynaptic) neuron by a specialized protein called the *dopamine transporter*, and after this, it is either destroyed or reused. If cocaine is present, the drug attaches to the dopamine transporter protein and blocks the normal recycling process, resulting in a buildup of dopamine in the synapse leading to an increase in postsynaptic activity. This increased neuronal activity contributes to the pleasurable effects of cocaine. Repeated exposure to cocaine causes addiction because the brain starts to adapt; that is, the pleasure pathways become less sensitive to normal stimulation and cocaine. This means that a user needs higher doses and more frequent use of cocaine to get the same high.

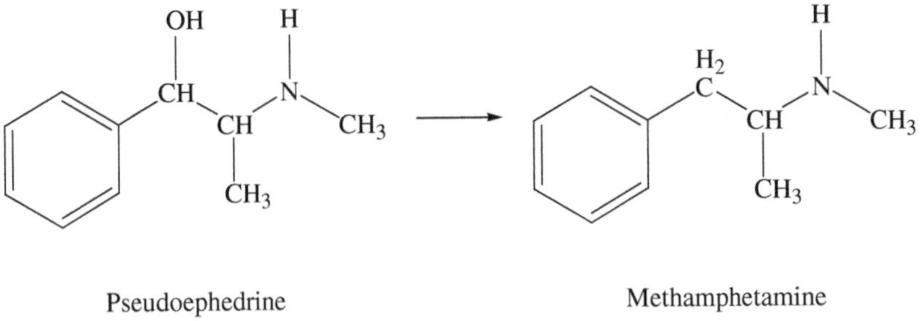

Pseudoephedrine   Methamphetamine

Scheme 5.8 The synthesis of methamphetamine from pseudoephedrine.

Cocaine Smuggling versus Methamphetamine Manufacture

Most of the cocaine illicitly sold in the United States has been smuggled in; however, the same is not true of methamphetamine. This is due to differences in the complexity of the two molecules, and in the availability of starting materials used for their synthesis (Scheme 5.8).

In recently published information on the conversion of an easily obtained starting material to cocaine, 15 reaction steps are detailed. More importantly, this synthesis requires an advanced knowledge of organic chemistry techniques, making it virtually impossible for a layman to perform. In addition, cocaine is a relatively structurally complex molecule. It is actually easier to grow the coca plant, extract the cocaine, and then smuggle it into another country than to try to make it illicitly. This is true even for cocaine used for legal medical purposes.

In the case of methamphetamine, however, its illicit production requires only one or two reaction steps using *pseudoephedrine* as a starting material, with little knowledge of organic chemistry required. Pseudoephedrine is a molecule commonly used in over-the-counter decongestants and fairly easy to obtain at almost any drug store. Compared to cocaine (see the structure in the previous story), methamphetamine is much less complex structurally, making it easier to make artificially. Because of this, it is much easier to illicitly synthesize methamphetamine than it is to smuggle it in from outside the United States.

What Basic Common Ingredient Is Needed to Make the Drugs Vicodin®, Percocet®, Oxycontin®, and Percodan®?

These pharmaceuticals are all derived from the same molecule, called thebaine, which is isolated from opium poppies. Industrial processes use thebaine to synthesize oxycodone, which is the active pharmaceutical agent found in Vicodin®, Percocet®, Oxycontin®, and Percodan®.

The opium poppy, *Papaver somniferum*, was first cultivated circa 3400 BC in lower Mesopotamia. For some time, the active ingredient of opium – morphine – was used as a medicinal agent. People soon realized that morphine was addictive, however, and began looking for nonaddictive alternatives. In 1874, English researcher C.R. Wright first synthesized diacetylmorphine; and in 1898, the German company Bayer (which also manufactures aspirin) began to market diacetylmorphine under the brand name Heroin. Bayer halted the production of heroin in 1913 and then turned to thebaine as a source of alternatives, hoping to develop a drug that would retain the analgesic effects of morphine and heroin, with less risk of leading to dependency. Chemists at Bayer first created oxycodone in 1916, and since then, thebaine-based dugs have become wildly popular (Figure 5.11).

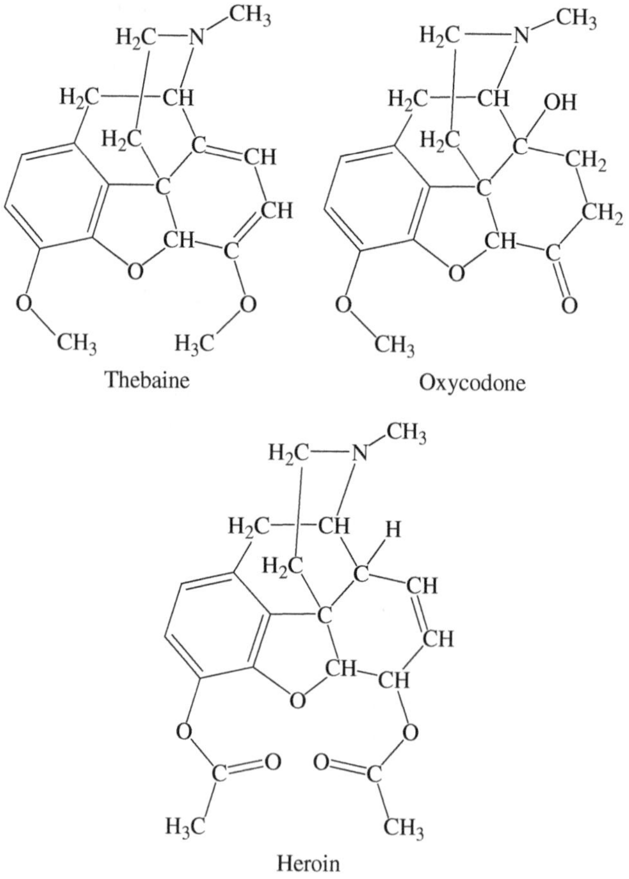

Thebaine

Oxycodone

Heroin

Figure 5.11 A comparison of the structures of opioids.

Oxycodone is called a semisynthetic drug because starting material used in its synthesis is derived from a natural source. After isolation and purification, the thebaine isolated from opium poppies is converted into oxycodone in a few synthetic steps. Although oxycodone could be made from simpler, petroleum-based starting materials the structure of thebaine is relatively complex and would require a large number of synthetic steps. Each step of an industrial synthesis of a drug increases costs due to increased processing. This makes the complete synthetic synthesis of oxycodone not commercially viable as a source of oxycodone. In short, it is cheaper to grow poppies, isolate thebaine, and then convert the thebaine into oxycodone.

The United Nations Single Convention on Narcotic Drugs, as well as other international drug treaties, allow for the legal growth and processing of opium poppies.[1] The process of manufacturing narcotic medical supplies in the United States has relied exclusively upon imports of opium gum from foreign countries. US policy which is more than 50 years old limits the production of narcotics, and the number of opium-producing nations from which opium gum is imported. Currently, India and Turkey are the US biggest suppliers of opium gum, supplying at least 80% of the nation's requirements. France, Poland, Hungary, Australia, and Yugoslavia supply the other 20%. This means that a huge number of US dollars are going to these countries to grow poppies as a source thebaine, when American famers could grow the poppies themselves. In fact, Canada as started to grow their own thebaine-producing poppies, instead of importing the product. Because opiate pain relievers represent a $500–600 million Canadian market, these poppies could yield Canadian farmers 3–5 times the average return per acre compared to other crops such as grain and oilseed. Currently only 200 acres of poppies have been approved by the Canadian government, it has been estimated that 25,000 acres of poppies could meet Canada's domestic needs for thebaine-based analgesics. The author wonders if it is time for the United States to consider doing something similar.

In 2009, there were 202 million prescriptions for opioid painkillers in the United States, 84.9% of which contained hydrocodone or oxycodone. This makes opioid painkillers one of the most prescribed classes of medication in the United States. The United States also has the dubious distinction of being the largest user of oxycodone in the world, with an estimated 55,927,804 g of the drug being distributed each year, which is more than half of the estimated annual worldwide consumption.[2]

---

1 Once during a presentation I was giving on this topic, a colleague of mine interrupted saying, "Do you expect me to believe there are huge fields of opium poppies?" The answer is quite simply, Yes!

2 If you ever want to truly understand the scope of this topic, ask a room full of people how many of them have ever had a prescription for a painkiller. My students are generally in their early twenties and usually 80–90% of them raise their hand.

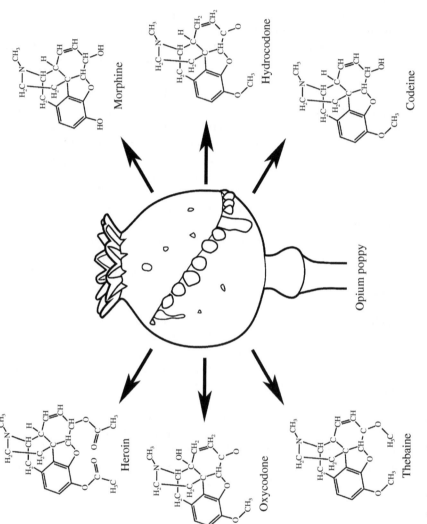

Figure 5.12 Molecules that come from the opium poppy.

Enormous problems have arisen due to the structural similarity of oxycodone and other opioid painkillers such as heroin. Oxycodone abuse is a continuing problem in the United States. For this reason, the government has classified it as a Schedule II controlled drug. Opioid painkillers have been shown to be the most commonly abused class of prescription drugs in the United States. A 2010 study reported that more than 5.1 million Americans admitted they had used opioid painkillers recreationally. It is therefore not surprising that there has been a fivefold increase in drug treatment admissions for opioid painkillers from 1998 (19,941) to 2008 (144,644). In addition, there has been a corresponding increase in emergency room visits involving opioid painkillers, with 305,885 reported cases in 2008.

The magnitude of the problem with opioid painkillers is demonstrated by the fact that there were over 12,000 unintentional opioid-related overdose deaths in 2007, more than the number from all other illegal drugs combined. In fact, opioid painkiller overdose is now the second leading cause of unintentional death in the United States, with the Center for Disease Control and Prevention now referring to opioid painkiller overdose as a national epidemic.

An excellent question that begs to be asked is why do doctors prescribe opioid painkillers when they know they are so addictive? The answer is that nothing else works as well. There have been many attempts to create nonopioid painkillers, but the results have been disappointing. A prime example is the drug Rofecoxib which was marketed under the brand name Vioxx®. Initially, Rofecoxib was approved by the FDA in 1999 as a treatment for arthritis and acute pain. However, in 2004, rofecoxib was removed from the market because of concerns about increased risk of heart attack and stroke associate with its long-term use. Although it was on the market for 5 years, it was estimated that rofecoxib was responsible for between 88,000 and 139,000 heart attacks of which 30–40% were probably fatal. As recently as January, 2016, the Portuguese pharmaceutical company, Bial, had to suspend human drug trials of a new painkiller after tests in France left one person brain-dead and five others hospitalized (Figure 5.12).

Drug Money Is Right

A typical US $1 bill is printed on a cotton/linen paper blend and lasts little more than 21 months. In this time, it can pick up quite a few nasty things. A test of 234 US bank notes found that 90% were contaminated with cocaine. A typical amount was 28.75 µg of cocaine per bill, which equates to extracting the drug from 100,000 bills to get a little under 3 g of cocaine. The data makes sense considering that the US Office of National Drug Control Policy reported that more than 2 million Americans used cocaine in 2007. That same year, the United Nations Office on Drugs and Crime reported that 6 million Americans admit to using cocaine, amounting to as much as 457 metric tons of this drug in a year.

The bills become contaminated if a drug user snorts with it. Contamination also results from drug transactions and exposure inside currency-counting machines at banks. The denominations of US currency most likely to be contaminated are $5, $10, $10, and $50 bills, probably because $1 is too little in value to be used to purchase cocaine.[3]

The frequency of money being contaminated depends on the geographic location where it circulates, with bills from urban areas much more likely to have drug residues. One hundred percent of paper money from large cities, such as Miami and Boston, were found to be contaminated but only 67–87% of that collected from smaller cities. Tests of other countries' currency showed similar contamination but to a lesser extent than bills from the United States. Money from Canada had the next highest rate of contamination, 85%, followed by Brazil with 80%. China and Japan had the lowest reported contamination, 20% and 12%, respectively.

In another study, tests were run to see what other kinds of drugs were present in 10 randomly collected US$1 bills from 5 cities. Heroin was detected on 7 bills, morphine on 3, methamphetamine on 3, and PCP (phencyclidine) on 2.

So now that you know your money is contaminated with cocaine, you may wonder what else may be lurking on paper currency. Recent testing of genetic material found on $1 bills showed that they contain almost 3000 different types of bacteria. Among these strains are bacteria that cause gastric ulcers, pneumonia, food poisoning, and staph infections.

Another interesting find was the toxic molecule bisphenol A (BPA) on bills. In the study, all but 1 of 22 bills tested were contaminated with BPA. It was theorized that the bills were likely contaminated owing to their proximity to receipts. Many receipts have a thermal ink coating along with BPA as a loose powder. These receipts could be rubbing BPA off onto dollar bills in wallets or through handling.

What Percentage of Americans Use Prescription Drugs?

Between 2005 and 2008, 47.9% of Americans had taken at least one prescription drug in the past month, 21.4% were using three or more prescription drugs per month, and 10.5% were using five or more. The most commonly prescribed types of drugs are antidepressants, painkillers, and cholesterol-lowering drugs. Although this information changes frequently, Lipitor, used for lowering cholesterol, is currently the single most prescribed drug.

---

3 I replicated this experiment in an upper-division laboratory course at the university where I teach. The presence of cocaine on several $1 bills was scientifically confirmed.

It is no surprise that the pharmaceutical industry is a big business, with $234.1 billion being spent on prescription drugs in the United States in 2008. In line with this is the fact that nearly three-quarters of doctor visits result in some form of drug therapy. Honestly, how many people do you know that have never had a drug prescribed to them?

Unfortunately, the overwhelming frequency of doctor-ordered medication in the United States brings with it the possibility for prescription medicine abuse. As of 2008, an estimated 6.2 million Americans (2.5% of the population) admitted to having abused prescription drugs in the past month. Of these, the most commonly abused drugs were pain relievers (4.7 million), tranquilizers (1.8 million), stimulants (904,000), and sedatives (234,000). During this same year, misuse of prescription opioid painkillers was responsible for 475,000 emergency room visits and caused 73.8% (14,800) of all drug overdose deaths in the United States. This is more overdose deaths than those from cocaine and heroin combined.

During the following year, 2009, the misuse of pharmaceuticals led to 1.2 million emergency room visits, representing more than all of the emergency room visits due to illegal drugs – such as heroin or methamphetamine – combined. Correspondingly, Americans were dying from prescription drug overdoses at the rate of almost 100 per day in 2007, which is roughly three times as many as in 1991.

The United States is in the midst of an epidemic of prescription painkiller abuse. Sales of prescription painkillers have markedly increased since 1999. In 2010, 2 million Americans admitted to misusing a prescription painkiller for the first time in the past year. This amounts to almost 5500 people per day. In 2010, more than 12 million people reported using prescription medication without physician approval, seeking the drugs' various side effects.

Are You Ready for Powdered Alcohol?

You should be because in March, 2015 the controversial powdered alcohol product call Palcohol® received approval from the Alcohol and Tobacco Tax and Trade Bureau (TTB). As part of the US Treasury Department, the TTB has the authority to review the formulation and labeling of distilled spirits products. The Palcohol® product would come in 100 ml pouches, with 5 ounces (150 ml) of water being added to the powder inside to create the equivalent of an alcoholic drink. The Arizona-based company that produces Palcohol®, Lipsmark LLC, has been given approval for four flavors of powdered alcohol: cosmopolitan, margarita, vodka, and rum. The company says that Palcohol® was dreamed up for people who love the outdoors but do not want to travel with heavy alcohol containers.

Because Palcohol® is an entirely new form of alcohol, there has been intense concern from legislatures about the possible misuse by people. Concerns include accidental overdose due to unfamiliarity with its potency, the possibly of its flavors being appealing to small children, it being easy to sneak into public events, the fact that people can snort the powder, and that it could be used to surreptitiously spike a drink. The TTB was quick to point out that although Palcohol® has received federal approval, states can also regulate alcohol sales within their borders. This is exactly what multiple states – including South Carolina, Louisiana, and Vermont – have done by preemptively banning powdered alcohol products. Other states – such as Nebraska, Pennsylvania, Virginia, and Massachusetts – have proposed obstructive legislation against powdered alcohol products.

So how does powdered alcohol work? Although the actual formulation of Palcohol® is a secret, the process of making powdered alcohols is quite well known. In fact, a US patent for powdered alcohol products was submitted by the General Foods Corporation in 1972. To put it simply, the liquid alcohol is absorbed by a solid substance which retains it original form. A good example of this is how a sponge can absorb water and still remain a solid. Now imagine a powder made up of a multitude of tiny sponges and you are getting close to what is going on. In the case of Palcohol®, the absorbent substance is most likely made up of a class of molecules called dextrins. Dextrins are made by breaking up starch, which is produced by many common foods such as potatoes, corn, and rice. Because dextrins contain multiple –OH groups they form a strong hydrogen bonding interaction with ethanol. Dextrins are commonly used as the absorbing agent in powdered alcohols because they do not increase the sweetness of the drink, are easily dissolved in cold water, and do not impart any flavor. In addition, dextrins are known to produce powders even with an alcohol content of 60%. Once the ethanol-containing dextrin powder is added to water, the water kicks the ethanol out of the dextrin and you have your alcoholic beverage (Figure 5.13).

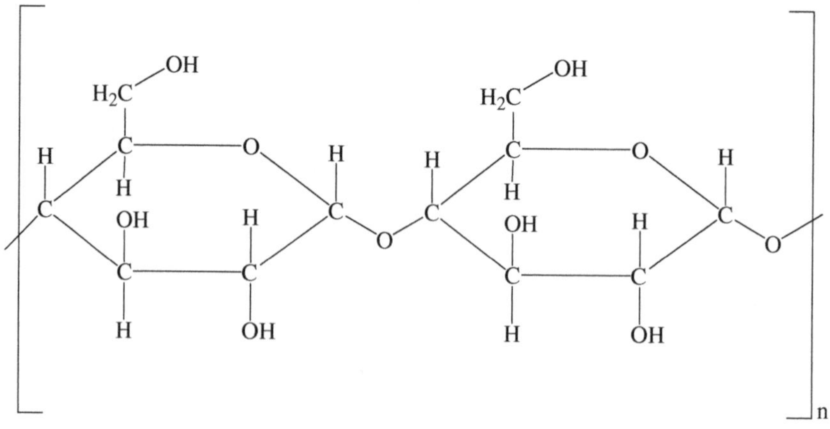

Figure 5.13 The structure of dextrin.

Ecstasy Is Ruining the Rain Forests

MDMA (3,4-methylenedioxy-methamphetamine), popularly known as ecstasy or Molly, is a synthetic drug that has structural similarities to both the stimulant amphetamine and the hallucinogen mescaline. MDMA produces feelings of increased energy, euphoria, empathy toward others, and distortions in sensory perception. MDMA was first synthesized in 1912 by Merck Pharmaceutical but was ignored by the scientific community for decades.

Drug experts say ecstasy emerged in the early 1970s as a substitute for methylenedioxy-amphetamine, or MDA, which was a popular psychedelic drug outlawed in the late 1960s. However, it was not until the early- to mid-1980s that ecstasy came into regular use on college campuses and in nightclubs. According to the 2011 National Survey on Drug Use and Health (NSDUH), there were 14.6 million lifetime users of ecstasy among people aged 12 and older. MDMA was placed on the Drug Enforcement Agency's Schedule I list of controlled substances in 1985 (Figure 5.14).

Besides its impact on the health of individuals and society, ecstasy has some other repercussions you might not have thought about, such as its effect on the environment. This starts to make sense when you consider the starting material used to make illegal MDMA. Typically, MDMA is made from *safrole oil*, a viscous, fragrant extract from some trees, often used in cosmetics and perfumes (Scheme 5.9).

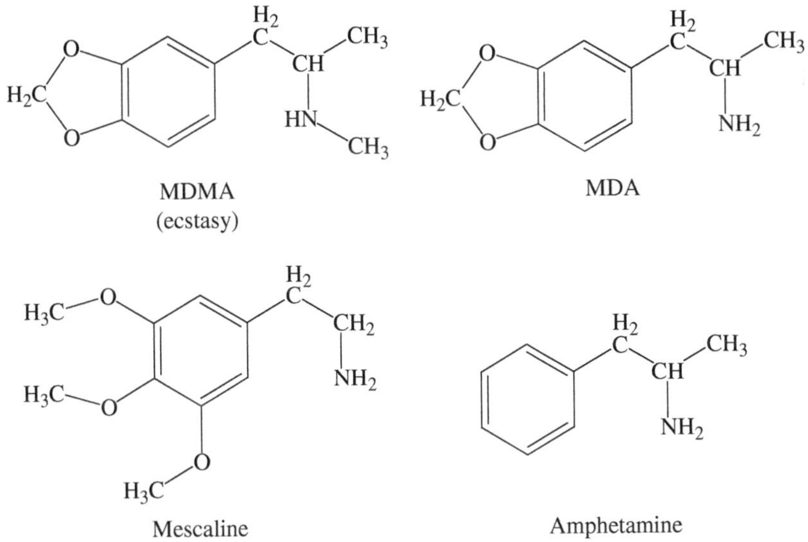

Figure 5.14 Comparison of molecules which are similar in structure to MDMA.

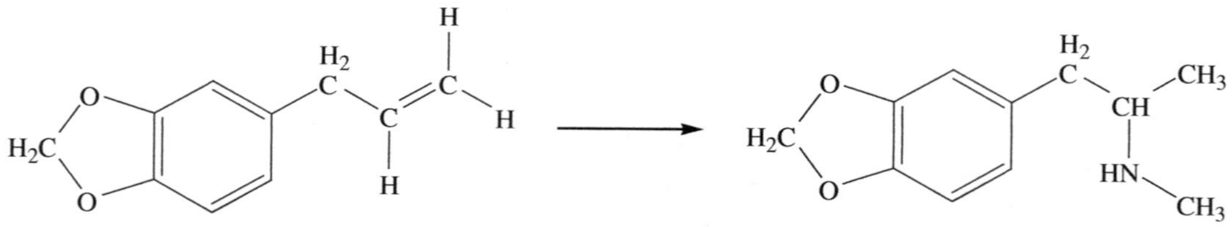

Safrole

Ecstasy

Scheme 5.9 The conversion of safrole to ecstasy.

Safrole oil is produced by a tree found in the western Cardamoms, part of southeast Asia's largest mainland contiguous rainforest. For many threatened species, including the Asian elephant, Indochinese tiger, and the Siamese crocodile, the Cardamoms is their last refuge. The source of the safrole is a tree locally known as "Mreah prew phnom," which experts think might be *Cinnamomum parthenoxylon*. These trees are harvested, and their oil-rich roots are shredded and boiled in large cauldrons. For the ease of smuggling, the extraction of safrole oil is usually done in distilleries in the forest itself, and the oil is transported to factories for conversion into ecstasy tablets. It takes an estimated four Mreah prew phnom trees to produce a 40-gallon barrel of safrole oil. Because the boiling process requires so much firewood, six trees of lesser value are burned in order to process a single safrole-containing tree.

In 2007, between 72 and 137 metric tons of ecstasy were produced globally, with between 11 and 23.5 million people using ecstasy at least once. With this level of demand for ecstasy, its manufacture is contributing to deforestation; it has been estimated that 500,000 trees per year are being destroyed to obtain the required safrole. In 2008, authorities were able to block the smuggling of 33 tons of safrole, capable of creating 245 million ecstasy tablets worth $7.6 billion.

How Are Moldy Bread, Migraine Headaches, LSD, and the Salem Witch Trials All Related?

Ergot (*Claviceps purpurea*) is a fungus that grows parasitically on rye and to a lesser extent on other species of grain. Ergot was a problem when rye was stored for long periods in poor conditions. In the Middle Ages, ergot infestation was the cause of mass poisonings which affected thousands of people at a time. The last major mass ergot poisoning was in southern Russia in 1926 and 1927. The ingestion of ergot has been shown to produce hallucinations, insanity, seizures, and nausea. In fact, ergot poisoning has been implicated, in part, for causing the Salem Witch Trials.

Lysergic acid diethylamide (LSD) was first made in 1938 by Swiss chemist Dr Albert Hofmann as part of a research program geared at developing pharmaceuticals based on molecules found in nature, called *natural products*. In the beginning of this type of research, chemists isolate and characterize natural products from medicinal plants and other sources known historically to have some form of biological activity. Organic chemistry allows for the generation of structural variations of these natural products, called *derivatives*. Chemists make a wide variety of derivatives in an effort to enhance certain pharmacological properties of a particular natural product.

Along with its hallucinogenic effects, ergot was known to be a vasoconstrictor. It could produce a burning sensation in the limbs called *St. Anthony's fire*, named after the patron saint of the victims of ergot poisoning. Because vasoconstrictors have medical uses, including treating migraines, Hofmann was making a series of derivatives of one of the natural products found in ergot, called *ergometrine*. He decided to start making derivatives with a core form of ergometrine called *lysergic acid*. It should be noted that as part of this project, he was successful in generating *methergrine*, a drug used to induce labor, and *hydergine*, which is used to treat dementia. Many other additional drugs were found or developed by investigating ergot natural products.

One of the derivatives generated by Hofmann was a product called *LSD*. After working on this compound, he accidentally got a small amount on his hand and later, being the good scientist he was, recorded the effects of the world's first LSD trip: "In a dreamlike state, with eyes closed (I found the daylight to be unpleasantly glaring), I perceived an uninterrupted stream of fantastic pictures, extraordinary shapes with intense, kaleidoscopic play of colors." He assumed the effects were caused by the compound he was working on. To test his theory, he gave himself a 250 µg dose, what he figured to be 1/100 of a typical dose for other pharmaceuticals. However, LSD is incredibly powerful, with the typical dose being about 10 µg. This means Hofmann actually gave himself over 25 times a typical LSD dose. The very next day, he experienced and recorded the first *bad* LSD trip. The effects were not so pleasant. In his description, "Everything in the room spun around, and the familiar objects and pieces of furniture assumed grotesque, threatening forms." Hoffman describes this discovery in his book *LSD: My Problem Child*. He passed away in 2008 at the age of 102 (Figure 5.15).

It was later determined that LSD is the most potent hallucinogenic substance known, with dosages measured in micrograms, or millionths of a gram. By comparison, cocaine and heroin dosages are measured in milligrams, or thousandths of a gram. Even compared to other hallucinogenic substances, LSD is 100 times more potent than psilocybin and 4000 times more potent than mescaline. The similarity of structures of ergometrine and LSD should be noted, which shows that even a slight change in chemical structure can produce drastic changes in their effects on humans.

LSD has become a common drug of abuse. The 2011 NSDUH indicated that 23 million people in the United States, 12 and older, have used LSD. DAWN ED reports that an estimated 3817 emergency department visits were associated with LSD in 2010. Compared with other illicitly manufactured drugs, such as methamphetamine and PCP, few clandestine LSD laboratories have been located or seized. In fact, only six LSD synthesis laboratories have been confiscated by the DEA since 1981. A limited number of chemists, probably less than a dozen, are believed to be manufacturing nearly all of the LSD available in the United States. This small number is not surprising given that the LSD synthesis is a difficult process to master. Unlike other illegally synthesized drugs, the makers must adhere to precise and complex production procedures. Most likely, the production recipe is passed on by personal instruction, which makes the number of people with the knowledge very small. Further evidence of the

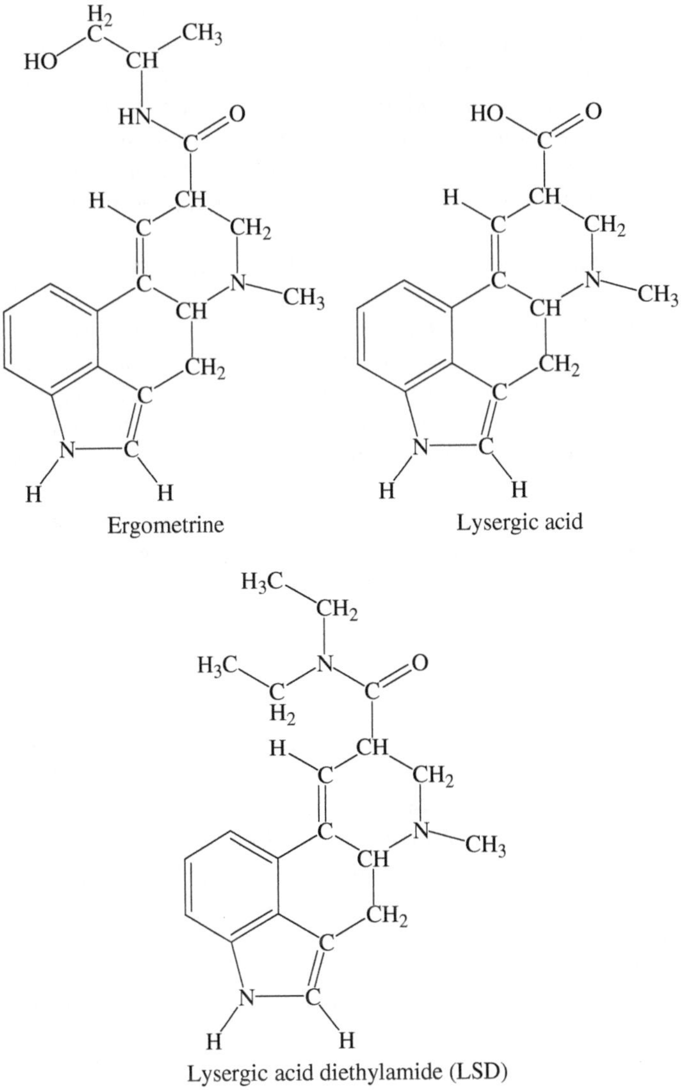

Figure 5.15 Comparison of the structures of ergometrine, lysergic acid, and LSD.

premise that most LSD manufacturing is done by a small fraternity of chemists comes from the fact that virtually all the LSD seized during the 1980s was of consistently high purity and sold in uniform dosages.

The illicit synthesis of LSD starts with ergotamine tartrate, a drug used mainly to treat migraines, typically obtained from sources located abroad, most likely Europe, Mexico, Costa Rica, and Africa. Although obtaining starting materials is one of the primary problems for illicit drug manufacturers, it is less of a problem for LSD. A mere 25 kg of ergotamine tartrate can produce 5 or 6 kg of pure LSD, which could be processed into 100 million dosage units. This is more than enough to meet what is believed to be the entire annual US demand for the hallucinogen. Pure LSD is a white crystalline material that is soluble in water. It is dissolved and diluted in a solvent for application onto paper or other materials. This is where the term "blotter acid" comes from. The paper is broken into single-dose tabs that are either ingested or applied sublingually (Scheme 5.10).

Not so commonly known is that LSD is actually an extremely toxic compound, on the same order as sodium cyanide. If this is true, however, then why are LSD overdose deaths virtually unheard of? The answer is in the vast difference between the effective dose of LSD, that is, how much it takes to get high, versus its lethal dose, or how much it would take to kill a person. It has been estimated that 14 mg of LSD could kill a human, but when you consider that it would take over 1400 doses to achieve this amount, you begin to see why LSD overdoses are so rare.

Nonetheless, there have been numerous reports of LSD overdose. In one case, several people were hospitalized following a dinner party, having snorted pure LSD under the mistaken impression that it was cocaine. Another example of LSD's toxicity occurred in 1962 during an experiment to determine whether an injection of LSD would induce "musth" in a bull elephant. While in *musth*, a normally cooperative elephant runs berserk for a period of about 2 weeks and may attack and destroy anything in his path. It had previously been shown that high doses of LSD given to animals would produce a violent reaction, and it was hoped that LSD could product musth in an elephant so that this behavioral condition could be studied. To test this, a 14-year-old male Indian elephant named Tusko, housed at the Lincoln Park Zoo in Chicago, Illinois, was injected with about 297 mg of LSD dissolved in 5 ml of water. For the roughly 3000 kg Tusko, this was considered to be far below a toxic dose. Unfortunately, Tusko immediately collapsed and went into seizures. The researchers were understandably quite surprised and distressed by the elephant's seriously bad reaction, and although multiple remedies were tried, Tusko died an hour and 40 minutes later.

Stories have been circulated about LSD causing a user to become permanently psychotic. Fact or fiction? One of the advantages of being a scientist is the ability to distinguish proven fact from rumor. When someone uses LSD or any other hallucinogen, there is a proven possibility of developing permanent hallucinations called *Hallucinogen Persisting Perceptive Disorder*. First described in 1955, the condition is often described by users as a drug trip that never ends. The altered perception may persist for weeks to indefinitely. There is no cure for this syndrome, and the subsiding of symptoms is unpredictable and sometimes never occurs. Furthermore, certain stressors such as anxiety or fatigue may cause symptoms to appear. It is believed that people who claim to suffer

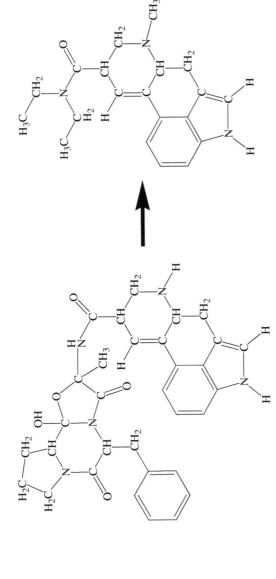

Ergotamine
(a migraine drug)

Lysergic acid diethylamide
(LSD)

Scheme 5.10 The conversion of ergotamine to LSD.

from LSD "flashbacks" might have this affliction. The causes and frequency of this disorder are not fully known but are under investigation. Other rumors about LSD involve people killing themselves after ingesting LSD. Unfortunately, this is not a rumor. It should always be remember that LSD is a powerful and dangerous drug.[4]

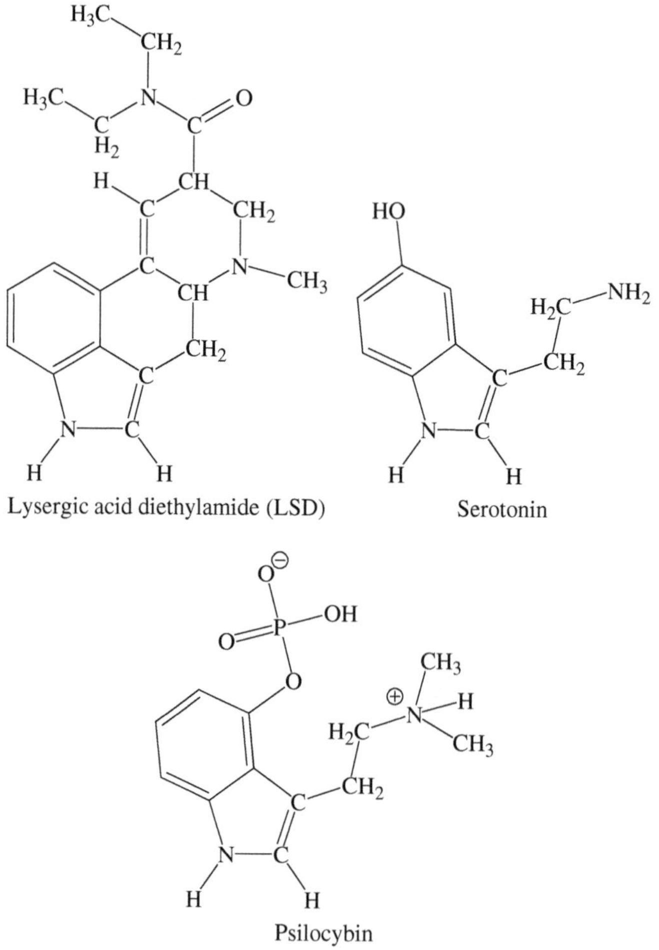

Lysergic acid diethylamide (LSD)

Serotonin

Psilocybin

Figure 5.16 A comparison of hallucinogenic molecules that mimic the structure of serotonin.

---

4 When I was a junior in high school, a group of students I knew went downtown to take LSD together. For one of them, it was his first time, and unfortunately, he had a bad reaction. The other members of the group, breaking LSD etiquette, abandoned him. After night fell, he wandered out to the highway that ran through town, stepped out in front of a car, and was killed. I make it a point to tell my students this story to show that LSD is a very powerful drug and should not be trifled with.

On a side note, in 1958, Hofmann also isolated the psychoactive substance *psilocybin* from Mexican magic mushrooms (*Psilocybe mexicana*). Spanish chroniclers and naturalists of the sixteenth century, who entered the country soon after the conquest of Mexico by Cortes, provided the first written evidence of the use of inebriating mushrooms during festivals, religious ceremonies, or for supernatural healing practices. Hofmann, after hearing that these mushrooms caused hallucinations, theorized that the mushrooms must contain a molecule causing this effect. The average active dose of psilocybin in human beings is about 10 mg, more than 100 times less active than LSD (Figure 5.16).

Thus far, researchers have been unable to discover what LSD does in the brain to cause hallucinations. Classic hallucinogens are thought to produce their perception-altering effects by acting on neural circuits in the brain that use the neurotransmitter serotonin. Although it is not true for all hallucinogens, LSD and psilocybin have distinct structural similarities to serotonin. LSD binds to serotonin receptors in neurons and triggers new patterns of neurotransmission that are perceived as hallucinations. Some of LSD's most prominent effects are in the prefrontal cortex, the area of the brain associated with mood, cognition, and perception.

Further Reading

What are the Dangers of Bath Salts?

ABC News 2012, *Face-eating cannibal attack may be latest in string of 'bath salts' incidents*, ABC News, Viewed 2 October 2016, http://abcnews.go.com/Blotter/face-eating-cannibal-attack-latest-bath-salts-incident/story?id=16470389

Banks, ML, Worst, TJ, Rusyniak, DE & Sprague, JE 2014, 'Synthetic cathinones ("bath salts")', *The Journal of Emergency Medicine*, vol. 46, pp. 632–642.

Boyle, L 2013, *Did all these festival-goers die from taking BATH SALTS? One death confirmed as fears others were duped into buying toxic street drug while believing it was 'molly'*, Dailymail.co.uk, Viewed 2 October 2016, http://www.dailymail.co.uk/news/article-2419654/Bath-salts-death-Matthew-Rybarczyk-died-toxic-drug-connected-deaths.html

Brown, H 2014, *Man charged with selling bad molly to Electric Zoo 2013 attendees (report)*, The Hollywood Reporter, Viewed 2 October 2016, http://www.hollywoodreporter.com/earshot/man-charged-selling-bad-molly-722500

Baumann, MH 2014, 'Awash in a sea of 'bath salts': implications for biomedical research and public health', *Addiction*, vol. 109, pp. 1577–1579.

Katz, DP, Bhattacharya, D, Bhattacharya, S, Deruiter, J, Clark, CR, Suppiramaniam, V & Dhanasekaran, M 2014, 'Synthetic cathinones: "A khat and mouse game"', *Toxicology Letters*, vol. 229, pp. 349–356.

myFOXDetroit.com Staff 2014, *Six high school students hospitalized from new 'Cloud 9' drug*, myFOXDetroit.com.

Office of National Drug Control Policy 2014, *Synthetic drugs (a.k.a. K2, spice, bath salts, etc.)*, The Whitehouse.

Sacco, LN & Finklea, K 2016, *Synthetic drugs: Overview and issues for congress*, Congress Research Service, Viewed 2 October 2016, http://www.fas.org/sgp/crs/misc/R42066.pdf

Smith, R 2015, Rep. *Charlie Dent holds public forum on synthetic drugs to promote new bill*, WFMZ-TV News.

US Food and Drug Administration 2014, *Lancaster man sentenced to 33 months in federal prison for sales of bath salts and synthetic marijuana*, US Food and Drug Administration, Viewed 2 October 2016, http://www.fda.gov/iceci/criminalinvestigations/ucm406718.htm

Valente, MJ, Guedes de Pinho, P, de Lourdes Bastos, M, Carvalho, F & Carvalho, M 2014, 'Khat and synthetic cathinones: a review', *Archives of Toxicology*, vol. 88, pp. 15–45.

What to do If You Want Your Skin to Turn Blue

Howgego, J 2012, *Chemistry behind the 'blue man' unlocked*, Chemistry World, Viewed 2 October 2016, http://www.rsc.org/chemistryworld/2012/11/chemistry-blue-man-argyria-colloidal-silver

Lansdown, ABG 2010, 'A pharmacological and toxicological profile of silver as an antimicrobial agent in medical devices', *Advances in Pharmacological Sciences*, vol. 2010, pp. 1–16.

Palamar, M, Midilli, R, Egrilmez, S, Akalin, T & Yagci, A 2010, 'Black tears (melanodacryorrhea) from argyrosis', *Archives of Ophthalmology*, vol. 128, pp. 503–505.

Wadhera, A & Fund, M 2005, 'Systemic argyria associated with ingestion of colloidal silver', *Dermatology Online Journal*, vol. 11, pp. 12.

The Flesh-rotting Street Drug

Katselou, M, Papoutsis, I, Nikolaou, P, Spiliopoulou, C & Athanaselis, S 2014, 'A "Krokodil" emerges from the murky waters of addiction. Abuse trends of an old drug', *Life Sciences*, vol. 102, pp. 81–87.

Office of Diversion Control 2013, *Desomorphine (Dihydrodesoxymorphine; dihydrodesoxymorphine-D; Street Name: Krokodil, Crocodil)*, Office of Diversion Control.

Thekkemuriyi, DV, Gheevarghese John, S & Pillai, U 2014, "Krokodil'—A designer drug from across the Atlantic, with serious consequences', *The American Journal of Medicine*, vol. 127, Clinical Communication to the Editor.

How does a Breathalyzer Detect a Blood Alcohol Level?

Martin, D 2002, Robert F. Borkenstein, 89, *Inventor of the breathalyzer*, The New York Times, Viewed 2 October 2016, http://www.nytimes.com/2002/08/17/us/robert-f-borkenstein-89-inventor-of-the-breathalyzer.html

Mothers against Drunk Driving 2016, *Drunk driving statistics*, Mothers against Drunk Driving, Viewed 2 October 2016, http://www.madd.org/drunk-driving/about/drunk-driving-statistics.html

National Center for Statistics and Analysis 2014, *Alcohol-impaired driving*, National Center for Statistics and Analysis, Viewed 2 October 2016, http://www-nrd.nhtsa.dot.gov/Pubs/812102.pdf

Susan 2012, *Breathalyzer — Markets reach $3.2 billion by 2018*, WinterGreen Blog, Viewed 2 October 2016, http://www.wintergreenblog.com/?p=119

How to Become a Brewery

Cordell, B & McCarthy, J 2013, 'A case study of gut fermentation syndrome (auto-brewery) with *Saccharomyces cerevisiae* as the causative organism', *International Journal of Clinical Medicine*, vol. 4, pp. 309–312.

Logan, BK & Jones, AW 2000, 'Endogenous ethanol 'auto-brewery syndrome' as a drunk-driving defense challenge', *Medicine, Science, and the Law*, vol. 40, pp. 206–215.

MedlinePlus 2016, *Bacterial gastroenteritis*, US National Library of Medicine, Viewed 2 October 2016, http://www.nlm.nih.gov/medlineplus/ency/article/000254.htm

How was a Painkiller Used to Free Hostages?

BBC News 2002, *Gas 'killed Moscow hostages'*, BBC News, Viewed 2 October 2016, http://news.bbc.co.uk/2/hi/europe/2365383.stm

Buffalo Field Campaign 2015, *Carfentanil-immobilizing bison*, Buffalo Field Campaign.

Krechetnikov, A 2012, *Moscow theatre siege: Questions remain unanswered*, BBC News, Viewed 2 October 2016, http://www.bbc.co.uk/news/world-europe-20067384

Riches, JR, Read, RW, Black, RM, Cooper, NJ & Timperley, CM 2012, 'Analysis of clothing and urine from Moscow theatre siege casualties reveals carfentanil and remifentanil use', *Journal of Analytical Toxicology*, pp. 647–656.

Stanley, TH, Egan, TD & Van Aken, H 2008, 'A tribute to Dr. Paul A. J. Janssen: Entrepreneur extraordinaire, innovative scientist, and significant contributor to anesthesiology', *Anesthesia & Analgesia*, vol. 106, pp. 451–462.

Wax, PM, Becker, CE & Curry, SC 2003, 'Unexpected "gas" casualties in Moscow: A medical toxicology perspective', *Annals of Emergency Medicine*, vol. 41, pp. 700–7005.

Wines, M 2002, *Chechens seize Moscow theater taking as many as 600 hostages*, The New York Times, Viewed 2 October 2016, http://www.nytimes.com/2002/10/24/world/chechens-seize-moscow-theater-taking-as-many-as-600-hostages.html

The Secret Ingredient in Coca-Cola ®

Elmore, BJ 2014, *Citizen Coke: The Making of Coca-Cola Capitalism*, W. W. Norton & Company, New York, NY.

Gootenberg, P 2004, 'Secret ingredients: The politics of coca in US–Peruvian relations, 1915–65', *Journal of Latin American Studies*, vol. 36, pp. 233–265.

Hellerman, C 2011, *Cocaine: The evolution of the once 'wonder' drug*, CNN, Viewed 2 October 2016, http://www.cnn.com/2011/HEALTH/07/22/social.history.cocaine/index.html

Why is Crack Cocaine so Addicting? Cocaine Smuggling versus Methamphetamine Manufacture

Estroff, TW 2001, *Manual of Adolescent Substance Abuse Treatment*, American Psychiatric Publishing, New York, NY.

Mans, DM & Pearson, WH 2004, 'Total synthesis of (+)-cocaine via desymmetrization of a meso-dialdehyde', *Organic Letters*, vol. 6, pp. 3305–3308.

National Institute of Drug Abuse 2016, *What is cocaine?* National Institute of Drug Abuse, Viewed 2 October 2016, http://www.drugabuse.gov/publications/research-reports/cocaine/what-cocaine

The Washington Times 2004, *Coca kick in drinks spurs export fears*, The Washington Times, Viewed 2 October 2016, http://www.washingtontimes.com/news/2004/apr/19/20040419-093635-4754r/

Volkow, ND 2010, *Letter from the Director*, National Institute of Drug Abuse, Viewed 10 September 2014, http://www.drugabuse.gov/publications/research-reports/cocaine/letter-director

What Basic Common Ingredient is Needed to Make the Drugs Vicodin®,
Percocet®, Oxycontin®, & Percodan®?

Glen, B 2014, *Company studies economic feasibility of thebaine poppy*, Producer,
Viewed 2 October 2016, http://www.producer.com/2014/04/company-studies-
economic-feasibility-of-thebaine-poppy/

Halvorsen, H & Lovli, T 2009, *Preparation of oxycodone*, US Patent WO
2009004491 A2.

International Narcotics Control Board 2015, *Narcotic drugs*, International
Narcotics Control Board, Viewed 2 October 2016, https://www.incb.org/
documents/Narcotic-Drugs/Technical-Publications/2014/Narcotic_Drugs_
Report_2014.pdf

National Institute of Drug Abuse 2011, *DrugFacts: Drug-related hospital
emergency room visits*, National Institute of Drug Abuse, Viewed 2 October
2016, http://www.drugabuse.gov/publications/drugfacts/drug-related-
hospital-emergency-room-visits

National Institute of Drug Abuse 2015, *DrugFacts: Prescription and
over-the-counter medications*, National Institute of Drug Abuse, Viewed 2
October 2016, http://www.drugabuse.gov/publications/drugfacts/prescription-
over-counter-medications

Novak, BH, Hudlicky, T, Reed, JW, Mulzer, J & Trauner, D 2000, 'Morphine
synthesis and biosynthesis – An update', *Current Organic Chemistry*, vol. 4,
pp. 343–362.

PBS 1998, *Opium throughout history*, PBS, Viewed 2 October 2016, http://www
.pbs.org/wgbh/pages/frontline/shows/heroin/etc/history.html

Schmidt, B 2015, *CDC: Drug overdose deaths at all-time high in U.S.*, NEWS 8000,
Viewed 2 October 2016, http://www.news8000.com/news/cdc-drug-overdose-
deaths-at-alltime-high-in-us/37089890

The Drug Abuse Warning Network 2010, *Highlights of the 2009 Drug Abuse
Warning Network (DAWN) findings on drug-related emergency department
visits*, The Drug Abuse Warning Network.

University of Pennsylvania School of Medicine 2011, *Opioids now most prescribed
class of medications in America*, Science Daily, Viewed 2 October 2016, http://
www.sciencedaily.com/releases/2011/04/110405161906.htm

US Drug Enforcement Agency 2016, *Thebaine Control Program*, US Drug
Enforcement Agency, Viewed 2 October 2016, https://www.dea.gov/druginfo/
concern_thebaine.shtml

Volkow, ND & Mclellan, TA 2011, 'Curtailing diversion and abuse of opioid
analgesics without jeopardizing pain treatment', *JAMA*, vol. 305, pp.
1346–1347.

Drug Money is Right

Biello, D 2009, *Cocaine contaminates majority of U.S. Currency*, Scientific American, Viewed 2 October 2016, http://www.scientificamerican.com/article .cfm?id=cocaine-contaminates-majority-of-american-currency

Carpenter, S 2016, *Report: Dollar bills, receipts tainted with BPA*, Chicago Tribune, Viewed 2 October 2016, http://www.chicagotribune.com/lifestyles/ sns-green-money-tainted-with-bpa-story.html

Hunter, A 2010, *BPA: Are your dollar bills contaminated?* CBS News, Viewed 2 October 2016, http://www.cbsnews.com/news/bpa-are-your-dollar-bills- contaminated/

Jenkins, AJ 2001, 'Drug contamination of US paper currency', *Forensic Science International*, vol. 121, pp. 189–193.

Park, M 2009, *90 percent of U.S. bills carry traces of cocaine*, CNN, Viewed 2 October 2016, http://www.cnn.com/2009/HEALTH/08/14/cocaine.traces .money/index.html

Singletary, M 2014, *Dirty Dollars: That paper money is teeming with germs*, The Washington Post, Viewed 2 October 2016, https://www.washingtonpost.com/ business/economy/dirty-dollars-that-paper-money-is-teeming-with-germs/ 2014/04/24/ae806962-cad2-11e3-93eb-6c0037dde2ad_story.html

What Percentage of Americans use Prescription Drugs?

Centers for Disease Control and Prevention 2010, *Prescription drug use continues to increase: U.S. Prescription Drug Data for 2007–2008*, Centers for Disease Control and Prevention, Viewed 2 October 2016, http://www.cdc.gov/nchs/ data/databriefs/db42.htm

Centers for Disease Control and Prevention 2011, *Vital signs: Overdoses of prescription opioid pain relievers – United States, 1999–2008*, Centers for Disease Control and Prevention, Viewed 15 September 2015, http://www.cdc .gov/mmwr/preview/mmwrhtml/mm6043a4.htm

Centers for Disease Control and Prevention 2014, *Therapeutic drug use*, Centers for Disease Control and Prevention, Centers for Disease Control and Prevention, Viewed 2 October 2016, http://www.cdc.gov/nchs/fastats/drug- use-therapeutic.htm

Centers for Disease Control and Prevention 2016, *Injury prevention & control: Opioid overdose*, Centers for Disease Control and Prevention, Viewed 2 October 2016, https://www.cdc.gov/drugoverdose/pubs/index.html#tabs-760094-4

Office of Applied Studies 2009, *National survey on drug use & health*, Substance Abuse and Mental Health Service Administration.

The Drug Abuse Warning Network 2010, *Highlights of the 2009 Drug Abuse Warning Network (DAWN) findings on drug-related emergency department visits*, Substance Abuse and Mental Health Service Administration, Viewed 15 September 2015, http://oas.samhsa.gov/2k10/dawn034/edhighlights.htm

What is Powdered Alcohol?

Adams, P 2014, *How to make powdered booze at home*, Popular Science, Viewed 2 October 2016, http://www.popsci.com/article/technology/how-make-powdered-booze-home

CBS News 2015, *"Palcohol" powdered alcohol wins federal approval*, CBS News, Viewed 2 October 2016, http://www.cbsnews.com/news/palcohol-powdered-alcohol-wins-federal-approval/

Ecstasy is Ruining the Rainforests

Banks, ML, Worst, TJ, Rusyniak, DE & Sprague, JE 2014, 'Synthetic cathinones ("bath salts")', *The Journal of Emergency Medicine*, vol. 46, pp. 632–642.

Campbell, S 2009, *Harvested to make ecstasy, Cambodia's trees are felled one by one*, Global Post.

Drug War Facts 2016, *MDMA or ecstasy (methylenedioxymethamphetamine)*, Drug War Facts 2016, Viewed 2 October 2016, http://www.drugwarfacts.org/cms/Ecstasy

IRIN News 2008, *Ecstasy tabs destroying forest wilderness*, IRIN News, Viewed 2 October 2016, http://www.irinnews.org/PrintReport.aspx?ReportId=79340

MacKinnon, I 2009, *'Ecstasy oil' factories destroyed in Cambodian rainforest*, The Guardian, Viewed 2 October 2016, http://www.guardian.co.uk/environment/2009/feb/25/ecstasy-cambodia

National Institute of Drug Abuse 2016, *DrugFacts: MDMA (ecstasy/molly)*, National Institute of Drug Abuse, Viewed 2 October 2016, http://www.drugabuse.gov/publications/drugfacts/mdma-ecstasy-or-molly

The National Newspaper 2011, *Cambodia's jungles threatened by ecstasy*, The Huffington Post, CBS News, Viewed 2 October 2016, http://www.huffingtonpost.com/2009/01/14/cambodias-jungles-threate_n_158005.html

How are Moldy Bread, Migraine Headaches, LSD, and the Salem Witch Trials all Related?

Griggs, EA & Ward, M 1977, *LSD toxicity: A suspected cause of death*, The lycaeum.

Hofmann, A & Ott, J 2013, *LSD: My Problem Child*, Oxford University Press, Oxford.

Klock, JC, Boerner, U & Becker, CE 1973, 'Coma, hyperthermia and bleeding associated with massive LSD overdose', *Western Journal of Medicine*, vol. 120, pp. 183–188.

US Drug Enforcement Administration 2007, *LSD manufacture – Illegal LSD production*, U.S. Department of Justice Drug Enforcement Administration, Viewed 15 September 2015, https:/fas.org/irp/agency/doj/dea/product/lsd/lsd-5.htm

West, LJ, Pierce, CM & Thomas, WD 1962, 'Lysergic acid diethylamide: Its effects on a male Asiatic elephant', *Science Magazine*, vol. 138, pp. 1100–1102.

6

Why Oil Is Such a Big Part of Our Lives

Petroleum doubtlessly plays a huge role in everyday life. It allows cars to run, heats homes, and often generates electricity. However, the commercial products created from oil are even more common. Every day, we use, touch, and sometimes even ingest products that were once deep underground. Although I focus on the major uses of oil in this chapter, its role in the manufacture of commercial products is so important that it is discussed throughout this entire book.

What Substance Is Used to Make 80% of All Pharmaceuticals?

I have to say that this question is easily my favorite to ask classrooms of students, mainly because very few people know the answer to this question. When I started teaching, my classes would average 400 students and I would regularly stump the entire room. Imagine that a room containing that many people could not come up with the answer. I even went so far as to perform an online poll and determined that less than 1% of the general public knows the answer. The fact that the general public knows so little about the chemistry around them started me telling these stories in my classes and has eventual led to this book.

Petroleum, also called crude oil, is a mixture of hundreds of different types of organic molecules containing hydrogen and carbon. Once petroleum is removed from the ground, it is sent to a refinery where its components are separated. Chemical processes convert most of the petroleum (45%) into gasoline. However, other parts are converted into jet fuel, kerosene, lubricants, and waxes. A total of 2.1% becomes what is called *petrochemical feedstock*, mainly methanol, ethylene, propylene, butadiene, benzene, toluene, and xylene. These and other petrochemical feedstock chemicals are used as the precursors for many chemical products in a variety of industries. Although 2.1% may not sound like much, if you consider that the world's daily petroleum

Strange Chemistry: The Stories Your Chemistry Teacher Wouldn't Tell You, First Edition. Steven Farmer.
© 2017 John Wiley & Sons, Inc. Published 2017 by John Wiley & Sons, Inc.

production is 90,000,000 barrels, you can see that the petrochemical industry is actually quite large.

Petroleum is the basis for almost all plastics, roughly 80% of pharmaceuticals, and just about any organic compound made synthetically. A short list of products made from molecules that originated from petroleum include cosmetics, fertilizers, detergents, synthetic fabrics, asphalt, paint, inks, glues, cleaning chemicals, insecticides, antifreeze, carpeting, dyes, upholstery foams, soaps, artificial flavors, perfumes, hairsprays, laundry bleach, deodorants, shampoos, soaps, dishwasher detergents, some vitamins, crayons, rubbing alcohol, golf balls, toothpaste, shaving creams, and soft contact lenses. The list actually goes on and on.

One widely used product made from petroleum is aspirin, which has a whopping world consumption of 40,000 tons annually. Originally, aspirin was made from the molecule salicylic acid isolated from the Willow tree. Imaging how many Willow trees would need to be grown to keep up with world's demand of aspirin. Using petrochemical sources allows for large-scale production of molecules without utilizing a natural source. Aspirin is just one of many petroleum-based pharmaceuticals, again giving you an idea of the vastness of the petrochemical industry. During the process of making gasoline from petroleum, the molecule *benzene* is isolated. Benzene is then converted into aspirin through a series of chemical reactions illustrated in Scheme 6.1. The reactions used to convert petrochemicals into pharmaceuticals and other commercial products are an important aspect in the study of organic chemistry.

The birth of the modern petrochemical industry started in 1792 when it was discovered that burning coal in vacuum produced a flammable gas which burned brightly. For a time, coal gas was a major source of illumination. Unfortunately, burning coal left behind a foul-smelling residue called *coal tar*, which has no apparent use. Ingenious scientists found that coal tar was full of interesting chemicals, one in particular was called *aniline*. At the time, Malaria was a significant problem and the molecule quinine, which was used as a treatment, could only be obtained from the bark of the cinchona tree found in Central America. In 1856, the then 18-year-old chemist William Henry Perkin attempted to synthetically make quinine using aniline isolated from coal tar. In a moment of serendipity, the aniline used by Perkin contained an impurity, and instead of forming quinine he ended up making *mauveine*, the first organic chemical dye. Prior to this dyes were isolated from natural sources including animals, plants, and minerals which made them all very expensive. Mauveine was extremely cheap because it could be made from molecules isolated from relatively worthless coal tar. Soon the new purple color, mauve, produced by mauveine dye became one of the most sought after colors on the dye market. With this, the organic chemistry industry was born. The idea of creating useful products using organic chemistry reactions and cheap molecule sources flourished. Soon many pharmaceuticals were being

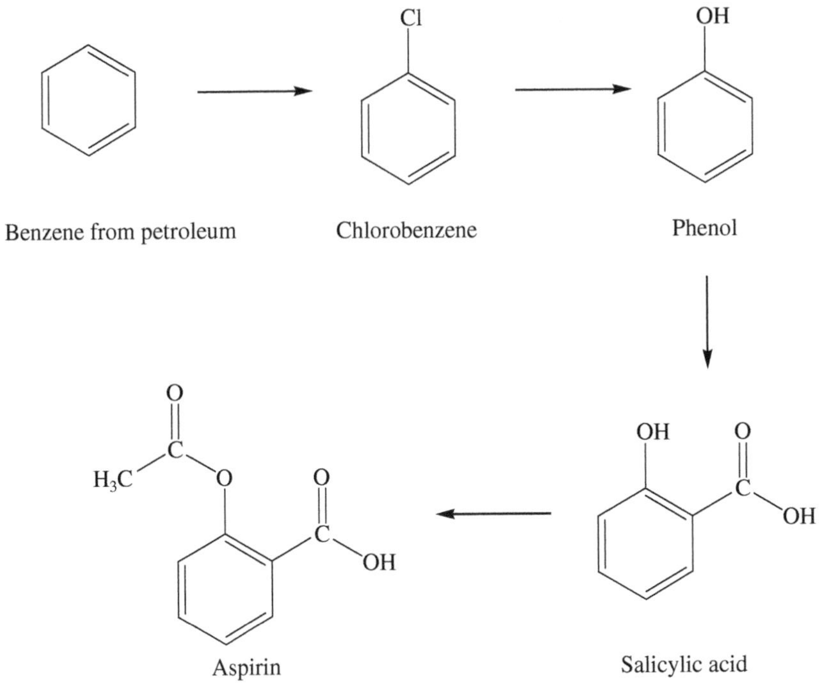

Scheme 6.1 The synthesis of Aspirin using Benzene obtained from petroleum.

made from coal tar derivatives; in fact, aspirin was one of the first drugs made industrially (Figure 6.1).

During WWII (World War II), coal tar was replaced with petroleum as the major source of cheap carbon-containing molecules. During WWI (World War I), TNT (trinitrotoluene) became the primary explosive; however, production was limited by the relatively small amount of toluene being produced by isolation from coal tar. A more abundant source of toluene was required to keep up with the munitions demands of WWII. New techniques were developed to synthesize toluene using the molecules found in petroleum, and the modern petrochemical industry was born. Technically, the term petrochemical refers to specific chemical compound that can be made from oil, natural gas, coal, or other similar sources. However, petroleum and natural gas are used to make approximately 99% of US petrochemicals. The basic petrochemicals are ethylene, propylene, butadiene, benzene, toluene, and xylene. Using the reactions in organic chemistry, these basic molecules can be converted as almost every carbon-containing product industrially synthesized in the United States. Just to give an idea of the scope of the petrochemical industry, the United States produced over 24 million tons of ethylene in 2013 (Figure 6.2; Scheme 6.2).

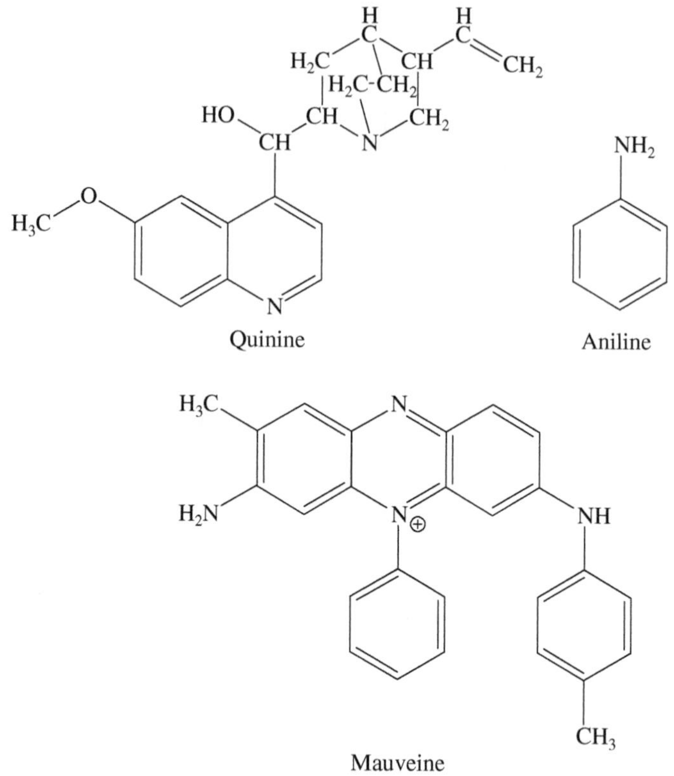

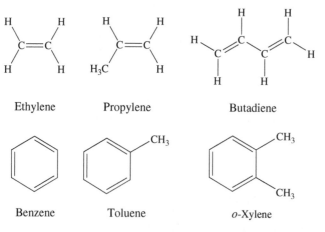

Quinine   Aniline

Mauveine

Figure 6.1 The structure of quinine, aniline, and mauveine.

Ethylene   Propylene   Butadiene

Benzene   Toluene   *o*-Xylene

Figure 6.2 The structure of some basic petrochemicals.

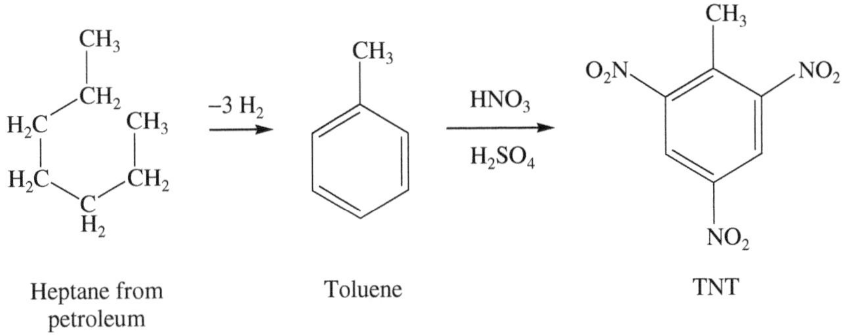

Scheme 6.2 The synthesis of TNT using heptane obtained from petroleum.

As you can imagine, petroleum is a nonrenewable resource, so if nothing is changed, a wide variety of industries could eventually die out. Fortunately, scientists are hard at work to find a renewable, cheap source of carbon feedstock for these industries. The current focus is on cellulose from wood and ethanol from corn (Figure 6.3).

Why Do Scientists Think Oil Comes From Fossilized Plants and Animals?

The idea that petroleum, a fossil fuel, comes from the fossilized remains of plants and animals arises from scientists' discovery of numerous organic molecules with complex structures that can be directly traceable to a biological source. This concept is called the *biogenic origin of petroleum*, and the organic molecules are called *biomarkers* or *geochemical fossils*.

Organic compounds from biological origins can be preserved within sedimentary rock layers for geological periods of time. *Organic geochemistry* is the study of these organic compounds in the geosphere. For petroleum in particular, the concept of its biogenic origin started when the father of organic geochemistry, Alfred Treibs, first isolated vanadium and nickel containing porphyrin molecules from crude oil in 1936. Plants and animals contain molecules called *porphyrins* – chlorophyll and hemoglobin, respectively – and it is thought that these compounds change under petroleum-forming conditions to create the porphyrin molecules that are found in crude oil. In addition to porphyrins, other biomarkers are commonly found: *algenan*, which comes from the cell walls of algae; *lignin*, a complex polymer found in wood; and *sporopollenin*, derived from the tough outer layer of plant spores and pollen (Figure 6.4).

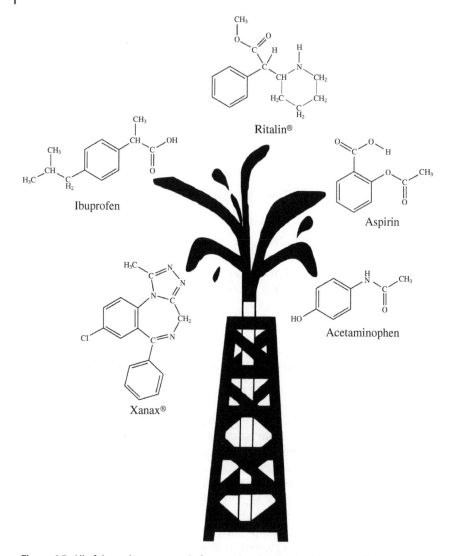

Figure 6.3 All of these drugs are made from molecules obtained from petroleum.

Coal is a carbonaceous material formed by compaction and hardening of plant remains that may have originally been peat in swamps. Oil is the fossilized remains of marine algae and zooplankton waste. This is why most coal fields are more inland on the continent, while oil fields tend to be closer to the oceans. Texas and Alaska both contain major oil fields. In addition, this explains why offshore drilling for oil is so lucrative.

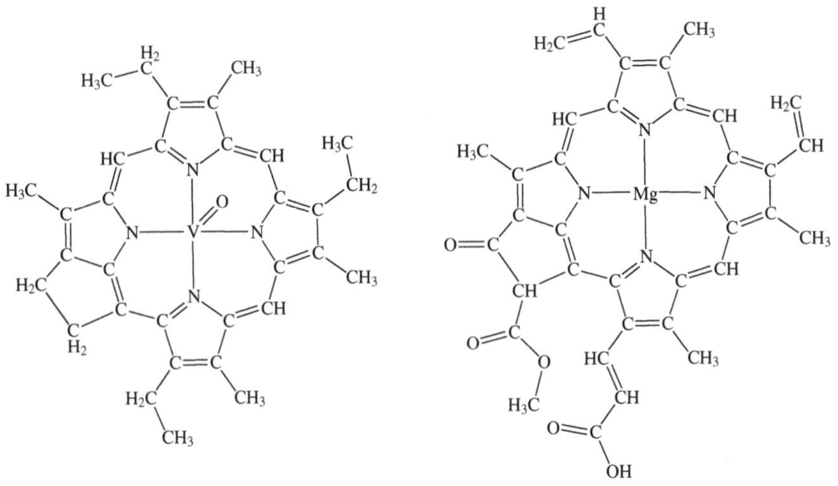

Vanadium porphyrin complex
isolated from crude oil

A magnesium chlorophyll complex

Figure 6.4 A comparison of a vanadium containing porphyrin found in crude oil and chlorophyll.

How Is Oil Made?

Much of the carbon that becomes petroleum originates from *phytoplankton* (algae). A large part of the organic material produced by phytoplankton is consumed by zooplankton, and the remains of planktonic organisms and zooplankton fecal pellets sink to the ocean floor. These remains are broken down by bacteria, releasing the organic compounds of which they are composed. Most of these organic compounds are recycled as nutrients for phytoplankton, but some of these organic compounds get absorbed into clay in the sediment. Organic compounds associated with life, for example, proteins, sugars, and lipids, get trapped in the sediment and turn into petroleum. As more layers form, the sediments and minerals are compacted to form sedimentary rock.

In the sediment, these organic compounds remain and undergo degradation through temperature, pressure, and microbial action, in a process called *diagenesis*. As they are broken down, structural changes occur, and the organic molecules combine together in a process called *polycondensation* to form kerogen in marine sediments. *Kerogen* is the polymeric organic material from which hydrocarbons are produced with increasing burial and heating. Kerogen is by far the most abundant form of organic carbon in the earth's crust, and by far the largest pool of organic matter on the planet. There is an estimated 10^{16} tons of carbon in kerogen, compared to roughly 10^{12} tons of carbon in the living biomass on earth (Figure 6.5).

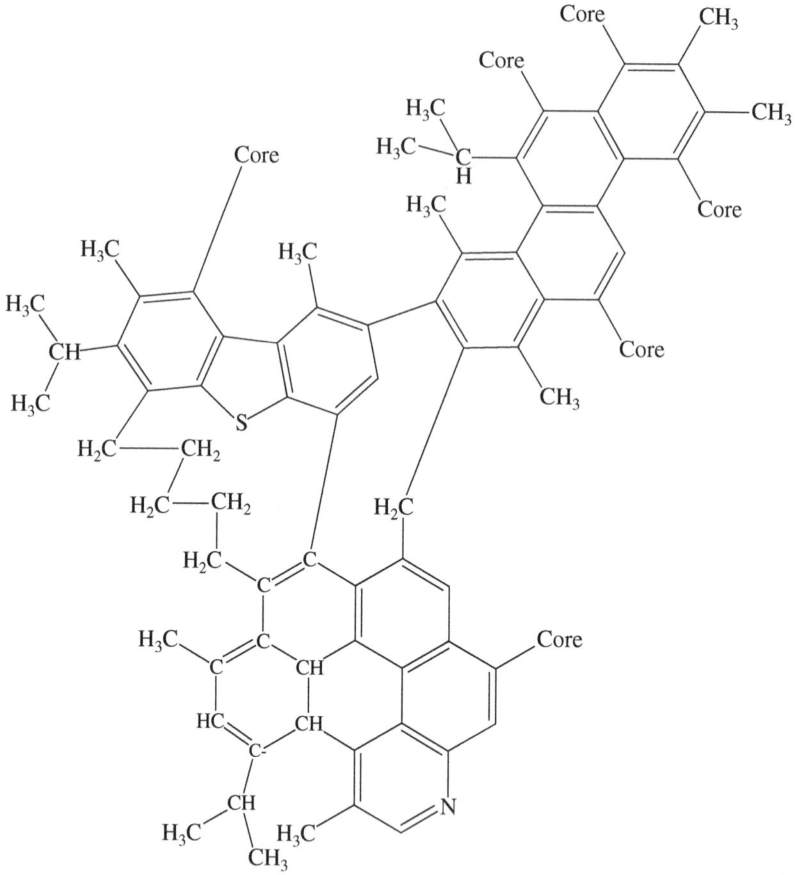

Figure 6.5 The structure of a kerogen fragment. The word "core" represents a connection to a larger molecular matrix.

With increasing burial depth and temperature, and catalyzed by the minerals in the rock, a process called *catagenesis* causes hydrocarbon chains to break off and be expelled from the kerogen. These chains form the liquid parts of petroleum in a process called *cracking*. This cracking process is very similar to the process used in oil refineries to increase the light hydrocarbon content of petroleum. At greater depths and temperatures, the oil molecules are cracked into smaller gas molecules such as methane (CH_4), carbon dioxide (CO_2), nitrogen gas (N_2), and hydrogen sulfide (H_2S) in a process called *metagenesis*. Hydrocarbon-rich liquids and gases evolved from kerogen during catagenesis and metagenesis are termed *petroleum*.

Petroleum forms under specific conditions defined as the "oil window." If conditions are too hot, metagenesis will convert all the petroleum into natural gas; if too cold, catagenesis will not occur and the organic material will remain as kerogen. The oil window lies between temperatures of approximately 60 and 120 °C, which explains why oil is found is very defined regions.

Where Is Most of the Carbon in the World?

The largest reservoir of carbon is in sedimentary rock, including calcium carbonate (limestone, $CaCO_3$) and dolomite ($CaMg(CO_3)_2$). Sedimentary rock accounts for about 99.9% of all carbon on the planet, with roughly 75% in the form of carbonates and 25% in organic compounds, such as kerogen. It is estimated that there is 5×10^7 Gt (a Gt is equal to 10^9 ton or about 10^{15} g) of carbon in sedimentary rock, while there is only 740 Gt in the atmosphere.

The Most Widely Recycled Material in the United States

It seems that when this question is posed, most people tend to think of products that they put in their recycling bin, such as plastic, glass, and aluminum cans. In fact, when thinking about large-scale recycling, one needs to look at industries. With some hints provided, many can come up with the second most-recycled material in the United States, which is steel.

The material holding the number one spot is *asphalt* pavement, the most widely recycled material in the United States, both in terms of gross tonnage and percentage. In addition, the asphalt industry remains the number one recycler in America. Roughly 99% of the asphalt pavement removed each year is reused in new roads. In 2010, about 62 million tons of recycled asphalt pavement was reused in roadway projects, with less than 1% of reclaimed asphalt pavement being sent to landfills. This is quite significant when compared to the 2.98 million tons of aluminum recycled each year, of which less than 20% is from to beverage cans.

The recycling of asphalt pavement saves US taxpayers an estimated $1.8 billion per year. American roadways are an asset valued at over $2.4 trillion. Over 92% of these roadways – more than 2.5 million miles – are surfaced with asphalt. This represents enough asphalt to cover the entire state of Nevada.

Asphalt pavement is a mixture of asphalt and crushed rock. The first recorded use of asphalt as a road material was in Babylon in 625 BCE. In 1870, the first asphalt pavement was laid in the United States in Washington, D.C., using natural asphalt obtained from Trinidad Lake. By 1907, the use of natural asphalt was abandoned and replaced by asphalt produced during petroleum refining. Asphalt is used as a road material because it is cheaper, more pliable, and creates less noise than concrete. In addition, asphalt can easily expand and contract with temperature changes while concrete cannot. For roads constructed of concrete, spaces must be incorporated to allow for contraction and expansion.

What Material Is Used to Make Asphalt?

Asphalt is a by-product of petroleum refining. At the refinery, crude petroleum undergoes a distillation process which separates it into fractions of different substances, each with its own characteristics.[1] Because asphalt is the heaviest component of crude petroleum, it does not boil off during distillation and remains a residue (Figure 6.6).

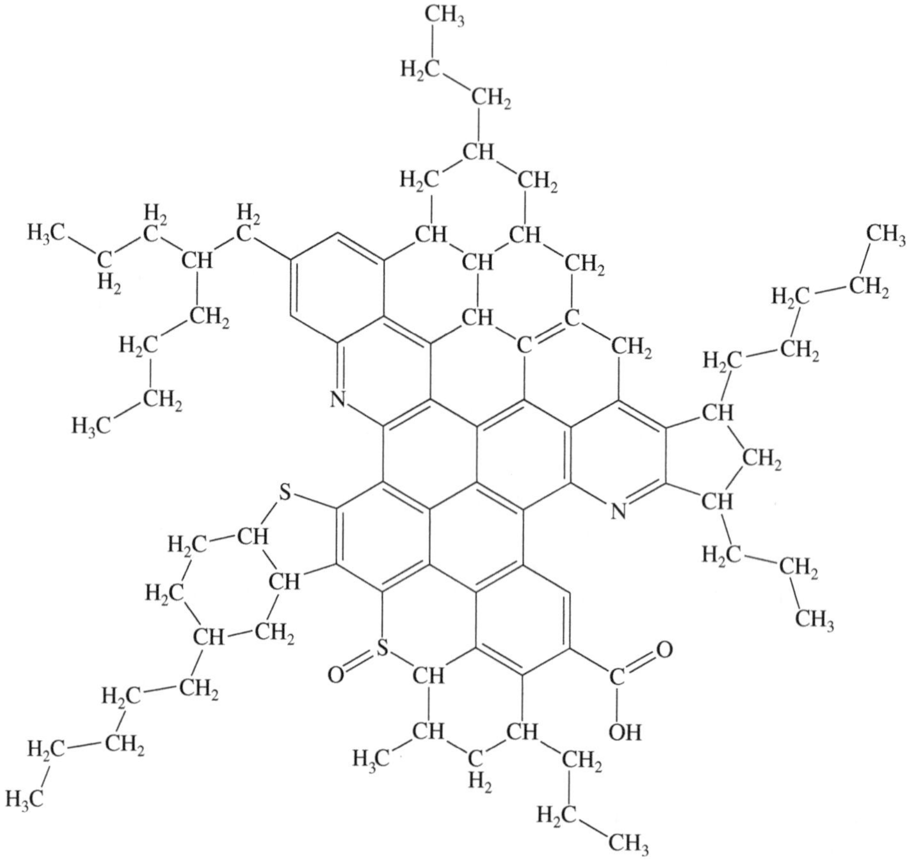

Figure 6.6 The average molecular structure of molecules found in asphalt.

---

1 Residing in the San Francisco Bay area, I actually live fairly close to two major petroleum refineries. They need to be located near the ocean because the petroleum is commonly shipped in using huge tanker ships. Major highways run right past both refineries so their sight has become quite familiar. If you want to know it is like, they smell horrible and sometimes at night you can see flames coming from the top of the processing towers.

Asphalt is a complex mixture of thousands of types of molecules containing up to 150 carbon atoms each. The composition of these molecules varies with the petroleum source, but they consist of hydrogen atoms along with a small number of sulfur, oxygen, or nitrogen atoms. The presence of sulfur-containing molecules is part of the reason why asphalt has a distinctive "eggy" smell. Asphalt also contains trace amounts of iron, nickel, and vanadium, which are believed to come from porphyrin biomarkers discussed previously. The large size of the molecules in asphalt allows them to have strong interactions and gives the material its semisolid character. This can be easily compared to the components of gasoline, which typically have eight carbons and therefore exist in liquid form.

Asphalt is described as a semisolid, but it is actually an extremely viscous liquid. It flows so slowly that the movement is imperceptible. Pitch (another term for asphalt) flow is so slow that it is competing for the title of the world's longest running laboratory experiment, the pitch drop experiment. This experiment started in 1927 at the University of Queensland, Australia, to show that asphalt is, in fact, a liquid. Asphalt was added to a glass funnel, and then experimentalists began to wait. It took a little more than 8 years for the first drop of asphalt to fall from the bottom of the funnel! The experiment has continued to this day; the ninth drop fell on 24 April 2014. In this experiment, researchers were able to show that asphalt is about 100 billion times more viscous than water. The experiment is expected to be able to continue for about another 100 years, so look for the tenth drop to fall around 2028.

Unfortunately, asphalt contains known carcinogens, and asphalt fumes is listed as a potential carcinogen by the US Center for Disease Control (CDC). A CDC study showed that workers in the asphalt industry experience nasal and throat irritation, headache, coughing, rashes, nausea, stomach pain, decreased appetite, headaches, and fatigue. Since July 1997, the asphalt industry has installed engineering controls that vent asphalt fumes away from workers, thereby reducing their exposure to asphalt fumes.

How Oil Helped to Save the Whales

Before the development of petroleum products, whale oil obtained from blubber was widely used as a fuel source in oil lamps. In addition, spermaceti, a waxy substance from sperm whales, was used to make candles. At one point, the world had a thriving whaling industry developed to supply whale oil. In 1846, the United States alone had a whaling fleet of 735 ships. At its height in 1856, the US whaling industry was producing 4–5 million gallons of spermaceti and 6–10 million gallons of whale oil annually. The whaling industry is aptly described in Herman Melville's novel *Moby Dick*.

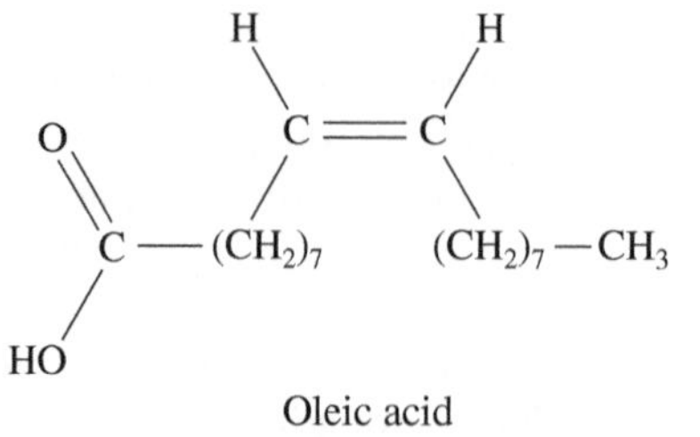

Figure 6.7 The structure of oleic acid, a fatty acid.

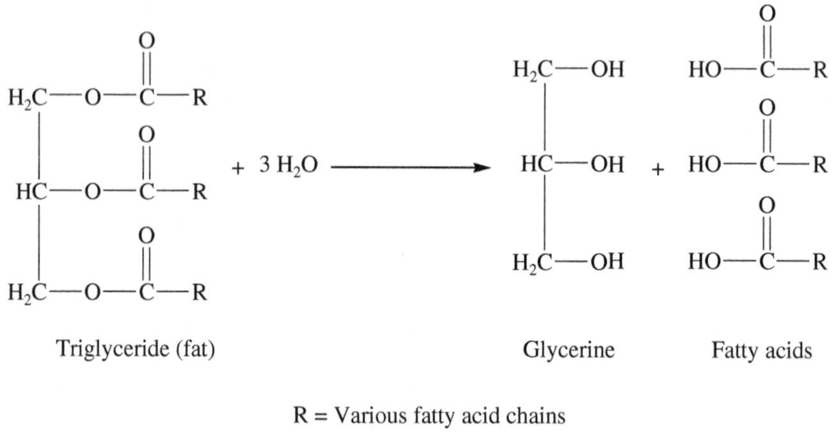

Oleic acid

Triglyceride (fat)   Glycerine   Fatty acids

R = Various fatty acid chains

Scheme 6.3 The decomposition of triglycerides to form glycerine and fatty acids.

The average sperm whale can produce between 25 and 40 barrels of oil, the composition of which varies with the species and method by which it is processed. Primarily, whale oil is made up of *triglycerides*, which are themselves made up of fatty acid molecules attached to a glycerin molecule. A wide variety of fatty acids is present, but the most common ones are oleic acid and its variants (Figure 6.7; Scheme 6.3).

Whale oil was a cheap albeit foul-smelling source of illumination. In 1849, Dr Abraham Gesner, a Canadian geologist, devised a method for the isolation of kerosene from petroleum. *Kerosene* is similar to gasoline, but it is composed of molecules with a greater mass and therefore has a higher boiling point. This makes kerosene less volatile and safer to use for lighting. In the late nineteenth century, whale oil was replaced by kerosene because it was cheaper, longer lasting, did not smell as bad when burned, and did not spoil. By the end of the 1850s, 30 kerosene plants were operating in the United States, and this more practical fuel began to drive whale oil out of the market. The 735-ship whaling fleet of 1846 shrank to 39 ships by 1876. Even after whale oil dropped in price to

40 cents-per-gallon, it could not keep up with the 7 cents-per-gallon price of kerosene. Of course, with Thomas Edison's invention of the incandescent light bulb, an even less costly and safer means to produce illumination, whale oil and kerosene were both driven from the market.

Despite the loss of the market for whale oil in the late 1930s, more than 50,000 whales continued to be killed each year because whale products were used in many other industries. Whales were an important source of the glycerin used to make nitroglycerine, the explosive used in WWI. In addition, during WWI, soldiers used whale oil to grease their feet to prevent trench foot and other problems associated with wet feet. A single battalion of over a thousand men would use up to 10 gallons of whale oil every day. In fact, because whale oil retains density under extremely high pressure and is resistant to freezing, it continued to be used in the automatic transmission fluid of cars until it was banned by the Endangered Species Act of 1973. During the early twentieth century, whale oil was used to make margarine until hydrogenation allowed for it to be replaced by vegetable oils. Even the collagen contained in tail flukes of whales was used to manufacture glue.

The continued hunting of whales forced some species to the verge of extinction. In response to growing concerns from scientists and the public, the US Congress passed the Marine Mammal Protection Act of 1972 (MMPA). The MMPA prohibits the taking of marine mammals and places a moratorium on the import, export, and sale of any marine mammal or marine mammal product within the country. This was followed by the 1973 Endangered Species Act (ESA). Under the ESA, the US government has the responsibility to protect endangered species, threatened species, and any critical habitat vital to the survival of these species. The hunting of whales virtually stopped in 1986 when the International Whaling Commission (IWC) banned commercial whaling so that stocks might recover. Almost overnight, the whaling industry disappeared. However, some commercial whaling still continues in Iceland and Norway despite objection from the IWC. In 2014, 897 whales were caught.

Further Reading

What Material is used to Make Roughly 80% of all Pharmaceuticals?

Processing 2013, *US ethylene production grows in 2013*, Processing, Viewed 24 September 2016, http://www.processingmagazine.com/processing-e-news/us-ethylene-production-grows-in-2013/

Wood, EJ 2005, *Aspirin: The Remarkable Story of a Wonder Drug*, Dairmuid Jeffreys, Bloomsbury, London.

Why do Scientists Think Oil Comes from Fossilized Plants and Animals?

Braun, RL & Burnham, AK 1993, *Chemical reaction model for oil and gas generation from type I and type II kerogen*, Lawrence Livermore National Laboratory, Viewed 24 September 2016, https://www.researchgate.net/publication/242102599_Chemical_reaction_model_for_oil_and_gas_generation_from_type_1_and_type_2_kerogen

Kvenvolden, KA 2006, 'Organic geochemistry – A retrospective of its first 70 years', *Organic Geochemistry*, vol. 37, pp. 1–11.

How is Oil Made? Where is Most of the Carbon in the World?

Broad, WJ 2010, *Tracing oil reserves to their tiny origins*, The New York Times, Viewed 24 September 2016, http://www.nytimes.com/2010/08/03/science/03oil.html?pagewanted=2&_r=0

Burlingame, AL, Haug, P, Belsky, T & Calvin, M 1965, 'Occurrence of biogenic steranes and pentacyclic triterpanes in an Eocene shale (52 million years) and an early Precambrian shale 2.7 billion year. A preliminary report', *Proceedings of the National Academy of Sciences of the United States of America*, vol. 54, pp. 1406–1412.

Fuel Chemistry Division 2014, *Petroleum*, American Chemical Society, Viewed 24 September 2016, http://www.ems.psu.edu/~pisupati/ACSOutreach/Petroleum_2.html

Hills, IR & Whitehead, EV 1966, 'Triterpanes in optically active petroleum distillates', *Nature*, vol. 209, pp. 977–979.

Orr, WL, Emery, KO & Grady, JR 1958, 'Preservation of chlorophyll derivatives in sediments of Southern California', *Bulletin – American Association of Petroleum Geologists*, vol. 42, pp. 925.

Treibs, A 1936, 'Chlorophyll und Häminderivate in organischen Mineralstoffen', *Angewandte Chemie*, vol. 49, pp. 682.

The Most Widely Recycled Material in the United States

Federal Highway Administration 1993, *A study of the use of recycled paving material*; Report No. FHWA-RD-93-147, Federal Highway Administration, Washington, DC.

National Asphalt Pavement Association 2011, *The Asphalt Paving Industry. A global perspective*, National Asphalt Pavement Association, Viewed 24 September 2016, http://www.eapa.org/userfiles/2/Publications/GL101-2nd-Edition.pdf

What Material is used to Make Asphalt?

Beyond Roads 2016, *The history of asphalt*, Beyond Roads, Viewed 24 September 2016, http://www.beyondroads.com/index.cfm?fuseaction=page& filename=history.html

How Oil Helped to Save the Whales

Edgerton, D 2016, *Not counting chemistry*, Chemical Heritage Magazine.
International Whaling Commission 2016, *Catches taken: under objection or under reservation*, International Whaling Commission, Viewed 24 September 2016, https://iwc.int/table_objection
Roberts, J 2016, *Whales in Space*, Chemical Heritage Magazine.
San Joaquin Valley Geology 2015, *How the oil industry saved the whales*, San Joaquin Valley Geology, Viewed 24 September 2016, http://www.sjvgeology .org/history/whales.html

7

Why Junior Mints® Are Shiny and Other Weird Facts about Your Food

I think everyone knows that many of the foods we eat have undergone some type of chemical process, but quite often these are hidden and not discussed. In addition, components of foods come from sources that are quite unexpected. These compounds are usually cloaked with clever names that hide their true origin, which might keep us from eating the food product. This chapter will explore the many strange things we commonly eat. I warn you that I may end up talking about some foods you regularly enjoy!

Why Is Gum Chewy?

The oldest known sample of gum, made from the resin of a birch tree, was over 9000 years old, found on the island of Orust in Sweden, and most likely stuck under a table. It was not until 1848, however, that John Curtis of Bangor, Maine, sold the first commercial gum, created primarily from the harvested sap of spruce trees. Another early gum was concocted from paraffin wax and sugar. None of these initial forms of gum was wildly popular, though. The true beginning of modern chewing gum came with the introduction of *chicle* (from the Aztec word for "sticky stuff"), a latex gum derived from the sapodilla tree (*Manilkara zapota*). Chicle has been used as a chewing gum in the Americas for hundreds of years. Ever heard of Chiclets®? They were the first candy-coated gum and given a name based on the chicle originally used to make them.

Legend has it that in 1869, a former President of Mexico, General Antonio López de Santa Anna, was exiled to America and happened to bring a sample of chicle with him. This is the same Santa Anna who had led the 1836 military victory over American defenders of the Alamo in the then-Mexican province of Texas. He gave the chicle sample to inventor Thomas Adams, who used it to create the first modern gum. The first flavored gums produced by Adams were Tutti Frutti and Black Jack licorice.

The sapodilla tree grows in the forests in Central America, and the ancient Mayans used its sap for a wide variety of purposes. Today, the sapodilla tree is

Strange Chemistry: The Stories Your Chemistry Teacher Wouldn't Tell You, First Edition. Steven Farmer.
© 2017 John Wiley & Sons, Inc. Published 2017 by John Wiley & Sons, Inc.

also grown in the Philippines, Mexico, Thailand, Vietnam, Puerto Rico, Florida, and California. Chicle is a milky fluid produced by the sapodilla tree whenever it is cut or attacked by insects. The chicle forms a protective layer over the damaged area and has even been shown to deter insects. Chicle can only be harvested from sapodilla trees that are at least 20 years old, with each tree yielding only about 1 kg of gum per tapping. Trees can only be tapped once every 3–4 years. Chicle harvesting itself is done by hand, and it is a dangerous and labor-intensive process. The chicle extractors, called *chicleros*, cut flesh along the length of the tree and allow the latex to run down the trunk.[1] Once the raw chicle latex is collected, they dry it, cook it, and form it into blocks to be sold to gum producers.

A natural latex, chicle is a polymer made when isoprene is polymerized to form poly(*cis*-1,4-isoprene). *Isoprene* is a volatile organic molecule that is produced and emitted by many trees, primarily as a deterrent against fungus and insect invaders. Numerous species of trees and plants contain enzymes that can cause the biological precursor to isoprene, isopentenyl diphosphate, to polymerize. In latex-forming trees, the enzyme is usually *cis*-prenyltransferase (Figure 7.1; Scheme 7.1).

After chewing gum became widely popular, chicle production could not keep up with demand. As of 2006, the gum economic sector had grown to $19 billion annually. The United States alone produced an estimated 145,000 tons of gum in 2012. Chew on that! The need for large amounts of low-cost, nonchicle gum base was further necessitated by an increase in US import and export taxes, which increased the price of chicle to gum manufacturers. Because of these

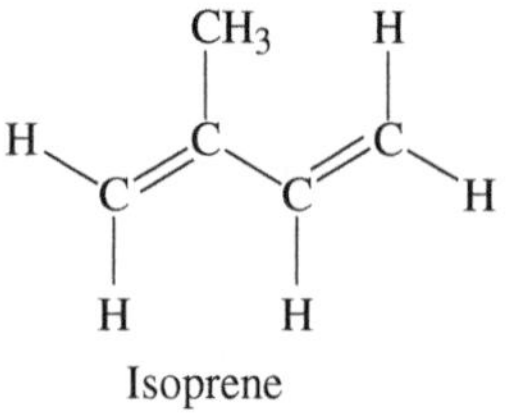

Figure 7.1 The structure of isoprene.

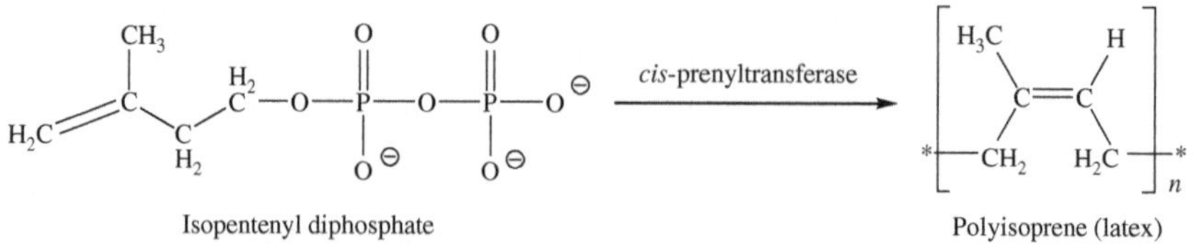

Scheme 7.1 The conversion of isopentenyl diphosphate into latex.

---

1 One of my students told me that as a child growing up in the Philippines, the chicleros would give her raw chicle by winding the flowing sap on a stick. She said it was a little bitter but very fun to chew.

factors, the 1940s saw the beginning of the use of synthetic, petroleum-based polymers as the gum base in chewing gum.

Because chicle is a natural polymer, it makes sense that gum makers would use a synthetic, petroleum-derived polymer as a substitute. The sticky stuff used in modern gums is called a *gum base* and its true composition is a trade secret. Flavoring and sugar are added to the gum base to create chewing gum. The composition of chewing gum is roughly 50% sugar, about 25% gum base, with flavorings, emulsifiers, humectants, and preservatives making up the rest. Some of the polymers used in modern gum bases are polyisobutylene, which is used in bicycle inner tubes; and polyvinyl acetate, which is also used in Elmer's® glue. Perhaps the people who ate glue as a child were on to something (Figure 7.2).

Despite the cost of chicle, a few modern gum manufactures still use it. At one point, the United States imported more than 7000 tons of chicle per year from Central America, but today, it is closer to 200 tons. Glee Gum®, Xylichew®, and Simply Gum® are few of the brands of gum that still use chicle as the gum base. On its web site, Glee Gum even offers a "Make Your Own Chewing Gum Kit" using chicle.

Have you ever noticed how hard it is to get gum off the bottom of your shoe? Many shoes are made with a polymer called Poly(styrene–butadiene–styrene). Gum base is also a polymer and the similarity of their structures allows for strong intermolecular forces to occur, causing them to stick. In short, the gum base sticks to the shoe polymer with roughly the same force as it sticks to itself. When an attempt is made to remove the gum, it sticks to the shoe while the rest pulls apart, creating the mess many of us have experienced. This effect also explains why gum sticks so well to asphalt. Part of the composition of asphalt is similar to a polymer so it to interacts strongly with gum base.[2]

Figure 7.2 A comparison of the structures of polymers typically found in gum base.

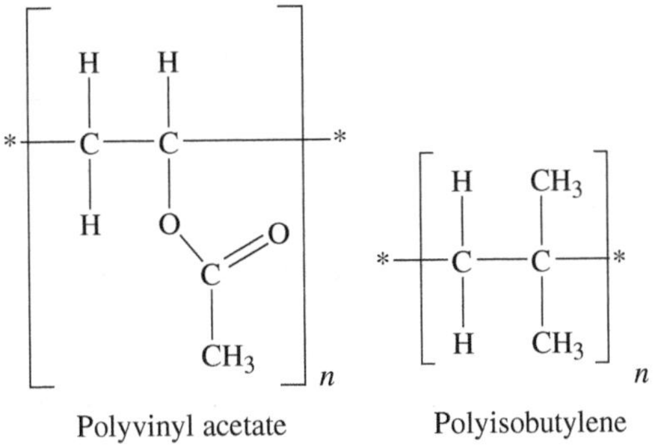

Polyvinyl acetate   Polyisobutylene

---

2 If you ever wondered what it is like to have a Ph.D. in chemistry, after rereading this story I decided to see if I could predict what else would stick to gum. Knowing that gum should stick to materials which a polymer-like structure, I was able to predict the outcome with 100% accuracy. Materials which stick to gum: paper, wax, wood, and most plastics. Materials which did not stick to gum: glass, tin cans, aluminum cans, and quartz.

Figure 7.3 You can thank chemistry for this situation.

If you have ever had gum stuck in your hair, peanut butter may have been used to get it out. Actually, it was the oil in the peanut butter that came to the rescue. The molecules in peanut butter oil have a structure vaguely similar to a polymer, which means that they can interact strongly with the molecules in gum base. With some work, the oil can break up the interaction between the gum base and the molecules found in hair. Because peanut butter oil is a liquid, the gum releases the hair and comes off relatively easy. If this ever happens to you, note that just about any oil can be used including olive oil or mineral oil (Figure 7.3).

The Problem with Gummi Bears

Gummi bears candy, along with many other commercial products such as marshmallows, Jell-O®, candy corn, toffees, and Peeps®, are made using *gelatin*. Food-grade gelatin (derived from Latin *gelatus*, meaning "stiff, frozen") is a translucent, colorless, flavorless protein extracted from collagen. *Collagen*, named from the Greek word for glue, *kolla*, is the most abundant protein in mammals and makes up roughly 30% of their total protein content. Found throughout the body, collagen makes up a major portion of many tissues such as tendons, skin, and dentin. The most abundant sources of the collagen used in gelatin production are pig skin (46%), bovine hides (29.4%), and pig and cattle bones (23.1%).

Gelatin has been known since antiquity and was first used as glue as far back as 6000 BC. The first English patent for gelatin production was granted in 1754, and the current worldwide production of gelatin is about 375,000 tons per year. Gelatin is one of the most versatile edible ingredients in commercial food manufacturing owing to its ability to be formed into gels with an elastic texture that melts in the mouth. This gives gelatin products a particular chewiness and excellent flavor release that is difficult to imitate. Gelatin is compatible with milk proteins and provides a smooth, even-textured consistency and creamy mouth feel. This makes gelatin well suited as an ingredient in the manufacture of a wide range of dairy products including yogurts, puddings, custards, whipped cream, and sour cream. Gelatin's unique fat-like melting properties are widely utilized for low-fat butters and margarine products, as they simulate the texture of the high-fat versions. Gelatin's whipping properties are also occasionally used for manufacturing aerated desserts such as mousses, chiffons, and soufflés. Lastly, gelatin is frequently an ingredient in certain meat products – including canned hams, terrines, patés, sausage, and luncheon meat – to enhance appearance, prevent drying out, bind juices, and prevent crumbling upon slicing by filling any cavities in the meat tissue. You may have noticed gelatin when opening a canned ham, in particular. In this case, it is used to cushion the ham to prevent damage to it during transport.

Because of gelatin's property of forming flexible plasticized films, it is recognized as one of the most versatile materials for "gel cap" pharmaceuticals. Gel cap pharmaceuticals are one of the most popular forms of drug dosage because they are more easily swallowed and digested, and the film prevents having to taste any of the drugs they encase.

Gelatin contains the amino acids glycine and proline in concentrations that are 10–20 times higher than in any other protein. Because these amino acids have been shown to play a major role in fortifying connective tissue, gelatin is often added to food supplements to maintain healthy joints, improve the elasticity of skin, and reduce wrinkles. In addition, gelatin is known to stimulate the growth of hair and nails. Gel-containing products applied externally, such as suntan lotions, shampoos, moisturizing creams, and facial creams, have many benefits. Gelatin can improve hair's manageability, shine, and body, and it also acts as an excellent skin moisturizer. Even for animals, gelatin is often added to pet food to increase the protein content, contribute to healthy joints, and promote a thick, healthy coat of fur (Figure 7.4).

Technically, all products made with gelatin are considered animal derived and should not be eaten by people following a vegan or kosher diet. Kosher gelatin is readily available but it is made with non-animal-based thickening agents

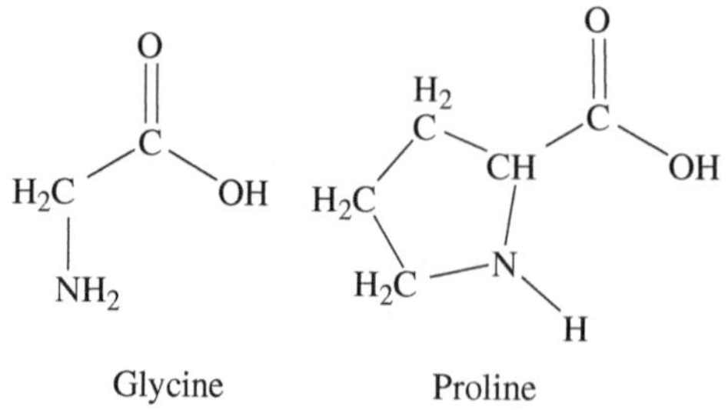

Figure 7.4 The structure of glycine and proline.

Glycine   Proline

such as guar gum, carrageenan, and xanthan. These Kosher alternatives are all polysaccharides and form gels in a similar manner to animal-derived gelatin.

Gelatin is a mixture of proteins created by the selective hydrolysis of collagen. In the gelatin-making process, collagen proteins are extracted from the skin and bones of animals by either an acid or alkaline bath. The proteins are then separated from the rest of the raw material, sterilized, and dried to form gelatin. Additional processing and purification steps may be added depending on the required end product.

The structure of collagen depends on the type and age of the animal. In fact, 28 different types of collagen have been identified. In general, collagen is made up of three protein chains, each roughly 1000 amino acids long, twisted together to form a helix structure – think of it like a rope. The helices are primarily held together by intermolecular hydrogen bonds between the protein chains. During the gelatin manufacturing process, heating causes these helices to fall apart, and the collagen loses its original structure. Because water also has a hydrogen bonding intermolecular force, it can interact with these protein chains. Upon cooling, the helixes are partially reformed, resulting in water being trapped in the mesh of protein chains. The protein chains' structure is changed, forming gelatin (Figure 7.5).

Because much of collagen comes from cows, there has been concern that gelatin may harbor the prions that cause BSE (bovine spongiform encephalopathy), also known as mad cow disease. A *prion* is an infectious agent made up entirely of protein material that can cause diseases that are similar to viral infections. Based on studies, the US Food and Drug Administration (FDA) considers the risk of BSE transmission through gelatin to be minimal as long as proper precautions are followed during manufacturing.

While we are on the subject, where do you think they get the collagen for cosmetic procedures such as collagen injections? This collagen usually comes from cows. With the concern about mad cow disease, patients have sought reassurance regarding the sources of medical collagen. Most medical collagen manufacturers use collagen from disease-free animals raised in "closed herds," or in countries that have never had a reported case of BSE, such as Australia, Brazil, and New Zealand. Most of the medical collagen used is from young

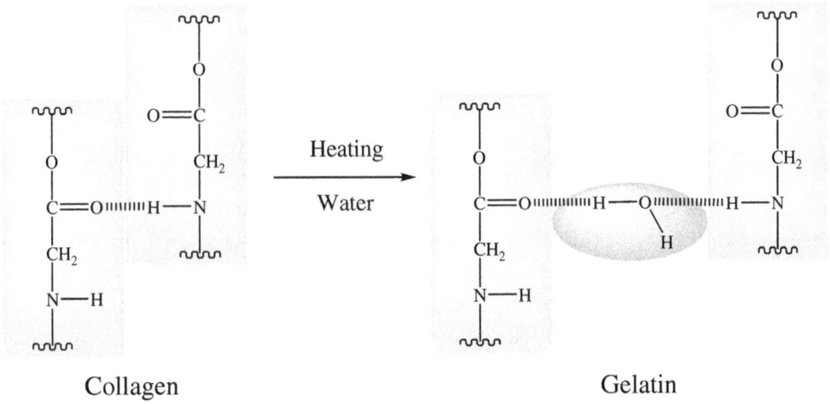

Figure 7.5 Showing how water becomes imbedded in collagen. Hydrogen bonds are dashed.

beef cattle (bovine) certified as disease free. Another commonly used collagen source is porcine (pig) tissue. In some cases, the patient's own fat is used.

What Is the Easiest Way to Peel a Tomato?

The easiest way to peel a tomato is to give it a sunburn of course! Unfortunately, it is not that easy. In the mid-western United States, reaction with *lye* (sodium hydroxide, NaOH) is the most common method for peeling tomatoes. In addition to tomatoes, products including peaches, sweet potatoes, white potatoes, apricots, mandarin oranges, and carrots have to be peeled as one of the first steps prior to canning. The lye solution peels by dissolving cell walls and weakening the cellulose network. The cells in the skin have thinner cell walls than the inside flesh of the fruit, so they are affected quicker. The difference of destruction rates causes a separation between the flesh and the skin.

Typical conditions include heating the vegetables in a 95 °C steam-jacketed kettle in a concentrated lye solution (18% NaOH) for approximately 30–75 seconds. Following the lye, the vegetables pass over disc or pinch rollers that mechanically eliminate the peel. There are several other alternatives to lye peeling, including steam peeling, freeze peeling, and peeling with calcium chloride ($CaCl_2$). However, these other methods tend to produce less desirable products than conventional lye peeling, with deteriorated peeling appearance, high loss in firmness, and lowered yields. Typically, lye peeling is the most industrially used method for processing tomatoes in the United States. However, due to pressures of cost and environmental regulations, some tomato processors have been forced to use steam peeling to reduce chemical contamination of water with lye.

Another Way to Eat Insect Parts!

A common food dye called *carmine* is a bright red pigment made from the molecule carminic acid, which is isolated from female cochineal insects (*Dactylopius coccus*). Cochineals are scale insect species indigenous to Peru, Mexico, Bolivia, Chile, and the Canary Islands. They feed on and are considered a parasite for wild cacti belonging to the genus *Opuntia*. Female cochineals are immobile, having neither wings nor legs, and it is believed that they produce carminic acid as an adaptation to deter predators. Artificial selection methods of both the cochineal insects and the cacti they grow on have been practiced for centuries by natives of Mexico and South American. As a result, strains of the insect have been produced, which are unique to specific geographic regions and vulnerable to slight shifts in climate, making them nearly impossible to relocate to other areas. Peru is the principal source of cochineal insects, producing roughly 200 tons in 2005. Chile, Peru, Bolivia, and Mexico also export cochineal insects (Figure 7.6).

The use of carmine for dyeing can be traced back to the Aztecs, Mayans, and Incas. Before the arrival of Europeans in the New World, kermes (kermesic acid, natural red 3) was the only red dye available in Europe and the Middle East. Oddly enough, it was also obtained from the dried bodies of female insects of the genus *Kermes*. Although these insects were found throughout most of Europe, kermes was very expensive and thus used only by the most privileged economic, governmental, and religious classes. In the later fifteenth and sixteenth centuries, after Spanish explorers invaded Central and South America, they discovered a high-quality red dye, carmine, used by the indigenous populations. Immediately, the global trade in cochineal insects became enormous, with exports to Spain ranging from 50 to 160 metric tons annually between 1575 and 1600. Only gold and silver were considered more important cargos (Figure 7.7).

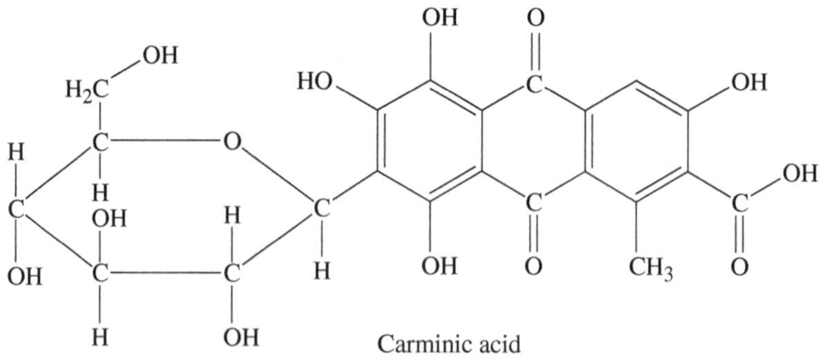

Figure 7.6 The structure of carminic acid.

Figure 7.7 The structure of kermesic acid.

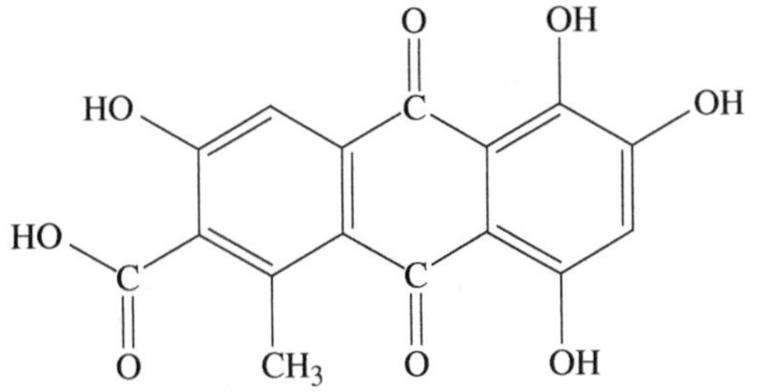

Kermesic acid

Production of carmine begins with harvesting cochineals through brushing them off the cactus pads and drying them in the sun or in ovens. The carminic acid is then extracted using water or a water/ethanol solution adjusted to be alkaline. The solution can be dried to form a solid called *cochineal* or sold as a liquid under the name "cochineal extract." It has been noted that even the purest cochineal contains microscopic insect body parts. One cochineal bug is estimated to contain up to 20% carminic acid, equating to roughly 70,000 insects being needed to make one pound of cochineal.

The dye, carmine, is formed via the addition of aluminum hydroxide to cochineal extract to create an aluminum salt of carminic acid. Carmine is used to color paints, and its rich red hue has attracted the eye of a wide variety of famous artists, including Van Gogh, Renoir, Vermeer, and Rembrandt, just to name just a few. Many brands of lipstick are tinted with cochineal extract. In the edible realm, it is used to color products such as yogurt, juice, meat products, jams, fruit syrups, gelatin deserts, bubble gum, ice cream, imitation crab meat, and sweet vermouth. Carmine dye is alternatively listed on food labels as crimson lake, cochineal, and natural red 4. The dye has been widely used for textiles as well, but it has been largely replaced by modern synthetic dyes that are cheaper and more readily available. However, cochineal dyes are regaining popularity because of fears over artificial food coloring.

Because they contain bugs, products containing carmine dye cannot be Kosher certified and are generally not considered vegan. As of 2014, the FDA has specified that food products intended for human use that contain cochineal extracts must have the common names "cochineal extract" or "carmine" listed on the label. A quick look in the local supermarket shows carmine listed in the ingredients of strawberry Yoplait® yogurt, most flavors of Activia® yogurt, and Good and Plenty® brand candy. In March of 2012, many Starbucks® customers were shocked to find that its Strawberries & Créme Frappuccino® mix had been switched to one that used cochineal extract to enhance the red color of the strawberry puree. The following month, Starbucks® announced that it would be transitioning the red dyes used in its products to lycopene, the molecule responsible for giving tomatoes their red color.

Why Is High Fructose Corn Syrup More Consumed than Sugar?

After the US FDA formally listed *high fructose corn syrup* (HFCS) as safe for use in 1983, it has become one of the most successful food ingredients in modern history. In fact, nowadays Americans consume more HFCS than sugar. So why is the use of HFCS so prevalent? In the 1980s, lobbying from United States sugar producers coerced the federal government to effect tariffs and sugar quotas resulting in the price of sugar in this country to be artificially higher than the global price. Literally, sugar is twice as expensive in the United States as it is in the rest of the world! Because of this, American companies started looking for a cheaper sugar substitute and came up with HFCS, a corn-based sweetener. During the 2000s, production of HFCS in the United States has averaged 9.2 million tons, utilizing about 4.7% of the country's total corn crop.

Some might wonder how a sweetener can be made from corn, when corn is not actually sweet. However, remember that corn is a high-starch food, and starch is a polymer of the sugar, glucose. The production of HFCS begins with enzymes being used to break starch into individual glucose molecules. This forms corn syrup, which is often used in food products. To form HFCS, a different enzyme is added to convert the glucose into a 50/50 mixture of glucose and fructose. The conversion of glucose to fructose is relatively simple, since their nonring structures are similar. This 50/50 mixture is made to mimic sucrose (common table sugar), a disaccharide that is made up of one glucose and one fructose molecule bonded together. If you are interested in seeing an example of these products, take a look at a bottle Karo® brand syrups, which usually contain corn syrup or a combination of corn syrup and HFCS (Figure 7.8; Schemes 7.2 and 7.3).

Sucrose (table sugar)

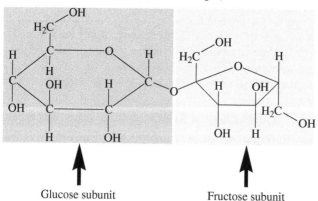

Figure 7.8 The structure of sucrose.

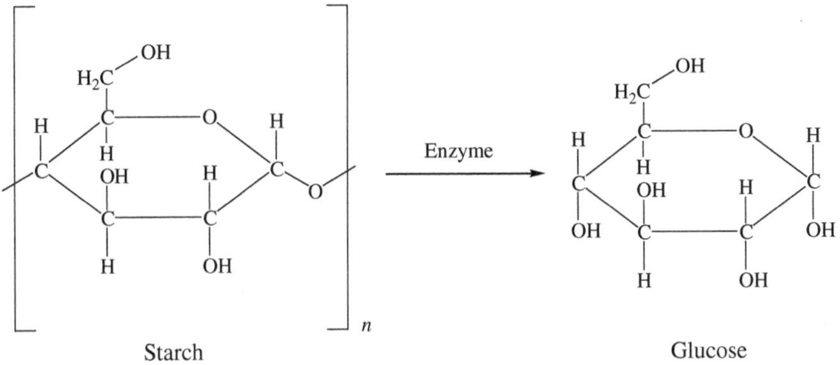

Scheme 7.2 The conversion of starch to glucose.

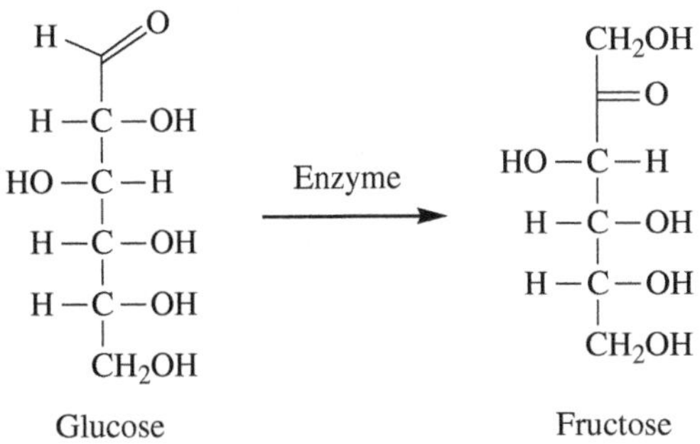

Scheme 7.3 The conversion of glucose to fructose.

HFCS is found just about anywhere that sugar is used, including in many soft drinks, sports drinks, and baked goods. HFCS in baked goods helps with browning and keeping the products soft, as in soft-bake cookies.

However, HFCS as a sugar substitute has been hotly debated due to the fact that it is a highly processed food. A 50/50 mixture of fructose and glucose is not the same as sucrose, so our bodies process them differently. High fructose diets have been implicated in the development of adult-onset diabetes and child-hood obesity, which you might notice have become issues in the United States. since the 1980s. Fructose is a natural sugar found in many fruits and some vegetables, and it has been part of the human diet for a long time. Therefore, it seems to make sense that HFCS itself is not the direct culprit. In fact, current thinking is that the obesity problem is caused by the simple fact that Americans are consuming more calories, which may be related to the cheapness of HFCS. One study showed that as a result of government subsidies, the consumer price of corn actually remained 25–30% below the cost of production between 1997 and 2005. This has allowed beverage manufacturers to increase serving sizes with only a marginal increase in consumer prices. In addition, the lower cost allowed companies to spend more on marketing campaigns. The majority of the HFCS produced in America goes into beverages. Between the 1970s and 1990s, there was a 123% increase in soft drink consumption in the United States. In 2000, the average American consumed more than 50 gallons of soft drinks per year, which equates to approximately one 16-ounce soft drink per day. Currently, HFCS-containing soft drinks make up 8% of the total energy intake in children and adults, which makes it America's real drinking problem.

Technically, because HFCS is made from corn, a natural grain product, and contains no artificial ingredients or color additives, it meets the FDA's requirements for use of the term "natural" in its labeling. However, the natural labeling is being used loosely. The fact that HFCS is a processed food has caused quite a bit of trouble for manufacturers. In May of 2006, a lawsuit was filed against Cadbury Schweppes for labeling 7-Up® as "All Natural," in spite of it containing HFCS. On 12 January 2007, Cadbury Schweppes agreed to remove that phrase from the 7-Up® label.

A completely different aspect of the HFCS discussion is its effect on bees. The fact that bees are attracted to HFCS was accidentally discovered when workers noticed that honey bees clustered and fed on HFCS spills during loading of the product into shipping tanks. Currently, HFCS has become a sugar alternative for feeding honey bees. It is used by commercial beekeepers to promote honey production and feed honey bees when sources of pollen and nectar are scarce. Oddly enough, the sweetener in common honey is a 50/50 mixture of fructose and glucose, just like HFCS. Because of this, the adulteration of honey with HFCS to increase its sweetness is beginning to become a problem.

On a funny side note, in the early 1980s, Coca Cola actually stopped making Coke® and instead came out with a variation called "New" Coke®. After extraordinary public outrage, including death threats, Coca Cola resumed production of Coke®. However, the product was called Coke Classic® and it tasted different than the original Coke®.[3] The difference was due to Coke Classic® being made with HFCS and not sugar. Since concerns about HFCS have come to the mainstream of public attention, many sodas offer products made with sugar and not HFCS – they are usually called "throwback" products. If you buy Coke® imported from Mexico, it is made with sugar because Mexico is not affected by US tariffs, and sugar is therefore cheaper in that country (Figure 7.9).

---

3 I distinctly remember my first taste of Coke Classic® because I instantly knew that something had changed. As a connoisseur of soda, I can say that sugar provides a superior product. I personally describe HFCS as having a grainy mouth feel, while sugar feels more glassy. If you have not already, I highly recommend trying your favorite soda with sugar as the sweetener.

You mean that's not where high fructose corn syrup comes from?

Figure 7.9 No it does not.

What Causes Rancid Butter to Stink?

Commercial butter is about 80% butterfat, 15% water, and 5% milk protein and other solids. Butterfat is primarily made up of a mixture of fat molecules called *triglycerides*. Triglycerides can be broken down into the molecule glycerine and three of several different types of fatty acid molecules.

Rancidification is the decomposition of fats and can occur by one of three main mechanisms: *hydrolysis* (addition of water), *oxidation* (reaction with the oxygen in air), or microbial degradation. For butter, hydrolysis is the initial cause of rancidification and the inevitable sour smell. Hydrolysis (from the roots *hydro*, meaning "water," and *lysis*, meaning "splitting") causes the fatty acid chains to split away from the glycerine backbone in triglycerides (Scheme 7.4).

Butterfat primarily contains the fatty acids oleic acid (~32%) and myristic acid (~20%). Because these fatty acids have high boiling points, 680 and 619.2 °F, respectively, they are relatively nonvolatile. Nonvolatile molecules generally have very little scent because they do not evaporate significantly, and substances in general must evaporate into a gaseous state in order to be detected by our noses. The smell of rancid butter comes from the fact that

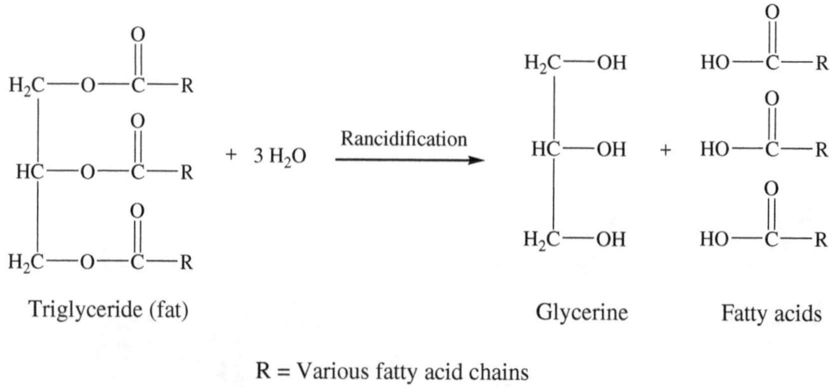

Scheme 7.4 The conversion of a triglyceride into glycerine and fatty acids.

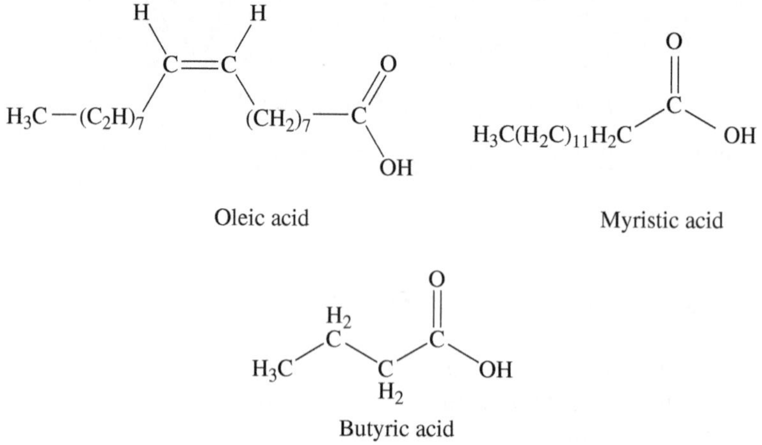

Figure 7.10 The structure of some fatty acids found in butter.

butterfat also contains small amounts of other fatty acids. The primary culprit in contributing to the bad smell is *butyric acid*, which comprises 3–4% of the fatty acids in butterfat and has a boiling point of only 326.75 °F. When butter goes rancid, butyric acid is liberated from the triglyceride by rancidification. Butyric acid is slightly volatile, quickly becoming a vapor that we smell and recognize as rancid butter. In short, butyric acid smells really bad. Published descriptions of butyric acid describe it as having a bad, unpleasant, or putrid odor characterized by odor qualities of buttery, sweaty, cheesy, and rancid (Figure 7.10).

Butyric acid actually plays a role in a number of foul odors. For example, it is the main odorant in human vomit due to the breakdown of triglycerides in our stomach. In addition, it is one of the many odorants in human fecal matter. Oddly enough, the decomposition of butter fats to form butyric acid is harnessed to impart favoring to fermented soybean, blue cheese, sour cream, and parmesan cheese. Even though butyric acid smells bad, it has an acrid taste with a somewhat sweet aftertaste.

Fatty acids have a group of carbon, hydrogen, and oxygen called a carboxylic acid. *Carboxylic acids* are a common element in molecules that have an odor described as sharp, acidic, or musty. Acetic acid is the main molecule in vinegar, and caprotic acid is the molecule most responsible for the smell of goats. In fact, human body odor is brought about, in part, when secretions from eccrine sweat glands and sebaceous glands are degraded into volatile carboxylic acids by microorganisms inhabiting the skin. Armpit odor has been shown to contain a wide variety of carboxylic acids, including acetic acid and butyric acid, although the molecule 3-methyl-2-hexenoic acid is believed to be most responsible for the odor (Figure 7.11).

The molecule isovaleric acid has been determined to be the major component in foot odor. Isovaleric acid is produced when the bacteria *Staphylococcus epidermidis* degrades the amino acid leucine, which is present in sweat. As you notice, many of these molecules are produced by the microbial degradation of molecules found in sweat. Deodorants work by killing these microbes through the action of an antimicrobial agent, such as triclosan, thus halting the degradation (Figure 7.12; Scheme 7.5).

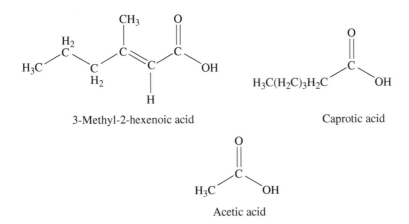

3-Methyl-2-hexenoic acid

Caprotic acid

Acetic acid

Figure 7.11 A comparison of carboxylic acids that smell bad.

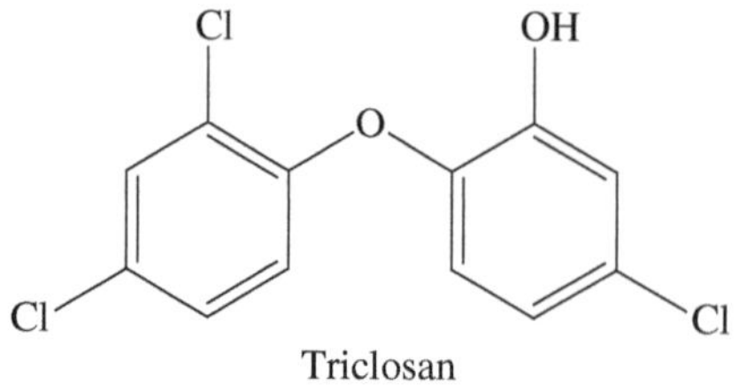

Figure 7.12 The structure of triclosan.

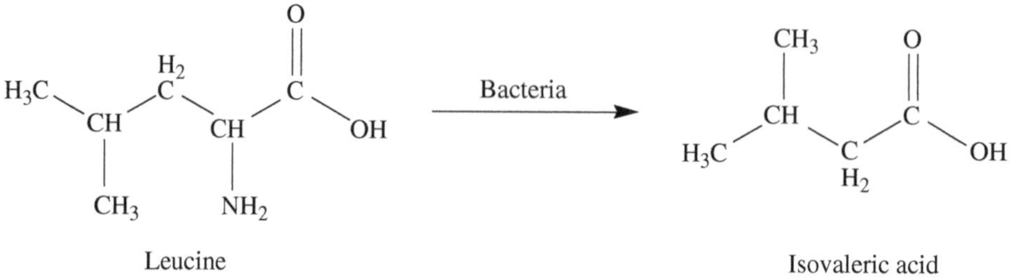

Triclosan

Leucine

Isovaleric acid

Scheme 7.5 The conversion of leucine into isovaleric acid.

Why Does Mint Make Your Mouth Feel "Cold?"

Our body is covered with nerve cells that function solely in providing the sensation of heat or cold. Usually, these nerve cells are activated by the presence of hot or cold objects. However, certain molecules can activate these cells to produce the same sensations. A prime example of this is how capsaicin from hot peppers stimulates heat-sensing nerve cells called *TRP-V1* to provide a hot sensation, without causing an actual change in temperature. Eugenol in clove oil also activates the TRP-V1 nerve cells, producing a warm sensation when a clove is in the mouth. Present in small amounts in many herbs and spices such as bay leaf, sage, basil, nutmeg, rosemary, and even marijuana, eugenol lends them a unique aroma and taste (Figure 7.13).

In a similar manner, the menthol molecule, present in peppermint; carvone, which is present in spearmint; and eucalyptol, found in eucalyptus oil, can provide a sensation of feeling cold by activating a nerve cell receptor called *TRPM8*. The TRPM8 nerve cell receptors become active at temperatures below 25 °C and are utilized by the body to detect innocuous cooling and assist with regulating body temperature. Because the TRPM8 nerve cells receptors are present in the skin, menthol will provoke a cooling sensation, whether eaten or applied to the skin. This refreshing cooling effect is used in many different types of mouthwashes, toothpastes, chewing gums, and candies. TRPM8 receptors are also present in the nose, so vapor inhalation of camphor, eucalyptol, and menthol provide a cold sensation in the nasal cavities (Figure 7.14).

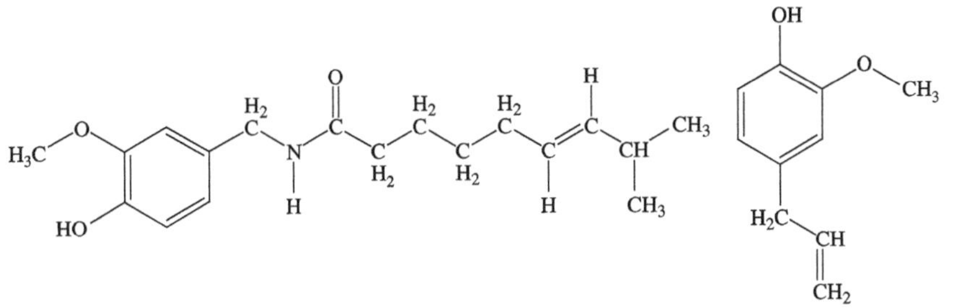

Capsaicin Eugenol

Figure 7.13 The structure of capsaicin and eugenol.

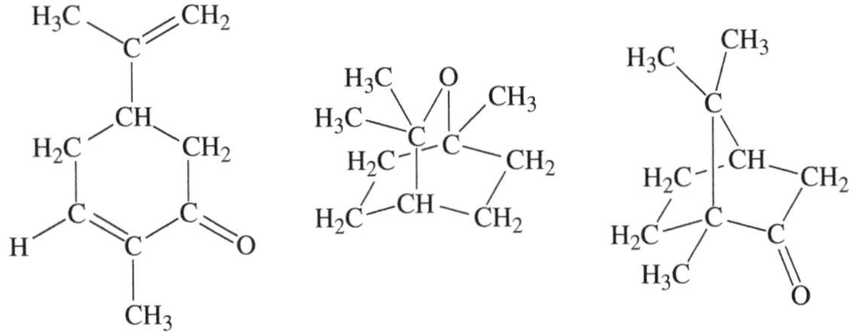

Carvone Eucalyptol Camphor

Figure 7.14 A comparison of molecules that activate TRPM8 nerve receptors.

In 1974, a super-cooling agent called *icilin* was first patented. Discovered by scientists, icilin is nearly 200 times as potent as menthol in activating the TRPM8 nerve cell receptors. A few human test subjects consumed 5–10 mg of icilin and reported a coldness in the mouth and chest and mild coolness on the inner surfaces of the arms and legs. The coolness lasted 30–60 minutes and was described as pleasant. Icilin is currently not available commercially and is mainly used for research of nerve cells. But think how interesting a candy made with icilin might be! (Figure 7.15).

Have you ever noticed how icing a pulled muscle helps relieve the pain? This is called *cooling-mediated analgesia,* and not surprisingly, the TRPM8 nerve cells play an important role. This analgesic effect is utilized in many over-the-counter preparations containing molecules that activate the TRPM8 nerve cell receptors. Some examples are topical analgesics such as Tiger Balm®, Bengay®, and Icy Hot®; decongestants such as Vicks® VapoRub™, and Mentholatum®; and cough drops such as Halls® Mentho-lyptus™. Because clove oil contains eugenol, it is often used to relieve toothache pain.

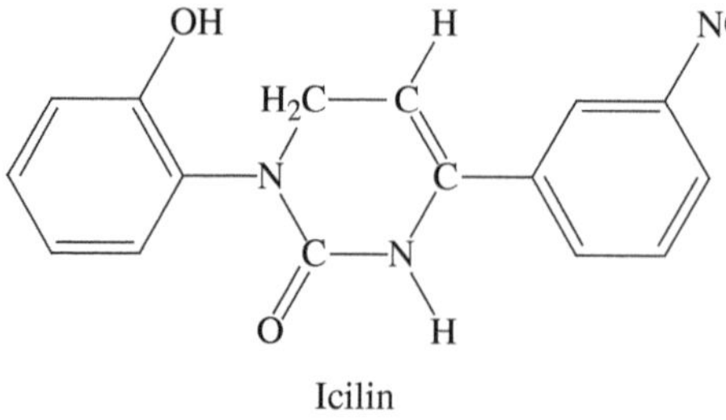

Figure 7.15 The structure of icilin.

Icilin

Even menthol cigarettes provide cooling and painkilling effect, which makes smoking a unique experience.[4]

It Is Probably Not Really Fresh Squeezed

Ever wonder how orange juice you buy in the store manages to taste fresh year round, as though oranges are eternally in season? Or have you ever wondered how a particular orange juice brand can consistently taste the same? The explanation for both these questions is the same: what you are drinking is not pure orange juice.

In 2012, the United States consumed 708,000 tons of orange juice, part of the world consumption of 2,127,000 tons. Research has shown that well over 40 compounds are responsible for the aroma and taste of fresh oranges. Unfortunately, the processing of orange juice often strips off many of these compounds, leaving what is essentially orange-colored sugar water. Compounds captured during processing, along with additional compounds obtained from orange peels, are later readded in what are called "flavor packs." Other juices, especially apple and berry, also use flavor packs.

Extensive research has been done regarding the compounds and ratios of these compounds used in the favor packs of different brands of orange juice, which give each brand a unique taste. Because flavor packs are not made from chemicals but rather from the captured natural compounds from the oranges, the FDA does not require that manufacturers list flavor packs in the ingredients label. Ethyl butyrate is one of the chemicals found in high concentrations in the flavor packs added to orange juice sold in North American markets. Ethyl butyrate is added because flavor engineers have discovered that it imparts a fragrance liked by Americans, and one they strongly associate with a freshly squeezed orange. The molecule limonene is another molecule

---

4 Out of curiosity, interest, or perhaps pure craziness, I decided to test if the cooling and painkilling effects of menthol would counteract the painful sensations from eating capsaicin. After eating part of a Habanero pepper, I sucked on a menthol-containing cough drop. Let's just say the experiment failed horribly.

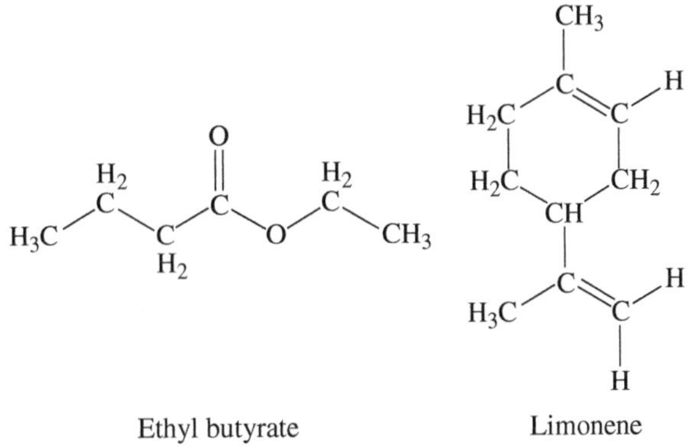

Figure 7.16 The structures of ethyl butyrate and limonene.

Ethyl butyrate

Limonene

strongly associated with the smell of oranges. It is odd to think that the orange juice we buy in stores probably does not actually smell like truly fresh squeezed juice (Figure 7.16).

There are two main ways that orange juice processing causes the loss of flavor compounds. The first is a deoxygenation process called *deaeration*. During the extraction of the juice from oranges, air is mixed in. Oxygen in air, if not removed, will cause reactions that can result in browning, changes in aroma, and loss of nutritional value. In order to prevent this and to give orange juice a longer shelf life, the air is removed using a vacuum. Specifically, the pressure in the space above the juice is reduced, causing the air to become less soluble in the juice and bubble out. Unfortunately, many of the flavor compounds are volatile so this vacuum process also removes them.

The other orange juice process that removes flavor compounds is concentration. Many orange juice processors condense their product to make it easier and cheaper to transport, store, and also freeze. Orange juice is heat intolerant so the water is removed under vacuum at temperatures of less than 80 °C, but once again, the volatile flavor compounds are removed along with the water, requiring them to be readded later.

In 2007, the US total citrus fruit production was 9.4 million metric tons, with a value of $3.1 billion. Roughly 40% of this crop is converted to concentrated orange juice. The concept of fresh-squeezed orange juice came about during World War II, in response to a desperate need for vitamin C by enlistees. The concentration process was patented in 1948, and since then, the annual US per capita consumption of fresh oranges dropped from 39.4 pounds to just 11.2 pounds in 2004. Correspondingly, the annual per capita consumption of orange juice increased from 0.48 gallons in 1948 to 4.77 gallons in 2004. This process has transformed the nation to one that drinks processed orange juice drinks instead of eating fresh oranges, with an estimated two-thirds of all Americans drinking orange juice for breakfast.

To be clear, frozen concentrated orange juice and orange juice concentrate are squeezed and concentrated in the country of origin. The concentrate is either frozen or sealed in an aseptic storage container. After being shipped to the country of use, it is reconstituted with water, flavor packs are added, and the product is packaged for sale. Orange juice that is labeled "Not from concentrate" is squeezed juice that is either lightly pasteurized and frozen or aseptically packed for shipment to the country where it will be consumed. However, this product still goes through aeration and requires the addition of flavor packs. To create orange juice that is labeled as "freshly squeezed," the whole fruit is shipped directly to the country of use and the juice is squeezed immediately before consumption.

Many orange juices are highly processed, which has led to a debate about what can be called "natural." According to the FDA, products that use flavor packs can still be called natural because the contents are obtained from oranges. Many consumer groups disagree, however. In 2012, a lawsuit was filed against the orange juice giant Tropicana. Lawyers claimed the Tropicana® juice should not be labeled "natural" because it is heavily processed and uses chemically engineered "flavor packs."

With the current obesity epidemic in the United States, there has been intense research on sources of sugar in the American diet. Shockingly, orange and other fruit juices have been implicated as a significant source of sugar and calories. In fact, many orange juices have almost as much sugar and more calories than the equivalent amount of Coca Cola® soda. In Britain, this is considered such a problem that the British government is calling for a tax on fruit juice and a label warning stating that the product has as much sugar as Coke®.

Why Are Viruses Added to Some Sandwich Meat?

On 18 August 2006, for the first time, the FDA approved a bacteriophage preparation that can be used in products for human consumption. Bacteriophages are viruses that infect and destroy only specific kinds of bacteria. The bacteriophages approved by the FDA act as antimicrobial agents against *Listeria monocytogenes*, a food-borne pathogen particularly dangerous for the elderly, newborns, and people with impaired immune systems.

Bacteriophages (bacteria-eater: from the Greek word phago meaning "to eat") are estimated to be the most abundant organisms in the Earth's biosphere and are found in high numbers in food, water, and inside or on our bodies. Studies have estimated that there are 1.5×10^7 bacteriophages per gram of ground soil, 2.0×10^7 bacteriophages per milliliter of fresh water, 1×10^8 bacteriophages per milliliter of human saliva, and an incredible 1×10^{31} bacteriophages total on earth. Fortunately, bacteriophages are harmless to humans because we are routinely exposed to them. Bacteriophages kill bacteria cells by landing on their

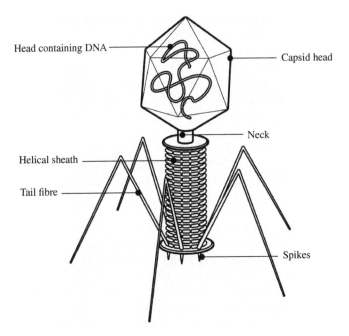

Head containing DNA

Capsid head

Neck

Helical sheath

Tail fibre

Spikes

Figure 7.17 The structure of a typical bacteriophage.

surface and injecting them with the DNA contained in their head. This converts the bacteria cell into a bacteriophage "factory". Eventually, the bacteria cell bursts and releases a new generation of bacteriophages. Bacteriophages are, in fact, considered quite important because, owing to their abundance, they destroy half of the world's bacterial mass every 48 hours (Figure 7.17).

The bacteriophage-containing food additive called ListShield® was introduced by a company called Intralytic®. The additive was approved for fruits, vegetables, and ready-to-eat poultry meat products, such as hotdogs and sandwich meat. It has been shown that the use of ListShield® in these products significantly reduces their contamination by *Listeria monocytogenes*, usually by 99–99.9%. The FDA also approved LISTEX®, developed by Micreos Food Safety®, initially for use on cheese. Later, the FDA assigned GRAS (generally recognized as safe) status to LISTEX® and approved it for use on all food products. These bacteriophages do not alter food flavor, aroma, or nutritional value. In addition, they are natural, considered "organic," and are certified both kosher and halal.

Listeria monocytogenes is a bacterium found in soil, water, vegetation, and livestock feces. Fresh or minimally processed fruits, vegetables, and meats can easily be contaminated with this bacterium, leading to human ingestion. Although *Listeria monocytogenes* can be killed by cooking and pasteurization, some ready-to-eat meats, such as hot dogs and deli meats, become contaminated after factory cooking but before packaging or even at the deli counter. Eating listeria-contaminated food causes *listeriosis*. Most other common foodborne pathogens, such as *Salmonella*, are rarely associated with fatalities; however, listeriosis has a mortality rate of 20–30%. Although listeriosis represents a small fraction of the hospitalizations caused by food borne pathogens (~3.8%), its importance is shown by the fact that it causes over 27% of the foodborne disease deaths.

Symptoms of listeriosis include fever, muscle aches, headache, confusion, convulsions, and even death. Listeriosis is particularly dangerous for pregnant women because it can cause miscarriages, premature delivery, or life-threatening infections of the newborn. Fortunately, listeriosis can be treated with antibiotics.

The CDC (Center for Disease Control) estimates that listeriosis causes roughly 1600 illnesses and 260 deaths each year in the United States. Outbreaks occur when a particular farm or food processing plant becomes contaminated with *Listeria monocytogenes* and their products are sold to the public. In 2011, the largest listeriosis outbreak in US history occurred when tainted cantaloupe from a single farm caused 147 illnesses, 33 deaths, and 1 miscarriage. More recently, in 2015, listeria-contaminated ice cream products sold by Blue Bell Creameries caused three deaths and the hospitalization of seven people. The contamination was traced to their Broken Arrow, Oklahoma, plant which was temporarily shut down (Figure 7.18).

Figure 7.18 Bacteriophages make a great sandwich condiment.

What Is Margarine Made From?

Margarine was first invented in 1869 in response to a request by Napoleon III for a wholesome butter alternative. Margarine's name comes from the Greek word for pearl, *margarite*, because some of the product appeared as lustrous, white, pearly drops. Originally, margarine was made from oil discharged from solidified fat. After World War II, hydrogenated vegetable oils were adopted as a cheaper source of margarine. By the late 1970s, roughly 60% of all edible oils and fats in the United States were partially hydrogenated.

Currently, most margarine is made out of partially hydrogenated soybean oil. The United States Department of Agriculture (USDA) estimates that over 8.5 million tons of soybeans were used in the United States during 2013, with approximately 85% of the crop being processed into soybean meal and vegetable oil. An estimated 75% of the soy oil used in the United States was hydrogenated to make shortening and margarine. Overall, soybean oil constitutes about half of worldwide edible vegetable oil production.

Hydrogenated soybean oils are popular in the commercial food industry because they reduce costs, enhance the texture of baked goods, extend shelf life of products, and replace animal fats, allowing people following vegetarian, vegan, kosher, or halal (adherent to Islamic law) diets to consume the goods.

Oils are made of two parts: glycerin and fatty acids. Fatty acids that contain one or more double bonds between carbon atoms are called *unsaturated* because each carbon could possibly bond to additional hydrogen atoms. Similarly, fatty acids that do not contain double bonds are called *saturated*. Due to the shape of the double bonds, unsaturated fatty acids are not straight but appear bent. Unsaturation produces kinks in the chains of the molecules that keep them from being tightly packed together. This affects the molecules' ability to interact and makes them more fluid. Vegetable oils, such as olive and soybean, contain fats with unsaturated fatty acids and are liquids. Fats containing greater amounts of saturated fatty acids are more linear; the molecular chains can pack more closely, and these substances tend to be solid. Animal fats, such as butter and lard, contain fats with saturated fatty acids, which explains why they are solids (Figure 7.19).

After extraction, soybean oil is a liquid and nowhere near the consistency of margarine. In order to make soybean oil into a solid, some of the double bonds are removed in a process called *hydrogenation*. Hydrogenation literally means the addition of hydrogen molecules. In this case, hydrogenation is a chemical process wherein the liquid is heated to high temperatures after adding a catalyst, typically nickel, and hydrogen gas is forcefully passed through. Filtration at the end removes the catalyst, and what is left is hydrogenated soybean oil. In spite of purification, minute amounts of the catalyst remain, which makes some brands of margarine an unusual source of nickel. Note, however, that only some of the double bonds are removed during the processing of soybean oil, which is

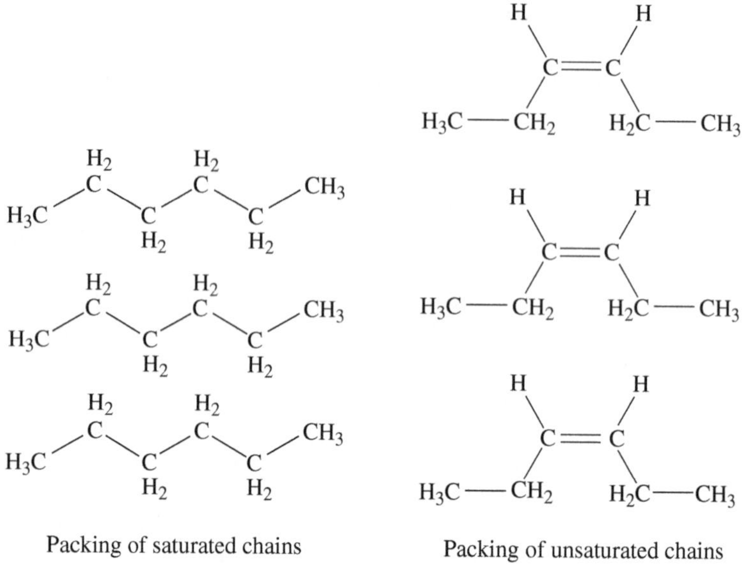

Packing of saturated chains   Packing of unsaturated chains

Figure 7.19 A comparison of the packing of saturated and unsaturated chains.

why the process is called *partial hydrogenation*. If all of the double bonds were removed, the product would be more like wax and lack the desired butter-like consistency. In looking at the structure of a typical wax, one can see that it is similar in structure to saturated fats (Figure 7.20; Scheme 7.6).

Double bonds in fatty acids contain two possible configurations, one with hydrogen atoms on the same side of the double bond, called *cis*, and the other with hydrogen atoms on different sides of the bond, called *trans* (Latin root meaning "across"). Naturally occurring unsaturated fatty acids have a cis-configuration; however, the partial hydrogenation process may lead to up to 45% of the fatty acids being transformed to the trans-configuration in a process called *isomerization*. Studies have shown a link between the consumption of trans-fatty acids (TFAs) and an increased risk of developing cardiovascular disease. In addition, there have been some claims that TFA may contribute to cardiovascular disease, cancer, neurological disorders, blindness, diabetes, obesity, liver disease, and infertility. Of all trans-fats consumed by humans, the major contributors of TFAs in the diet are partially hydrogenated oils (Figure 7.21).

In 2003, the US FDA issued a regulation requiring manufacturers to list TFA on the "Nutrition Facts" panel of foods and some dietary supplements. The FDA estimated that TFA labeling would prevent from 600 to 1200 cases of coronary heart disease and 250–500 deaths each year. In 2006, New York prohibited restaurants from serving foods with a TFA content exceeding 0.5 g per serving. In 2008, California followed suit.

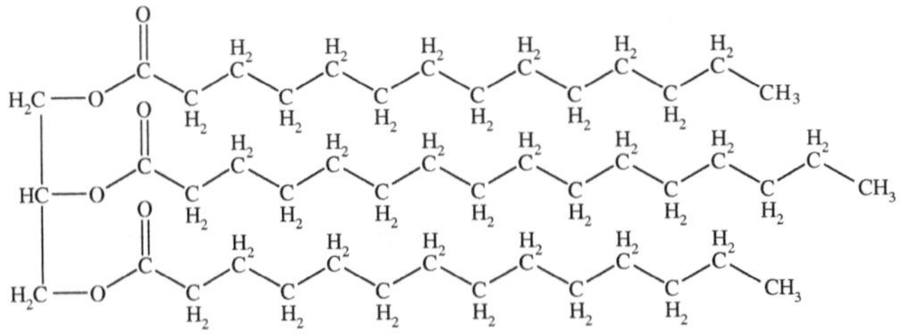

A saturated fat

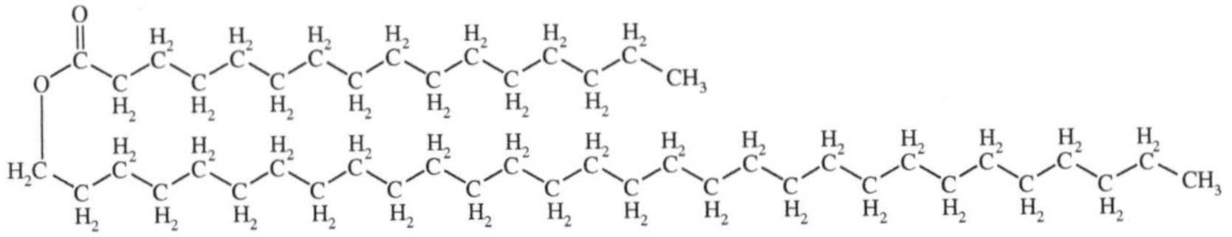

Triacontanyl palmitate (a major component of beeswax)

Figure 7.20 A comparison of the structure of a saturated fat and a wax.

Hydrogenation

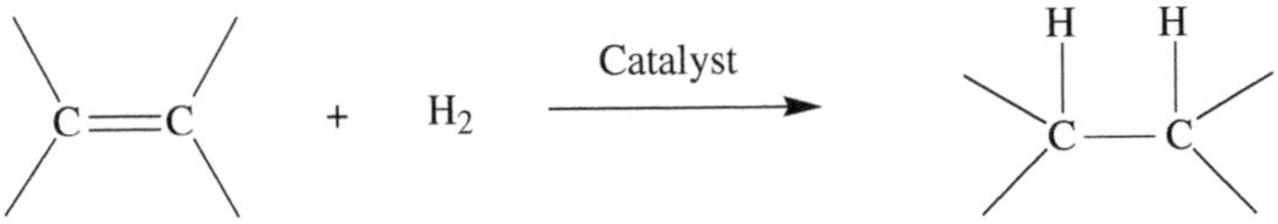

Unsaturated   Saturated

Scheme 7.6 The removal of a double bond using hydrogenation.

Figure 7.21 The cis- and trans-configurations of a double bond.

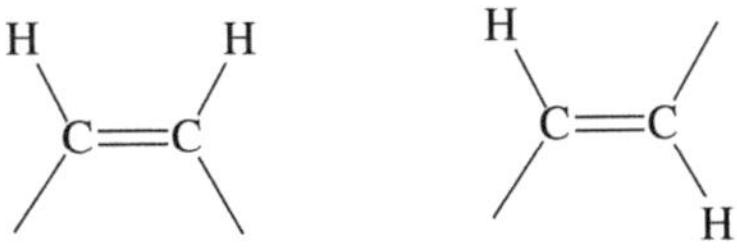

cis-Configuration   trans-Configuration

Why Are Junior Mints® Shiny?

Confectioner's glaze is made from *shellac*, a resin secreted by the female Lac (*Kerria lacca*), an insect indigenous to Southeast Asia and Mexico. Shellac is a hard, brittle, resinous solid, commonly used as a varnish by the application of multiple layers. Even though a few allergic reactions have been reported for shellac-containing products, the material is generally regarded as nontoxic and is recognized as safe by the FDA. This allows the use of shellac in food products, where it already plays a major role as a coating material for citrus fruits and confectionaries.

Shellac also goes by the names glaze, pure food glaze, natural glaze, beetle juice, or Lac-resin. In addition, it can be mixed with an alcohol solution to create confectioner's glaze. This is what gives the shiny coating to candies such as jellybeans, Jelly Bellies®, Mike and Ikes®, Hot Tamales®, Sugar Babies®, See's® Candies, chocolate raisins, Nestle Raisinets®, Junior Mints®, Reese's Pieces®, and Milk Duds®. In case you are interested, M & M's® candies do not contain confectioner's glaze.

Moreover, shellac is sprayed onto fruits and vegetables including avocados, lemons, grapes, bananas, cucumbers, tomatoes, melons, oranges, limes, passion fruits, peaches, and apples, to replace the natural wax coating that is often removed during washing. Shellac on produce extends its freshness and shelf life, gives it a shiny finish, protects it against insects and fungal contamination, slows ripening, and prevents the loss of water. Other waxes, such as beeswax, can be used for these same purposes. As it is almost impossible to tell which kind of wax is used, if you want to avoid eating shellac, choose unwaxed fruits and vegetables instead.

In addition to being used as a confectioner's glaze, shellac can be used as a pharmaceutical glaze for tablets and capsules. The glaze improves pharmaceuticals' appearance, extends shelf life, protects from moisture, and aids with swallowing. In addition, shellac is resistant to stomach acids, making it useful for time-released, delayed action, or colon-targeting pills.

As mentioned above, shellac is produced naturally by the Lac beetle. During a certain season of the year, these insects swarm in huge numbers onto trees and feed on the sap. They alter the sap chemically and then secrete it to form a crust that covers them during their larval stage. Think about this the next time you have a handful of jellybeans. Harvesters collect the encrusted branches and scrape off the secretion, which is then ground, sifted, rinsed, and bleached. After this, it is melted, filtered, and combined with solvent (usually denatured alcohol) to produce orange-colored shellac. Confectioner's glaze is produced with a different solvent and additional bleaching. It has been estimated that it takes 300,000 Lac bugs to produce 1 kg of secretions, with approximately 20,000 tons of shellac being produced annually. That's a lot of bugs and secretions!

Shellac is a natural polymer made up primarily of the polyhydroxy acids consisting of aleuritic acid, butolic acid, shellolic acid, and jalaric acid. The carboxylic acid and alcohol groups on these molecules can combine via an ester linkage to form a specific type of polymer called a *polyester* (Figures 7.22–7.24).

Figure 7.22 The structures of aleuritic acid and shellolic acid.

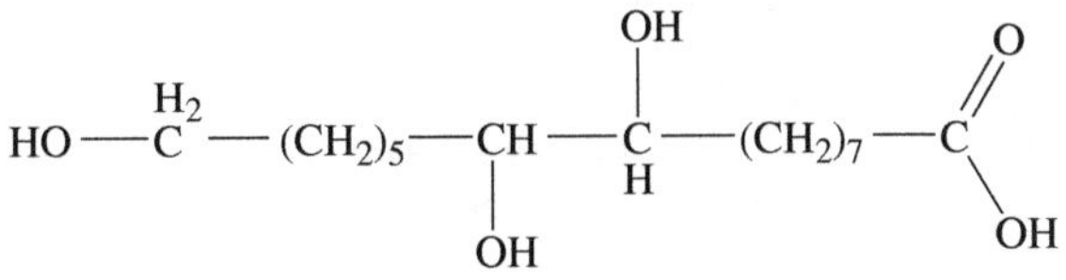

Aleuritic acid

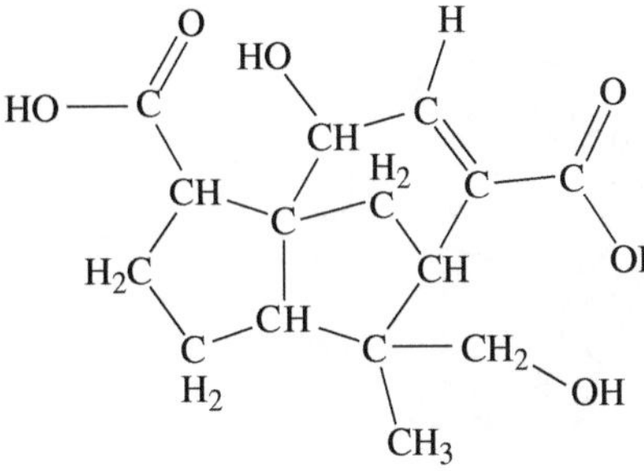

Shellolic acid

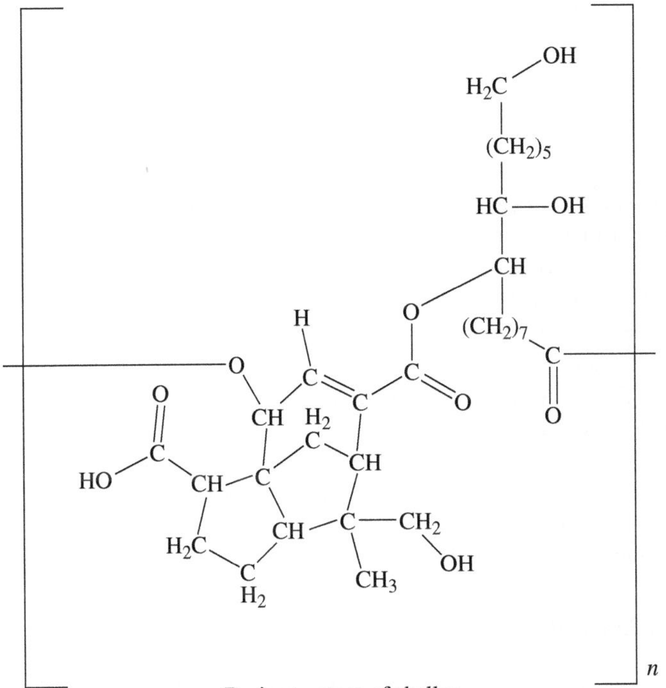

Basic structure of shellac

Figure 7.23 The structure of the polymer that forms shellac.

Figure 7.24 Why they call it confectioner's glaze.

Further Reading

Why is Gum Chewy?

Emsley, J 2004, *Vanity, Vitality, and Virility: The Science Behind the Products You Love To Buy*, Oxford University Press, Oxford, NY.

Epping, J, van Deenen, N, Niephaus, E, Stolze, A, Fricke, J, Huber, C, Eisenreich, W, Twyman, RM, Prufer, D & Schulze-Gronover, C 2015, 'A rubber transferase activator is necessary for natural rubber biosynthesis in dandelion', *Nature Plants*, vol. 1, pp. 15048.

Mathews, JP 2009, *The Chewing Gum of the America's Chicle*, The University of Arizona Press, Tucson, AZ.

Stanley, H 2008, *Materials science to the rescue: easily removable chewing gum*, www.scienceinschool.org, Viewed 20 September 2016, http://www.scienceinschool.org/print/594

Wardlaw, L 1997, *Bubblemania*, Aladdin Paperbacks, New York.

The Problem with Gummi Bears

Duconseille, A, Astruc, T, Quintana, N, Meersman, S & Sante-Lhoutellier, V 2015, 'Gelatin structure and composition linked to hard capsule dissolution: A review', *Food Hydrocolloids*, vol. 43, pp. 360–376.

PB Gelatins 2016, *PB Gelatins*, Viewed 20 September 2016, http://www.pbgelatins
.com/

Shoulders, MD & Raines, RT 2009, 'Collagen structure and stability', *Annual Review of Biochemistry*, vol. 78, pp. 929–958.

Singh, K & Mishra, A 2014, 'Gelatin nanoparticle: preparation, characterization and application in drug delivery', *IJPSR*, vol. 5, pp. 2149–2157.

What is the Easiest Way to Peel a Tomato?

Das, DJ & Barringer, SA 2006, 'Potassium hydroxide replacement for lye (sodium hydroxide) in tomato peeling', *Journal of Food Processing and Preservation*, vol. 30, pp. 15–19.

Garcia, E & Barrett, DM 2006, *Peelability and yield of processing tomatoes by steam or lye*, Viewed 20 September 2016, http://ucanr.edu/datastoreFiles/234-584.pdf

Vimpex International 2015, *Canned mandarin oranges*, Vimpex International, Viewed 20 September 2016, http://cannedmandarinoranges.com/

Another Way to Eat Insect Parts!

Dapson, RW 2007, 'The history, chemistry and modes of action of carmine and related dye', *Biotechnic & Histochemistry*, vol. 82, pp. 173–187.

Jaslow, R 2012, *Starbucks Strawberry Frappuccinos dyed with crushed up cochineal bugs*, report says, CBS News, Viewed 20 September 2016, http://www.cbsnews
.com/news/starbucks-strawberry-frappuccinos-dyed-with-crushed-up-cochineal-bugs-report-says/

Korte, A 2016, *Scientific analysis reveals Vincent van Gogh's true colors*, American Association for the Advancement of Science, Viewed 20 September 2016, http://www.aaas.org/news/scientific-analysis-reveals-vincent-van-gogh-s-true-colors

O'Hanlon, G 2013, *Palette of Rembrandt van Rijn*, Natural Pigments, Viewed 20 September 2016, http://www.naturalpigments.com/art-supply-education/rembrandt-van-rijn-color-palette/

US Food and Drug Administration 2015, *CFR - Code of Federal Regulations Title 21*, *US Food and Drug Administration*, Viewed 20 September 2016, http://www
.accessdata.fda.gov/scripts/cdrh/cfdocs/cfcfr/cfrsearch.cfm?fr=73.100

US Food and Drug Administration 2016, *Guidance for industry: Cochineal extract and carmine: Declaration by name on the label of all foods and cosmetic products that contain these color additives, small entity compliance guide*, US Food and Drug Administration, Viewed 20 September 2016, http://www.fda.gov/ForIndustry/ColorAdditives/GuidanceCompliance RegulatoryInformation/ucm153038.htm

Why is High Fructose Corn Syrup More Consumed Than Sugar?

Barker, RJ & Lehner, Y 1978, 'Laboratory comparison of high fructose corn syrup, grape syrup, honey, and sucrose syrup as maintenance food for caged honeybees', *Apidologie*, vol. 9, pp. 111–116.

Morgan, RE 2013, 'Does consumption of high-fructose corn syrup beverages cause obesity in children?', *Pediatric Obesity*, vol. 8, pp. 249–354.

The Corn Refiners Association 2016, *About high fructose corn syrup*, The Corn Refiners Association, Viewed 20 September 2016, http://corn.org/products/sweeteners/high-fructose-corn-syrup/

US Department of Agriculture 2016, *U.S. sugar production*, Viewed 20 September 2016, http://www.ers.usda.gov/topics/crops/sugar-sweeteners/background.aspx

Yu, Z, Lowndes, J & Rippe, J 2013, 'High-fructose corn syrup and sucrose have equivalent effects on energy-regulating hormones at normal human consumption levels', *Nutrition Research*, vol. 8, pp. 1043–1052.

What Causes Rancid Butter to Stink?

Ara, K, Hama, M, Akiba, S, Koike, K, Oskisaka, K, Hagura, T, Kamiya, T & Tomi, F 2006, 'Foot odor due to microbial metabolism and its control', *Canadian Journal of Microbiology*, vol. 52, pp. 357–364.

Cometto-Muniz, JE & Abraham, MH 2010, 'Structure–activity relationships on the odor detectability of homologous carboxylic acids by human', *Experimental Brain Research*, vol. 207, pp. 75–84.

Leffingwell & Associates 2016, *Odor & flavor detection thresholds in water (in parts per billion)*, Leffingwell & Associates, Viewed 20 September 2016, http://www.leffingwell.com/odorthre.htm

Mallia, S, Escher, F & Schlichtherle-Cerny, H 2008, 'Aroma-active compounds of butter: A review', *European Food Research and Technology*, vol. 226, pp. 315–325.

Senese, F 2015, *What is the chemical structure of butter?*, General Chemistry Online, Viewed 20 September 2016, http://antoine.frostburg.edu/chem/senese/101/consumer/faq/butter-composition.shtml

Yoon, K, Chang, Y & Lee, G 2013, 'Characteristic aroma compounds of cooked and fermented soybean (Chungkook-Jang) inoculated with various Bacilli', *Journal of the Science of Food and Agriculture*, vol. 93, pp. 85–92.

Why Does Mint Make Your Mouth Feel "Cold?"

Andersson, DA, Chase, HWN & Bevan, S 2004, 'TRPM8 activation by menthol, icilin, and colds differentially modulated by intracellular pH', *The Journal of Neuroscience*, vol. 24, pp. 5364–5369.

Eccles, R 1994, 'Menthol and related cooling compounds', *Journal of Pharmacy and Pharmacology*, vol. 46, pp. 618–630.

Ecogreen Oleochemicals 2016, *Sugar alcohols*, Viewed 20 September 2016, http://www.dhw-ecogreenoleo.de/Sugar-Alcohols_130924.pdf

Janssens, A & Voets, T 2011, 'Ligand stoichiometry of the cold and menthol activated channel TRPM8', *The Journal of Physiology*, vol. 589, pp. 4827–4835.

Knowlton, WM, Daniels, RL, Palkar, R, McCoy, DD & McKem, DD 2011, 'Pharmacological blockade of TRPM8 ion channels alters cold and cold pain responses in mice', *PLoS ONE*, vol. 6, pp. e25894.

Kuhn, JPF, Witschas, K, Kuhn, C & Luckhoff, A 2010, 'Contribution of the S5-Pore-S6 domain to the gating characteristics of the cation channels TRPM2 and TRPM', *Journal of Biological Chemistry*, vol. 285, pp. 26806–26814.

Lillian, T 2011, *Dude, you got some gum?*, The Science Creative Quarterly, Viewed 20 September 2016, http://www.scq.ubc.ca/dude-you-got-some-gum

Matt, S 2010, *Why does mint make your mouth feel cold?*, Mental_floss, Viewed 20 September 2016, http://www.mentalfloss.com/blogs/archives/52885

McKemy, DD, Neuhausser, WM & Julius, D 2002, 'Identification of a cold receptor reveals a general role for TRP channels in Thermosensation', *Nature*, vol. 416, pp. 52–61.

Vogt-Eisele, AK, Weber, K, Sherkheli, MA, Vielhaber, G, Panten, J, Gisselmann, G & Hatt, H 2007, 'Monoterpenoid agonists of TRPV3', *British Journal of Pharmacology*, vol. 151, pp. 530–540.

Wei, ET & Seid, DA 1983, 'AG-3-5: a chemical producing sensation of cold', *Journal of Pharmacy and Pharmacology*, vol. 35, pp. 110–112.

It's Probably Not Really Fresh Squeezed

AgResearch Magazine 2004, *New ways to make condensed O.J. Taste more like fresh*, United States Department of Agriculture, Viewed 20 September 2016, http://www.ars.usda.gov/is/AR/archive/sep04/juice0904.htm

Donaldson-James, S 2012, *California woman sues OJ Giant Tropicana over flavor packs*, ABC News, Viewed 20 September 2016, http://abcnews.go.com/Health/california-woman-sues-pepsicos-tropicana-alleging-deceptive-advertising/story?id=15394357&page=2

Flores, A 2004, *Making orange juice taste even better*, US Department of Agriculture, Viewed 20 September 2016, http://www.ars.usda.gov/is/pr/2004/040915.htm

García-Torres, R, Ponagandla, NR, Rouseff, RL, Goodrich-Schneider, RM & Reyes-De-Corcuera, JI 2009, 'Effects of dissolved oxygen in fruit juices and methods of removal', *Comprehensive Reviews in Food Science and Food Safety*, vol. 8, pp. 409–423.

ICE Futures U.S. 2012, *Frozen concentrated orange juice*, ICE Futures U.S., Viewed 20 September 2016, https://www.theice.com/publicdocs/ICE_FCOJ_Brochure.pdf

Matthews, RF 1994, *Frozen concentrated orange juice from Florida oranges*, University of Florida, Viewed 20 September 2016, http://www.ultimatecitrus .com/pdf/fcoj.pdf

Men's Health 2014, *5 Juices with more sugar than soda*, Time Magazine, Viewed 20 September 2016, http://time.com/103898/5-juices-with-more-sugar-than-soda/

Mintec 2012, *Orange juice: commodity factsheet, Chartered Institute of Procurement and Supply*, Mintec, Viewed 20 September 2016, http://www.cips .org/Documents/Knowledge/Categories-Commodities/Mintec/MintecCFS_ Orange_juice.pdf

Nikolas, K 2012, *Tropicana orange juice's all-natural claims challenged in court*, Digital Journal, Viewed 20 September 2016, http://www.digitaljournal.com/ article/325843

Why are Viruses Added to Some Sandwich Meat?

Ashelford, KE, Day, MJ & Fry, JC 2003, 'Elevated abundance of bacteriophage infecting bacteria in soil', *Applied and Environmental Microbiology*, vol. 69, pp. 285–289.

Centers for Disease Control and Prevention 2016, *Listeria (Listeriosis)*, Centers for Disease Control and Prevention, Viewed 23 September 2016, http://www.cdc .gov/listeria/

Clokie, MRJ, Millard, AD, Letarov, AV & Heaphy, S 2011, 'Phages in nature', *Bacteriophage*, vol. 1, pp. 31–45.

Federal Resister 2011, *Food additives permitted for direct addition to food for human consumption; bacteriophage preparation*, US Food and Drug Administration, Viewed 23 September 2016, https://www.federalregister.gov/ articles/2011/03/23/2011-6792/food-additives-permitted-for-direct-addition-to-food-for-human-consumption-bacteriophage-preparation

Hatfull, GF 2015, 'Dark matter of the biosphere: The amazing world of bacteriophage diversity', *Journal of Virology*, vol. 89, pp. 8107–8110.

Hennes, KP & Simon, M 1995, Significance of bacteriophages for controlling bacterioplankton growth in a mesotrophic lake, *Applied and Environmental Microbiology*, vol. 61, pp. 333–340.

Pride, DT, Julia Salzman, J, Haynes, M, Rohwer, F, Davis-Long, C, White, RA, Loomer, P, Armitage, GC & Relman, DA 2012, 'Evidence of a robust resident bacteriophage population revealed through analysis of the human salivary virome', *The ISME Journal*, vol. 6, pp. 915–926.

Walker, K 2006, *Use of bacteriophages as novel food additives*, Michigan State University, Viewed 23 September 2016, http://www.iflr.msu.edu/uploads/files/ Student%20Papers/USE_OF_BACTERIOPHAGES_AS_NOVEL_FOOD_ ADDITIVES.pdf

What is Margarine Made From?

Filip, S, Fink, R, Hribar, J & Vidrih, R 2010, 'Trans fatty acids in food and their influence on human health', *Food Technology and Biotechnology*, vol. 48, pp. 135–142.

Kala, AL, Joshi, V & Gurudutt, KN 2011, 'Effect of heating oils and fats in containers of different materials on their trans fatty acid content', *Journal of the Science of Food and Agriculture*, vol. 92, pp. 2227–2233.

Krettek, A & Thorpenberg, S 2008, *Trans fatty acids and health: A review of health hazards and existing legislation*, European Parliament, Viewed 23 September 2016, http://www.europarl.europa.eu/RegData/etudes/etudes/join/2008/408584/IPOL-JOIN_ET%282008%29408584_EN.pdf

McGreevy, P 2008, *State bans trans fats*, Los Angeles Times, Viewed 23 September 2016, http://articles.latimes.com/2008/jul/26/local/me-transfat26

Nash, AM, Mounts, TL & Kwolek, WF 1983, *Determination of ultratrace metals in hydrogenated vegetable oils and fats*, US Department of Agriculture, Viewed 23 September 2016, http://naldc.nal.usda.gov/download/26204/PDF

United States Department of Agriculture 2016, *Record U.S. peanut exports on strong shipments to China and Vietnam*, United States Department of Agriculture, Viewed 23 September 2016, http://apps.fas.usda.gov/psdonline/circulars/oilseeds.pdf

Why are Junior Mints Shiny?

Farag, Y 2010, *Characterization of different shellac types and development of shellac-coated dosage forms*, Ph.D. Thesis, University of Hamburg, Viewed 21 September 2016, https://www.chemie.uni-hamburg.de/bibliothek/2010/DissertationFarag.pdf

Hagenmaier, RD 2004, 'Fruit coatings containing ammonia instead of morphine', *Proceedings of the Florida State Horticultural Society*, vol. 117, pp. 396–402.

Rajkumar Shellac Industries 2010, *Welcome to our Raj Kumar Shellac Industries*, Rajkumar Shellac Industries, Viewed 23 September 2016, www.shellac-india.com

Scientific Analysis Laboratories 2011, *Morpholine*, Scientific Analysis Laboratories.

8

The Radioactive Banana and Other Examples of Natural Radioactivity

It may be a common misconception that radioactivity is something manmade and inherently unnatural. This mistaken belief could be caused by the fact that the uranium used to make nuclear bombs and for powering nuclear reactors is the product of a complex refining process. In actuality, radioactivity is very natural, and every living thing gives off radiation. Sunshine, for instance, is caused by a nuclear reaction, and what could be more natural than the warmth and light from Earth's closest star?

In fact, nuclear reactors are found in nature and they may be thought of as just another source of natural radiation that humans have learned to harness. Granted, nuclear reactor incidents such as the 2011 Fukushima Daiichi nuclear disaster are devastating, but nuclear reactors remain a viable source of energy. The lack of understanding about radioactivity leads to fear, but given that we are surrounded by it constantly, more knowledge will help us feel more at ease.

Radiation is defined as the emission of various particles from a nuclei undergoing *radioactive decay*. There are many types of radioactive decay giving rise to radiation, but for the purposes of this discussion, we will address only the two most common types experienced in everyday life. When most people talk about "radioactivity," they are typically referring to alpha particle decay. An *alpha particle* is made up of two protons and two neutrons. Compared to other types of radiation, alpha particles are quite dangerous because of their relatively large mass. As is the case with any collision, greater mass of the colliding entities usually causes more damage. Whenever you hear of someone dying because of radiation exposure, alpha particles are usually the cause.

Nuclei are made up of protons and neutrons. When you think about it, protons should not want to be in close proximity to each other due to their positive charges repelling each other. Despite this, protons and neutrons are held together by nuclear forces, which can overcome the repulsion of the positive charges. However, if a nucleus gets too large, the increased number of protons creates instability. This is why elements with nuclei larger than uranium do not occur naturally. They are so large that they very quickly break apart into smaller, more stable elements. In general, larger nuclei tend to

Strange Chemistry: The Stories Your Chemistry Teacher Wouldn't Tell You, First Edition. Steven Farmer.
© 2017 John Wiley & Sons, Inc. Published 2017 by John Wiley & Sons, Inc.

undergo alpha particle decay to make their nuclei contain fewer protons and thereby increase their stability. Two of the most commonly known radioactive elements, uranium and plutonium, both undergo alpha particle decay.

Another way that a nucleus can gain stability is by having equal numbers of protons and neutrons, since having more protons than neutrons increases the forces of repulsion among the positively charged protons. Other types of radioactive decay are targeted to the end result of getting the number of protons to equal that of the neutrons.

A *beta particle* (β) is really just an electron, but one specifically created by radioactive decay. During beta particle decay, a neutron is converted to a proton. Because an electron has so little mass, beta particles are much less dangerous than alpha particles, albeit still hazardous. Beta particle decay tends to occur with smaller nuclei that have more neutrons than protons.

The number of protons in a nucleus is called the *atomic number*, and this characterizes each different element. For example, a nucleus containing six protons is that of a carbon atom, while a nucleus with 92 protons characterizes a uranium atom. If you change the number of protons, you change the element, which is actually part of what makes nuclear chemistry so exciting. All of the types of radioactive decays discussed so far alter the number of protons in a nucleus, thereby changing the element. Much like the dreams of ancient alchemists, the application of nuclear chemistry allows scientists to convert one element into another. In fact, nuclear chemists have actually changed lead into gold using a particle accelerator, but the costs involved are prohibitive.

With numbers of protons and neutrons being important in radioactivity, scientists have come up with an easy method for keeping track. The number of protons (atomic number) is shown as a subscript in front of the element's symbol. The number of protons plus the number of neutrons is called the *mass number* and is shown as a superscript in front of the element's symbol. For example, $^{39}_{19}K$ is called potassium-39. Its nucleus contains 19 protons and 20 neutrons and is considered relatively nonradioactive. However, potassium-40 ($^{40}_{19}K$) has 19 protons and 21 neutrons in its nucleus and is considered radioactive. Atoms with the same number of protons but different number of neutrons in their nucleus are called *isotopes* of the same element. Different isotopes of the same element can have vastly different properties, as we will soon find out.

One other concept commonly referred to when discussing radioactivity is half-life. *Half-life* is the amount of time taken for half of a given radioactive sample to decay to another form, and it can be used to estimate how long the sample will be radioactive. If a 4-g sample of a radioactive isotope has a half-life of 2 days, only 2 g would remain after 2 days, 1 g would remain after 4 days, and so on. Possible half-life times range from minutes to billions of years.

Where Does the Helium We Use in Balloons Come From?

The helium we use for commercial purposes, such as in helium balloons, actually comes from underground. In 1903, it was found that a natural gas well in Dexter, Kansas, contained relatively large amounts of helium (1.8%), leading to the discovery that other wells in Texas, Oklahoma, and Kansas were also rich in helium, with concentrations ranging from 0.3% to 2.7%. Similar to natural gas and oil, helium is trapped in pockets under the ground and accessed by drilling. Initially, the United States was the world's only supplier of helium. Currently, the United States still provides about 75% of the world's helium producing about 687 kton of helium every year with an estimated value of $730 million.

Helium is generated underground through the radioactive decay of heavy elements such as uranium and thorium. One radiation product of this decay is alpha particles that are basically helium atoms without electrons. Scheme 8.1 is uranium-238 undergoing alpha particle decay to form thorium-234.

Once created, these alpha particles can quickly gain electrons from their surroundings to become helium gas, which becomes trapped in rock layers along with natural gas (Scheme 8.2).

Gases used for commercial purposes – such as oxygen and nitrogen – are typically isolated from air in a process called *cryogenic air separation*. In this process, air is partially condensed by cooling it to a low temperature, and then a distillation is performed to separate the component gases. This process is utilized to obtain the nitrogen used in liquid nitrogen, the oxygen for powering rockets as liquid oxygen, and the neon for making neon lights. Dry air is made up of nitrogen (~78%), oxygen (~21%), and argon (~1%), along with very small amounts of neon, helium, krypton, and xenon. Helium comprises about 0.0005% of air, an amount too small to be easily isolated. The earth's atmosphere contains very little helium, as it is one of the few gases that can obtain a velocity great enough to escape the earth's gravitational pull. Literally, the helium you release from a balloon will most likely find its way into outer space. Most

$$^{238}_{92}U \rightarrow {}^{234}_{90}Th + {}^{4}_{2}He^{2+}$$

Scheme 8.1 The alpha particle decay of uranium-238.

$$^{4}_{2}He^{2+} + 2\,e^- \rightarrow {}^{4}_{2}He_{(g)}$$

Scheme 8.2 An alpha particle gaining two electrons to become helium gas.

other gases are large enough to be held by gravity and therefore remain as part of the earth's atmosphere. In light of this, helium is a nonrenewable resource, and the earth's supply will eventually run out. The most recent estimates say that the supply of helium will not become an issue until 2060.

Helium is vital for applications requiring very cold temperatures and is also used in welding, creating inert atmospheres, and in deep sea diving. Actually, only a small amount is used in the helium balloon industry. The largest amount, roughly 32%, is used for cryogenic applications. Helium is the only substance known to exist in a liquid form at the extremely low temperature of −269 °C (about 4 K or −452.2 °F). It should be noted that this is roughly 4 °C above absolute zero, the coldest temperature possible. Helium's uniquely low boiling point is used to cool the superconducting magnets that are crucial to many instruments such as magnetic resonance imaging (MRI) machines used in hospitals. A superconducting magnet is an electromagnet made of superconducting wire, typically a niobium–titanium alloy. The superconducting wire has an electrical resistance approximately equal to zero when it is cooled to a temperature close to absolute zero by immersing it in liquid helium. This means that once a current is applied to the superconducting wire, it will continue to flow and generate a magnetic field, as long as the coil is kept at liquid helium temperatures. A typical electromagnet needs a constant application of current to maintain a magnetic field. Superconducting magnets are used because they produce stronger magnetic fields and are cheaper to operate than a typical electromagnet.

The second most common use of helium, roughly 18%, is as a shielding gas for welding. Exposure of a weld to oxygen, nitrogen, and moisture contained in air can create a variety of problems, including porosity (holes within the weld) and excessive spatter. Helium, due to its inert nature, does not cause these problems so replacing air with a helium atmosphere during welding has been shown to produce a higher quality bond.

Understanding how helium is obtained helps to explain a historical question that needs to be asked. Why was the Hindenburg airship filled with highly flammable hydrogen instead of the inert helium? In case you do not know, the Hindenburg airship disaster occurred on 6 May 1937. After making a trans-Atlantic crossing, the Hindenburg burst into flames and was completely destroyed killing 36 of the 97 persons aboard. At that time, the United States had a monopoly on the production of helium. In fact, in 1934, a plant in Amarillo, Texas, was the sole producer of commercial helium in the world. In order to retain helium for national defense purposes, the United States Congress passed the Helium Control Act of 1927, which placed a ban on foreign helium sales. This foreign sales ban on helium forced the German government to find an alternate lifting gas for the Hindenburg, hydrogen (Figure 8.1).

Figure 8.1 A clown holding balloons filled with radioactive decay (helium).

Who Was the First Person to Win Two Nobel Prizes?

Those who know the answer to this trivia question are on top of their game, since few people seem to guess correctly. In 1903, Marie Curie received the Nobel Prize in physics along with her husband, Pierre Curie, for their joint research on the radiation phenomena. She was the first woman to win a Nobel Prize, and even more remarkable is the fact that this was only the third year that Nobel Prizes were given out. In fact, only one other woman has ever been awarded the Nobel Prize in physics: Maria Goeppert-Mayer, who received the prize in 1963 for her discoveries regarding the nuclear shell structure.

In 1911, Marie Curie received the Nobel Prize in chemistry, partly for her discovery of the elements radium and polonium and also for her study of radium compounds. With this award, she became the first person to win a second Nobel Prize. Still to this day, she is the only person to win Nobel Prizes in different scientific fields. Curie was also the first woman to receive the Nobel Prize in chemistry. The next woman to achieve this feat was Marie Curie's daughter,

Irène Joliot-Curie, who, in 1935, became the second woman in history to be awarded the Nobel Prize in chemistry, for discovering artificial radioactivity.

The discovery of polonium and radium started when Curie found that some substances, such as the uranium-containing ore *pitchblende*, were far more radioactive than what was expected in light of their uranium content. She theorized that this excess radioactivity was due to an unknown radioactive element that had escaped detection upon prior chemical analyses. After a difficult chemical separation, she established the existence of a new element which they named *polonium* after Marie's native land, Poland. After further chemical separation, they determined that pitchblende contained yet another radioactive element, which they named *radium*.

As some may know, Marie did most of her research with her husband, Pierre Curie. This begs the question, why did he not also win a second Nobel Prize along with his wife? Unfortunately, Pierre died in 1906 from being run over by a horse-drawn wagon, and Nobel prizes are not given out posthumously.

To give you an idea of how impressive an achievement it is to win two Nobel prizes, there are only three other individuals who have done so: John Bardeen, Linus Pauling, and Frederick Sanger. Bardeen received two Nobel Prizes in physics for work on transistors and semiconductors. Pauling received a Nobel Prize in chemistry for research into the nature of the chemical bond and a Nobel Peace Prize for his activism. Sanger won two Nobel Prizes in chemistry for his work on the structure of proteins and his studies of the biochemistry of nucleic acids, with particular regard to recombinant DNA.

In addition to winning two Nobel Prizes, Marie Curie also has numerous additional accolades, contributing to her reputation as a pillar among scientists. In addition to helping discover the elements polonium and radium, she coined the term "radioactivity" and, in 1903, became the first woman in France to receive a Ph.D. of any kind. In 1908, she became the first woman to hold a professorship position at the Science University of Paris. In addition, during World War I, she realized that portable X-ray machines could help wounded soldiers by allowing surgeons to locate bullets and broken bones. She put her research on hold and helped to raise funds to create about 200 fixed radiological installations and furnish 18 cars with X-ray equipment to offer radiological service close to the battlefields. It was estimated that during the winter of 1917–1918 alone, her vehicles and posts took over a million X-rays.

Unfortunately, at the time of Curie's research, the damaging effect of radiation was not completely understood. She lived her life surrounded by radioactivity and was known to carry bottles of the polonium and radium in her coat pocket and store them in her desk drawer. On 4 July 1934, she eventually died from aplastic anemia related to overexposure to radiation. In fact, after more than 100 years, her laboratory notebooks are still dangerously radioactive and are contained in lead-lined boxes. Anyone wishing to view them must do so in protective clothing and only after signing a waiver of liability.

Where Is the Radioactive Material in YOUR House?

There are two major types of smoke detectors: ionization and photoelectric. *Ionization smoke detectors* contain a small amount (about 0.28 µg) of the radioactive material americium-241. *Americium-241* is manmade and first discovered during the Manhattan project, a national program that developed atomic weapons during World War II. During its radioactive decay, americium-241 emits alpha particles and low-energy gamma rays to form neptunium-237. Its half-life is 432.2 years. Americium-241 in the form of americium dioxide is embedded in a gold foil matrix within an ionization chamber of the smoke detector. The matrix is thick enough to completely retain the radioactive material but thin enough to allow the alpha particles to pass through. The average radioactive activity in a smoke detector source is about 1 µCi (microcurie) or one-millionth of a curie. As long as the radiation source stays in the detector, exposure would be less than about 1/100 of a millirem per year, which is essentially harmless. For comparison, a lethal dose of radiation would require an exposure on the order of 450,000 mrem (millirem). Even swallowing the radioactive material from a smoke detector would not lead to significant internal absorption because americium dioxide is insoluble and would pass through the digestive tract without delivering a significant radiation dose. Needless to say, however, it is not a good idea to tamper with smoke detectors (Scheme 8.3).

How does the smoke detector work? The ionization chamber in a smoke detector is made up of two metal plates. One of the plates has a positive charge, and the other a negative charge. When alpha particles are emitted from the radioactive source, they ionize the oxygen and nitrogen molecules in air by causing an electron to be removed. The result is oxygen and nitrogen molecules with a positive charge and electrons with a negative charge. In a smoke-free ionization chamber, these positive and negative ions migrate to the charged plates, creating a small electrical current that is measured by the smoke detector. When smoke particles are present, however, they interact with these ions, making them neutral again. With fewer ions available to migrate, the current decreases and this triggers the alarm.

You can easily tell what kind of smoke detector you have because those containing radioactive material are required by law to have a warning label on them. Because they contain radioactive material, it is not a good idea to throw them away in the trash. Ionization smoke detectors contain less the 1 µCi of americium-241, making them exempt from Nuclear Regulatory

$$^{241}_{95}\text{Am} \rightarrow {}^{237}_{93}\text{Np} + {}^{4}_{2}\text{He}$$

Scheme 8.3 The alpha particle decay of americium-241.

Commission (NRC) regulations. This means that there are no federal laws that prohibit disposal of these detectors in the normal municipal waste stream. However, there are a number of state and local regulations and/or laws that do prohibit such disposal. Many states and municipalities offer periodic recycling opportunities for smoke detectors. Check with your state or local solid waste disposal authority for the most current local guidance.

Which Elements Were First Detected in Radioactive Fallout from a Nuclear Bomb?

Einsteinium, named after Albert Einstein, is a synthetic element with an atomic number of 99. *Fermium*, named after Italian physicist Enrico Fermi, is a synthetic element with an atomic number of 100. These two elements, by the way, were the first ever to be named after eminent scientists. Both fermium and einsteinium were discovered in the debris of the first successful hydrogen bomb explosion. This 10-megaton bomb test was called Ivy Mike and took place on 1 November 1952, at Enewetak Atoll in the Pacific Ocean. Airplanes were flown though the explosion cloud and airborne residues were captured on filter paper for analysis. Fewer than 200 atoms of einsteinium were recovered. The discovery of fermium required more material, as the yield was expected to be at least an order of magnitude lower than that of einsteinium. After the nuclear explosion, large amounts of material drifted down as fallout. Researchers at the University of Berkeley at California tested tons of soil, ash, and coral debris from nearby Pacific islands and found around 200 atoms of the element fermium. News of the discovery of einsteinium and fermium was kept secret until 1955, due to cold war tension.

What is commonly thought of as a nuclear explosion involves nuclear fission. *Nuclear fission* is a nuclear reaction where a large element breaks up into smaller elements with the production of energy. In reality, there are only two specific isotopes that can undergo nuclear fission $^{235}_{92}U$ and $^{239}_{94}Pu$. Uranium-235 can absorb a neutron to become uranium-236 that immediately undergoes fission to create barium-141, krypton-92, and three neutrons. During the reaction, a minute amount of mass is converted to energy. The three created neutrons can subsequently initiate three new fission reactions and a chain reaction is started. The process repeats until all the uranium-235 is used up, and an immense amount of energy is released. Nuclear fission reactions are 10,000,000 times more powerful than a typical chemical explosion, such as TNT (Schemes 8.4 and 8.5).

$$^{235}_{92}U + {}^{1}_{0}n \rightarrow {}^{236}_{92}U$$

Scheme 8.4 The absorption of uranium-235 to form uranium-236.

$$^{236}_{92}U \rightarrow {}^{141}_{56}Ba + {}^{92}_{36}Kr + 3 {}^{1}_{0}n$$

Scheme 8.5 The fission of uranium-236.

So why do natural uranium deposits not spontaneously explode? The answer comes from the fact that only 0.72% of natural uranium is uranium-235 while most is uranium-238. There is simply not a high enough concentration of uranium-235 to sustain the required chain reaction. In order for a sustained nuclear reaction to occur, the uranium sample must be *enriched* with uranium-235, which is not easy to do. Uranium used in power plants has 3–4% uranium-235 and is called reactor grade. However, the uranium used in nuclear weapons contain 90% uranium-235 and is called weapons grade. A large part of the Manhattan project, which produced the first atomic weapons during World War II, was spent developing methods for the enrichment of uranium.

Nuclear explosions are the most powerful man-made sources of neutrons, with a density of about 10^{29} neutrons/(cm^2 s). With the large number of neutrons produced, scientists expected there to be elements with atomic numbers higher than uranium produced during a nuclear explosion. As mentioned previously, nuclei with an abundance of neutrons tend to emit beta particles to try and even out the proton-to-neutron ratio. Each beta particle decay creates a proton and increases the atomic number by one. Because uranium-238 is present in the explosion, it can collide with as many as 15 neutrons to form uranium-253. The uranium-253 can then undergo the loss of seven beta particles to form element 99, einsteinium. Fermium, element 100, was created in a similar manner (Schemes 8.6 and 8.7).

The most common isotope of einsteinium, ^{253}Es, has a half-life of 20.47 days. Einsteinium is highly radioactive and actually produces a visible glow. Because of the short half-life, all of the einsteinium made during the formation of the earth has decayed to other elements. Einsteinium-253 is currently made artificially in some nuclear reactors by the intense neutron irradiation of plutonium or californium. The yield is roughly on the milligram per year scale. Fermium has 18 isotopes with mass numbers from 242 to 259. In particular, fermium-257 has a half-life of 100.5 days and can be obtained in microgram quantities from the neutron bombardment of plutonium (Figure 8.2).

$$^{238}_{92}U + 15 {}^{1}_{0}n \rightarrow {}^{253}_{99}Es + 7 {}^{0}_{-1}\beta$$

Scheme 8.6 The formation of einsteinium by neutron bombardment of uranium-238.

$$^{238}_{92}U + 17 {}^{1}_{0}n \rightarrow {}^{255}_{100}Fm + 8 {}^{0}_{-1}\beta$$

Scheme 8.7 The formation of fermium by neutron bombardment of uranium-238.

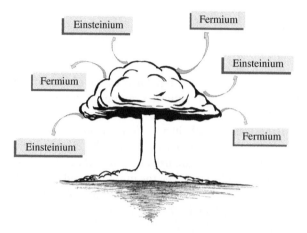

Figure 8.2 Even an atomic bomb can cause chemical reactions.

Radioactivity in Wristwatches, Exit Signs, and H-Bombs

Tritium, a radioactive isotope of hydrogen that contains two neutrons and one proton, is used as a component in nuclear weapons and as a tracer in biological studies. Colorless, odorless, tasteless tritium gas (T_2) is the radioactive equivalent of hydrogen gas (H_2) and is used on a practical basis to create visible light without a continuous external power source. Tritium undergoes radioactive beta decay to form ^{3}He by emitting a beta particle and an antineutrino from its nucleus (Scheme 8.8).

These ejected beta particles, which are essentially electrons, act as an energy source and interact with a phosphor material that converts the electrons into visible light. The lack of an external electrical source makes tritium lights very popular for use in emergency exit signs, watches, nighttime gun sights, and novelty key chains. Because tritium has a half-life of 12.33 years, most tritium lights have a useful life span of over 10 years. Tritium signs usually appear unlit in bright conditions but radiate a glowing green light in dark conditions. In addition, they should be labeled with the common radiation symbol and have an NRC sticker. During disposal, the tritium is captured and recycled to make new lighting devices. It has been estimated that there are more than 2 million tritium "EXIT" signs in the United States. If you are interested, tritium-containing light sources are available on the internet.

Although radioactive, contained tritium is relatively safe. The emitted electrons are relatively low energy, so they have very little penetrating power.

$$^3_1H \rightarrow \, ^3_2He + \, ^0_{-1}\beta$$

Scheme 8.8 The beta particle decay of tritium.

They typically travel through less than 6 mm of air and can stopped by our skin, keeping any real damage from occurring. However, ingesting or breathing in tritium is a different story. Once inside the body, it has the capability of being incorporated into many different biological molecules that use hydrogen, including DNA. As with most ionizing radiation sources, tritium ingestion has been shown to produce the typical radiogenic effects of cancer, genetic alteration, and reproductive abnormalities.

Because tritium is radioactive, the US NRC made it a requirement that no device containing tritium should be disposed of in normal waste streams and instead turned over to a licensed disposal. Currently, tritium is accountable to the US Department of Energy in amounts down to 0.01 g.

In nature, ordinary hydrogen (^1_1H) comprises over 99.9% of all hydrogen. Deuterium (^2_1H) comprises about 0.02%, and tritium (^3_1H) comprises about 10^{-16} percent. The tritium in nature is produced in the upper atmosphere by the interaction of atmospheric nitrogen with cosmic radiation. Once produced, the tritium becomes incorporated into water, which falls to earth as rain. Because such a small amount of tritium is present in nature, it must be made artificially to be used for commercial applications.

Tritium is made artificially by bombarding lithium-6 with a neutron to form a tritium atom and an atom of helium. It has been estimated that the United States has produced 225 kg of tritium, much of which has decayed, leaving a current inventory of about 75 kg (Scheme 8.9).

In actuality, tritium production in the United States stopped in 1988, and from 1988 until 2006, the nation had no source for tritium production. Starting in 2006, the Tritium Extraction Facility at the Savannah River Site started maintaining United States tritium stockpiles by recycling tritium from existing nuclear warheads, as well as extracting tritium from lithium rods irradiated in nuclear reactors.

Hydrogen bomb use tritium to make a nuclear explosion through the process of *fusion*. In this case, tritium and deuterium (both are isotopes of hydrogen) react to form helium and a neutron. The helium atom (^4_2He) has two protons, two neutrons, and weighs 4.00150 atomic mass units (amu). However, the mass of two protons together with two neutrons is 4.03188 amu. Therefore, during the process of combining two protons and two neutrons together to form a helium atom, a small amount of mass is lost. In accordance with Albert Einstein's famous equation $E = mc^2$, a small amount of mass converts to a large amount of energy, which is given off during this reaction (Scheme 8.10).

So why don't we use fusion for energy like we do with nuclear reactors? The problem is that both deuterium and tritium are positively charged, and their

$$^6_3Li + {}^1_0n \rightarrow {}^7_3Li \rightarrow {}^3_1H + {}^4_2He$$

Scheme 8.9 The formation of tritium by the neutron bombardment of lithium-6.

$$^{2}_{1}H + ^{3}_{1}H \rightarrow ^{4}_{2}He + ^{1}_{0}n$$

Scheme 8.10 The fusion reaction that creates a nuclear explosion.

positive charges repel with such force that immense temperatures ($\sim 1 \times 10^8$ K) are required to overcome it. This temperature is so high, generating and maintaining it is too difficult to make this a viable energy source. This is why the idea of *cold fusion* is so interesting. If we could cause this reaction to occur at lower temperatures, hydrogen fusion could provide an abundant energy source.

In a hydrogen bomb, a "kick start" is required to achieve these high temperatures. H-bombs contain fissionable material (uranium or plutonium), which are typically found in an atomic bomb. The fission reaction is easily initiated and produces a tremendous amount of energy and heat. These temperatures are high enough to start the hydrogen fusion reaction. So for a hydrogen bomb, a fission nuclear reaction is required to start a fusion nuclear reaction.

The Earth Is One Giant Nuclear Reactor

It is commonly known that the earth contains a certain amount of residual heat, clearly demonstrated when a volcano erupts and lava pours forth. In addition, the earth has a temperature gradient, with temperatures increasing the closer we get to the core. The mass of the core amounts to approximately 32.5% of the earth as a whole and mainly consists of a molten iron–nickel alloy. The temperature at the earth's core is predicted to be as high as 7000 K, well above the melting points of minerals and metals, and the reason it is molten. The fact that the earth has a molten core explains many planetary processes such as mantle convection, plate tectonics, and the magnetic field.

The earth's internal heat also helps to warm the planet as much of this energy is dissipated, even though you may not feel it. It has been estimated that the earth constantly radiates 44.2 TW of energy (TW stands for a unit of energy called the *terawatt*, which is equal to 1 trillion [10^{12}] watts). To give you an idea about how much energy this is, the world's average energy consumption is only 12 TW. This means if we were to capture even a fraction of the earth's radiated energy we could easily power the entire planet.

About half of the earth's total heat radiation comes from residual heat from the planet's formation (called *accretion*), including the energy generated by the impacts of meteorites and asteroids. The gravitational effects of the sun have also been theorized to be a source of some of Earth's heat radiation. The sun's gravity causes shifts in the earth's shape that generates heat. This is similar to how a wire will become hot if you repeatedly bend it. The other half of the heat radiation comes from the decay of radioactive isotopes inside the earth and

$$^{238}_{92}U \rightarrow {}^{234}_{90}Th + {}^{4}_{2}He$$

Scheme 8.11 The alpha particle decay of uranium-238.

is called *radiogenic heat*. Quite literally, there is an intense nuclear reaction going on in the center of the earth, generating an immense amount of heat. In particular, decays of uranium-238, thorium-232, and potassium-40 generate most of the radiogenic heat produced (Scheme 8.11).

Uranium-238 undergoes spontaneous alpha particle decay to form thorium-234. Many of the produced alpha particles eventually become helium atoms as previously discussed. After the Earth was formed roughly 4.5 billion years ago, a specific amount of uranium-238 was also formed. Since uranium-238 has a half-life of 4.5 billion years, it decays very slowly, so the current amount is almost half as much as existed in the earth when it was forming. Similarly, thorium-232 has a half-life of 14 billion years and potassium-40 has a half-life of 1.3 billion years, so much of these elements formed during the creation of the earth still remains. It is thought that the total amount of radiogenic heat dissipated by the earth is 19 TW, 84% of which is produced by uranium-238 and thorium-232 decay.

Scientists know the amount of radiogenic decay by estimating the number of geoneutrinos emitted by the earth. A *geoneutrino* is emitted during certain types of radioactive decay, and it can pass through the earth virtually unaffected. Geoneutrinos from ^{232}Th and ^{238}U radioactive decay are detectable because they have the ability to undergo inverse beta decay, which generates energy. The energy produced by this decay is detected as flashes of light by large underground liquid scintillator detectors. One of these detectors, the Kamioka Liquid Scintillator Anti-Neutrino Detector (KamLAND), is located roughly 1 km under Mount Ikenoyama near Kamioka, Japan. The detector is basically a large balloon filled with 1000 tons of ultra-pure liquid scintillator made up of 80% dodecane, 20% pseudocumene (1,2,4-trimethylbenzene), and a small amount of the fluorophore PPO (2,5-diphenyloxazole). This scintillator liquid is surrounded by very sensitive detectors that can detect a single photon of light formed. To give you an idea of how difficult these interactions are to detect, only 88 of these decay events were detected in 1.2 years.

Are Nuclear Reactors "Natural"?

In 1972, a natural nuclear reactor was discovered when French geologists noticed a slight discrepancy while assaying uranium samples from the Oklo Uranium mine in western Africa. Fissionable uranium-235 has a natural abundance of 0.720% in natural uranium, but the samples from the Oklo mine had a uranium-235 abundance of 0.717%. Although this difference seems miniscule, it was enough to alert scientists that something odd had happened. All of the uranium-235 was created at the birth of the universe, and it has been steadily decaying ever since. Because of this, uranium samples in the earth's crust, on the moon, and even in meteorites all have a uranium-235 abundance of 0.720%. This fact was actually used to estimate the age of the universe.

It was eventually determined that the isotope abundance in the Oklo mine was off because the uranium-235 had been used up in a nuclear fission reaction. As previously discussed, a stray neutron causes a uranium-235 nucleus to split, which gives off more neutrons, thus causing other uranium-235 atoms to fission in a nuclear chain reaction. Overall, this process causes the uranium-235 to be used up. It was proven that this is what occurred at the Oklo mine, in part because of the presence of certain products produced during fission, including neodymium and ruthenium.

Currently, there are no working natural nuclear reactors because the current levels of 0.720% uranium-235 in deposits are too low to cause a fission reaction. However, because uranium-235 has a half-life of 0.70 billion years, the amount 2 billion years ago was much higher. Roughly 1.9 billion years ago, when the Oklo natural reactor was started, the uranium-235 abundance was approximately 3%, which is the same as the enriched uranium used to fuel most modern nuclear power reactors.

It was estimated that the Oklo reactor operated for 150,000 years and consumed 6 tons of uranium while releasing an average of 15,000 MW of energy per year. Although this sounds impressive, its continual average output was about 100 kW, which is enough to power roughly 35 US homes. As a comparison, a typical modern nuclear reactor produces roughly 1000 MW of continual output.

There are numerous factors that allowed the Oklo site to become a nuclear reactor, whereas other uranium mines did not. In particular, the Oklo site was saturated with groundwater, which acted as a neutron moderator. A neutron moderator is required to slow down the ejected neutrons, giving them a better chance of reacting with a uranium atom to cause the fission reaction. Billions of years ago, the uranium at the Oklo site became flooded with groundwater, which allowed a nuclear chain reaction to occur. The energy produced by the nuclear reaction produced enough heat to boil the water off. This in turn caused the nuclear reaction to stop. Eventually, the groundwater would again flood the uranium, and the reaction would start once more. Evidence showed that the Oklo natural nuclear reactor proceeded in this start–stop manner at intervals of about 2 hours and 30 minutes for its lifetime.

Considering that the age of the earth has been estimated at about 4.5 billion years, why did the nuclear reactor not start up until about 2 billion years ago? The answer is oxygen. As the nuclear reaction progresses, many of the elemental isotopes generated as part of the fission reaction acted as a neutron poison, absorbing neutrons and stopping the reaction. After running for a short time, these neutron poisons would accumulate and cause the reaction to stop. These product isotopes tend to be insoluble in water unless oxygen is present. The oxygen allows the product isotopes to form oxides, which are water soluble. At the Oklo site, the water would continually wash away the neutron poisons so the reaction could continue. The nuclear reaction started 2 billion years ago because that is when oxygen appeared on Earth after what is called the "great oxidation event." This period saw the appearance of photosynthesizing bacteria, for example, cyanobacteria, which converted sunshine, CO_2, and water into carbohydrates and oxygen eventually creating earth's oxygen-rich atmosphere.

Are Your Gemstones Radioactive?

Have you ever touched something that was once in a nuclear reactor? If you answered no, you should think again. The use of man-made sources of radiation to improve the quality of gems has been around for almost 100 years, and *gemstone irradiation* has become a common practice in recent years. Certain gemstones are exposed to radiation to enhance, deepen, or change their color. In particular, topaz $(AI_2[SiO_4](OH,F)_2)$ is the most commonly treated stone. Typically orange, topaz becomes blue as a result of exposure to radiation. Because of this, the blue topaz industry has grown dramatically through the years and now accounts for $675 million of the worldwide gemstone market. It is estimated that the topaz market sells 2.25 million stones annually, with nearly 100% of them being created by laboratory irradiation.

Irradiation of gemstones is risky because it is impossible to predict how the process will affect a stone. Typically, numerous gemstones must be irradiated in order to obtain a few stones that are altered to the desired color. Gemstone irradiation usually takes place in one of three ways: in a nuclear reactor through neutron bombardment, in a particle accelerator through electron bombardment, or by exposure to gamma rays in an irradiator. *Gamma rays* are high-energy photons generated by radioactive decay.

Electron beam irradiators use a particle accelerator to create a narrow stream of electrons, which is directed at the gemstones. With higher electron energy, there is a probability that the gemstones will become radioactive. However, the half-lives of the produced radioactive elements tend to be short, so the radioactivity decays rapidly to background. For example, high-energy electron beam irradiation of topaz can create a radioactive fluorine-18 isotope, which has a half-life of 1.8 hours. Although the gemstone becomes radioactive, the emissions decay to background levels within a day or two.

The nuclear reactors used to irradiate gemstones with neutrons are typically designed for research rather than electrical power production. Although neutron irradiation produces a darker and richer color in the gemstone than electrons or gamma rays, there is a price, because the process creates some radioactivity in the gemstone. Although many of the radioactive isotopes produced have short half-lives, measured in hours, a few of those created have longer half-lives, measured in months or years.

Because a neutron is electrically neutral, it enters the positively charged nucleus relatively easily. Depending on the type of nucleus, this can alter the neutron-to-proton ratio and cause it to become unstable and radioactive. When a nucleus has too many neutrons, it typically undergoes radioactive decay (typically beta particle emission) to correct this imbalance. Consequently, relatively more induced radioactivity is generated during irradiation in a nuclear reactor. In theory, a neutron-irradiated sample of pure topaz would stop being radioactive after a few weeks because the neutron-activated isotopes of the main constituents of topaz (fluorine, aluminum, oxygen, and silicon) have half-lives of only seconds to hours. However, topaz is likely to contain trace element impurities such as tantalum, scandium, and manganese, the neutron-activated isotopes of which (Ta-182, Sc-46, and Mn-54) have half-lives of months or years. In short, this process of irradiation can make the gems slightly radioactive for some time.

All gemstones that are made radioactive through artificial irradiation fall under NRC regulations. These include blue topaz, irradiated diamonds, irradiated beryl, and some irradiated red tourmaline. The NRC regulates the initial distribution of these gemstones, requires distributers to be licensed, and requires the stones to be set aside to allow any radioactivity to fade. Moreover, a distributer must conduct radiological surveys to show that the stones are safe before they can be placed on the market. After this initial period, subsequent distributors, jewelers, and consumers do not need a license to own or sell them. To date, there have been no reported cases of anyone being harmed by wearing an irradiated gemstone. In fact, the NRC did a study which showed that a large (six-carat) blue topaz stone gives off less than half as much radiation as having false teeth.[1]

Gamma irradiators typically use sources containing cobalt-60 to supply gamma rays that will interact with the crystal structure of the gemstone and create the color change. Because the gamma rays are too weak to affect nuclei, no gemstone irradiated with cobalt-60 gamma rays will become radioactive and are not under NRC authority. These include treated pearls, yellow sapphire, treated quartz, and the majority of irradiated red tourmaline and irradiated kunzite.

Gamma rays cause electrons to be released from their normal location in the gem, forming *color centers*. A color center can be a "hole color center," wherein an electron is missing from a normally occupied position, or an "electron color center," formed by the presence of one extra electron in a different location. The

---

1 I tested one of my wife's large topaz necklaces with a Geiger counter. Although it was minuscule, it did give off a measurable amount of radiation.

resulting color change depends on multiple factors, including where the electrons relocate and the charges of the atoms near them. These factors affect the way the gemstone absorbs light, thus creating a distinct color. Topaz is colorless in its pure mineral form; however, irradiation can cause a blue color to appear. Just one electron displaced from its regular position can vary the specific wavelengths of light energy absorbed. Because white light is a combination of all the colors of the rainbow, we see the color of the remaining light reflected after absorption. In the specific case of blue topaz, the freed electron type absorbs light at red end of the spectrum, causing us to see the gemstone reflect the complementary blue color.

Through a similar process of irradiating diamonds, it is possible to induce black, green, blue-green, deep yellow, orange, pink, and red colors. When pearls are irradiated, dark gray colors result. Quartz can be irradiated to obtain a smoky color. Amethysts can be made by irradiating clear quartz. Even though this treatment sounds "artificial," it has been theorized that the color of many natural gemstones, including topaz, maxixe beryl, amethysts, and yellow diamonds, are attributed to natural radioactivity. In fact, the radiation from uranium impurities in zircons is thought to produce red hues.

While we're discussing radiation exposure, did you know that some airport baggage scanners irradiate luggage with neutrons from the spontaneous fission of californium-252? Although exposure to the source itself is highly unlikely, you have to remember that certain elements can become radioactive after neutron bombardment. Extensive studies have shown that exposure to this radiation is negligible but present. Luggage passed under scanners using californium-252 was found the most radioactive less than 1 hour after irradiation. The worst exposure came from the ingestion of food in the luggage less than 1 hour after irradiation. This is due, in part, to the presence of radioactive sodium-24 from salted food. The half-life of sodium-24 is 15 hours, so the potential radioactivity could remain for days. Furthermore, the application of cosmetics less than 1 hour after irradiation is another way we are exposed to radioactivity, due in part to the maganese-54 found in some cosmetic formulations. Because maganese-54 has a half-life of 313 days, trace radioactivity could continue for years.

Radon: The Radioactive Gas in Your Home

Most people know extremely little about the true sources of radiation in their world. Posing this question to a large group of people generally results in…silence. Once set thinking about it, some individuals venture a guess of cell phones, mainly because the radiation they emit has been a source of some recent concern. Others speculate that the answer might be dental X-rays, which is actually tied for second in terms of contributors to our total radiation exposure (about 11%). It simply does not occur to most people that we are being exposed to radiation from an invisible, odorless gas seeping out of the ground!

Radon is an odorless, tasteless, radioactive gas which comes from the natural decay of uranium in the ground. Because uranium is found virtually everywhere in the earth's crust, radon is also found almost everywhere. As the gas forms, it moves through the ground and into the air above it. It can actually move through small cracks in the foundation of a home, get trapped, and build up over time. In general, most people receive more radiation exposure from radon than from all other sources. The average American receives about 360 mrem of radiation exposure every year, and of this, about 200 mrem (~56%) comes from radon.

Radon as a problem in homes became evident in 1984, when a nuclear plant worker discovered radioactivity on his clothing when passing through radiation detectors on the way into his work station. It was later determined that the source of this radiation was radon gas.

As you can imagine, radioactivity is generally considered a bad thing. Radon is no exception. Problems occur when increased amounts of radon get trapped in a room, such as a basement, and are then inhaled over a long period of time. How big of a problem is radon gas? It has been identified as the second leading cause of lung cancer in smokers, causing an estimated 21,000 deaths per year, and it is the number one cause of lung cancer in nonsmokers. Radon was first identified as a possible cancer-causing agent when it was noticed that uranium miners were dying of lung cancer at higher rates than normal. Because uranium is one of the direct sources of radon gas, these miners were being directly exposed over long periods of time. The deadly nature of this high exposure to radon was confirmed when it was shown that rodents exposed to high radon gas levels developed lung tumors at a significantly higher rate than the unexposed control group.

Regarding radon's link to lung cancer, because it is a gas, radon can be breathed into the lungs, where it undergoes radioactive decay. Radon-222 has a half-life of 3.8 days and emits an alpha particle as it decays to polonium-218. The alpha particles produced cause damage to the DNA in lung tissue and over long periods of time, this can lead to lung cancer (Scheme 8.12).

The amount of radon in the air is measured according to radioactive emissions due to radon decay and is expressed in picocuries per liter of air (pCi/l). Radon levels can be measured through home testing kits. The average national level of radon outdoors is 0.4 pCi/l, and it is estimated that two-thirds of all homes exceed this level. The Radon Act 51, passed by Congress, sets this number as the target radon level for indoors and recommends that corrective

$$^{222}_{86}Rn \rightarrow \ ^{218}_{84}Po + \ ^{4}_{2}He$$

Scheme 8.12 The alpha particle decay of radon-222.

Have you checked it for Radon yet?

Figure 8.3 Even cavemen are concerned about radon.

measures be taken if radon levels exceed 4 pCi/l. Lung cancer risk has been shown to be 16% higher with exposure to each 2.7 pCi/l increase in the amount of radon.

Despite all of these health concerns, radon exposure is actually used as an alternative health treatment. The idea is that small doses of radiation which comes from a brief exposure to radon may have health benefits. The radon is provided from old uranium mines most commonly found in Montana. The treatment typically involves sitting in the mine and breathing in the radon-rich air for a short period of time. The websites for some of these health mines claim benefits for sufferers for a wide variety of conditions including diabetes, gout, cancer, lupus, and colitis (Figure 8.3).

The Radioactive Banana

Bananas are radioactive because they contain the element potassium. Potassium contains various isotopes, with the majority of potassium atoms being the benign potassium-39 isotope; however, a small number (0.012%) are the radioactive potassium-40 isotope. Potassium-40 typically undergoes radioactive beta decay to form calcium-40, with the very long half-life of 1.3

$$^{40}_{19}K \rightarrow {}^{0}_{-1}\beta + {}^{40}_{20}Ca$$

Scheme 8.13 The beta particle decay of potassium-40.

billion years. The average banana weights about 150 g and contains around 450 mg of potassium, which means that it undergoes about 14 radioactive decays each second. The miniscule amount of radiation that one could receive from a banana ($\sim 1 \times 10^{-5}$ rem) has been called a "banana equivalent dose," a nonstandard unit of radiation exposure used to exemplify that small doses of radiation present a correspondingly small risk. However, before you stop eating bananas, you should know that you would have to be in the presence of more than 41 million bananas to receive the expected lethal dose of 400–450 rem of radiation from potassium. Although the amount of radiation emitted by bananas seems insignificant, cargo-loads of bananas produce enough emissions to set off Radiation Portal Monitors at our nation's land and sea ports of entry (Scheme 8.13).

Interestingly, the second greatest source of radiation exposure to humans is our own body. This is due to the presence of potassium and other radioactive isotopes in our tissues. The effect, called *internal radiation*, is caused mainly by two elements, potassium and carbon. Potassium is common in our bodies, making up roughly 0.2% of the body's total weight. This means that a typical human body contains about 120 g of potassium of which 0.144 g are radioactive potassium-40. Our bodies cannot tell the difference between potassium-38 and potassium-40 so the radioactive element is distributed in our tissues. The other major source of internal radiation is carbon, comprising about 23% of the mass of a typical human body. Most carbon atoms are carbon-12, but a tiny percentage (0.0000000001%) is the radioactive isotope carbon-14, which

Figure 8.4 The radioactive banana.

also undergoes radioactive beta decay. On average, Americans receive a dose of 40 mrem per year from internal radiation, making up about 11% of our total annual radiation exposure. Before you get too worried about having radioactive elements in your body, it has been estimated that there is a 5000 times greater chance of dying for some other cancer cause than from potassium-40 exposure (Figure 8.4).

Further Reading

Where Does the Helium We Use in Balloons Come From?

Air Products and Chemicals, Inc. 2006, *Safetygram #7 Liquid nitrogen*, Brookhaven National Laboratory, Viewed 10 September 2015, http://www.bnl .gov/esh/shsd/PDF/Compressed_gas/SafetyGram_LN2.pdf

Lallanilla, M 2013, *World helium shortage expected to balloon drastically*, NBC News, Viewed 15 September 2016, http://www.nbcnews.com/science/science-news/world-helium-shortage-expected-balloon-drastically-f6c08963426

Mohr, S & Ward, J 2014, 'Helium production and possible projection', *Minerals*, vol. 4, pp. 130–144.

Texas State Historical Association 2016, *Helium production*, Texas State Historical Association, Viewed 15 September 2016, http://www.tshaonline.org/ handbook/online/articles/doh02

Universal Industrial Gases Inc. 2015, *Air: Its composition and properties*, Universal Industrial Gases Inc., Viewed 15 September 2016, http://www.uigi.com/air.html

U.S. Department of the Interior Bureau of Lan Management 2015, *About helium*, U.S. Department of the Interior Bureau of Lan Management, Viewed 15 September 2016, http://www.blm.gov/nm/st/en/prog/energy/helium_program/ about_helium.html

U.S. Geological Survey 2012, *Helium*, U.S. Geological Survey, Viewed 15 September 2016, http://minerals.usgs.gov/minerals/pubs/commodity/helium/ mcs-2012-heliu.pdf

Ward, DE & Pierce, AP 1973, "Helium', United States mineral resources', *US Geological Survey*, vol. 820, pp. 285–290.

The First Person to Win Two Nobel Prizes

Greer, E & Tolmachova, M 2011, 'Marie Curie: Pioneering discoveries and humanitarianism', *Helvetica Chimica Acta*, vol. 94, pp. 1893–1907.

O'Carroll, E 2011, *Marie Curie: Why her papers are still radioactive*, The Christian Science Monitor, Viewed 15 September 2016, http://www.csmonitor.com/ Technology/Horizons/2011/1107/Marie-Curie-Why-her-papers-are-still-radioactive

Pasachoff, N 1996, *Marie Curie: The discovery of radium*, American Institute of Physics, Viewed 15 September 2016, https://www.aip.org/history/exhibits/curie/brief/03_radium/radium_5.html

Scott, A 2011, 'Women in chemistry: Why it's Marie Curie every time', *Chemical Week*, vol. 173, pp. 38.

Where is the Radioactive Material in YOUR House?

Environmental Protection Agency 2016, *Radiation protection*, Environmental Protection Agency, Viewed 16 September 2016, http://www.epa.gov/radiation/sources/smoke_alarm.html

US Fire Administration 2009, *Disposal of fire/smoke detectors*, Federal Emergency Management Agency, Viewed 16 September 2016, http://www.usfa.fema.gov/downloads/pdf/techtalk/techtalk_v1n2_1209.pdf

World Nuclear Association 2014, *Smoke detectors and americium*, World Nuclear Association, Viewed 16 September 2016, http://www.world-nuclear.org/information-library/non-power-nuclear-applications/radioisotopes-research/smoke-detectors-and-americium.aspx

Which Elements were First Detected in Radioactive Fallout from a Nuclear Bomb?

Ghiorso, A, *et al.* 1955, 'New elements einsteinium and fermium, atomic numbers 99 and 100', *Physical Review*, vol. 99, pp. 1048–1049.

Morss, LR 1992, *Transuranium Elements: A Half Century*, American Chemical Society, Washington, DC.

Zielinski, S 2012, *What is enriched uranium?*, Smithsonian.com, Viewed 11 December 2016, http://www.smithsonianmag.com/science-nature/what-is-enriched-uranium-17091828/

Radioactivity in Wristwatches, Exit Signs, & H-Bombs

Agency for Toxic Substances and Disease Registry 2009, *Section 4: Public health implications of estimated tritium doses*, Agency for Toxic Substances and Disease Registry, Viewed 16 September 2016, http://www.atsdr.cdc.gov/hac/pha/PHA.asp?docid=37&pg=2

Department of Energy 2008, *Tritium handling and safe storage*, Department of Energy, Viewed 16 September 2016, http://energy.gov/sites/prod/files/2013/06/f1/DOE-HDBK-1129-2008.pdf

Dingwall, S, Mills, CE, Phan, N & Taylor, K 2011, 'Human health and the biological effects of tritium in drinking water: Prudent policy through science – Addressing the ODWAC new recommendation', *Dose Response*, vol. 9, pp. 6–31.

Folkers, C 2006, *Tritium: Health consequences*, Nuclear Information and Resource Service, Viewed 16 September 2016, http://www.nirs.org/factsheets/tritiumbasicinfo.pdf

Morss, LR 1992, *Transuranium Elements: A Half Century*, American Chemical Society, Washington, DC.

Savannah River Site 2008, *Tritium extraction facility*, Department of Energy, Viewed 16 September 2016, http://www.srs.gov/general/news/factsheets/tef.pdf

Savannah River Site 2012, *Defense programs*, Department of Energy, Viewed 16 September 2016, http://www.srs.gov/general/programs/dp/index.htm

Tritium Disposal Company 2016, *Rules and regulations regarding tritium signs*, Tritium Disposal Company.

US Environmental Protection Agency 2015, *Radionuclide basics: Tritium*, US Environmental Protection Agency, Viewed 16 September 2016, https://www.epa.gov/radiation/radionuclide-basics-tritium

The Earth is One Giant Nuclear Reactor

Araki, T 2005, 'Experimental investigation of geologically produced antineutrinos with KamLAND', *Nature*, vol. 436, pp. 499–503.

Asakura, K 2015, 'Study of electron anti-neutrinos associated with gamma-ray bursts using KamLAND', *The Astrophysical Journal*, vol. 806, pp. 1–5.

Dumé, B 2005, *Geoneutrinos make their debut*, Physics World, Viewed 16 September 2016, http://physicsworld.com/cws/article/news/22737/1/0507162

International Energy Agency 2016, *Key world energy statistics*, International Energy Agency, Viewed 10 December 2016, http://www.iea.org/publications/freepublications/publication/KeyWorld2016.pdf

McDonougha, WF & Suna, S 1995, 'The composition of the Earth', *Chemical Geology*, vol. 120, pp. 223–253.

Are Nuclear Reactors "Natural"?

Bentridi, S, Gall, B, Gauthier-Lafaye, F, Seghour, A, Pape, A & Medjadi, D 2013, 'Criticality of Oklo natural reactors: Realistic model of Reaction Zone 9', *IEEE Transactions on Nuclear Science*, vol. 60, pp. 278–293.

Karam, A 2005, *The natural nuclear reactor at Oklo: A comparison with modern nuclear reactors*, Idaho State University, Viewed 16 September 2016, http://www.physics.isu.edu/radinf/Files/Okloreactor.pdf

Sankaran, AV 2010, "Oklo phenomenon' of natural nuclear reactors – recent insights into its evolution and extinction', *Current Science*, vol. 98, pp. 749–750.

U.S. Energy Information Administration 2015, *How much electricity does an American home use?* U.S. Energy Information Administration, Viewed 16 September 2016, http://www.eia.gov/tools/faqs/faq.cfm?id=97&t=3

U.S. Energy Information Administration 2016, *Status of U.S. nuclear outages*, U.S. Energy Information Administration, Viewed 16 September 2016, http://www.eia.gov/nuclear/outages/

Are Your Gemstones Radioactive?

Ashbaugh, C 1988, 'Gemstone irradiation and radioactivity', *Gems & Gemology*, vol. 24, pp. 196–213.

American Gem Trade Association 2008, *The essential guide to the U.S. Trade in irradiated gemstones*, American Gem Trade Association, Viewed 16 September 2016, http://www.agta.org/education/docs/brochure-irradiated-gemstones.pdf

Ohio University 2016, *Irradiated gemstones*, Ohio University, Viewed 16 September 2016, https://www.ohio.edu/riskandsafety/radiationsafety/irradiated.htm

Salama, S 2011, *Study on properties of some treated gemstones*, Ph.D. Thesis, Benha University Physics Department, Viewed 16 September 2016, http://www.iaea.org/inis/collection/NCLCollectionStore/_Public/44/096/44096830.pdf

Schneider S, *et al.* 2001, *Systematic radiological assessment of exemptions for source and byproduct assessment*, U.S. Nuclear Regulatory Commission, Viewed 16 September 2016, http://www.nrc.gov/reading-rm/doc-collections/nuregs/staff/sr1717/nureg-1717.pdf

U.S. Nuclear Regulatory Commission 2016, *Background on irradiated gemstones*, U.S. Nuclear Regulatory Commission, Viewed 16 September 2016, http://www.nrc.gov/reading-rm/doc-collections/fact-sheets/irradiated-gemstones.html

Radon: The Radioactive Gas in Your Home

Abdel-Wahab, M, Youssef, S, Aly, A, el-Fiki, S, el-Enany, N & Abbas, M 1992, 'A simple calibration of a whole-body counter for the measurement of total body potassium in humans', *International Journal of Radiation Applications and Instrumentation. Part A. Applied Radiation and Isotopes*, vol. 43, pp. 1285–1289.

Air Chek Inc. 2009, *What is a safe and acceptable level of radon gas?*, Air Chek Inc., Viewed 2 October 2016, http://www.radon.com/radon/radon_levels.html

National Cancer Institute 2011, *Radon and cancer*, National Cancer Institute, Viewed 2 October 2016, http://cancer.gov/cancertopics/factsheet/Risk/radon

US Environmental Protection Agency 2016, *Radiation protection*, Viewed 2 October 2016, http://www.epa.gov/rpdweb00/radionuclides/radon.html

US Nuclear Regulatory Commission 2015, *Backgrounder on biological effects of radiation*, United States Nuclear Regulatory Commission, Viewed 2 October 2016, http://www.nrc.gov/reading-rm/doc-collections/fact-sheets/bio-effects-radiation.html

The Radioactive Banana

IAEA, European Commission 1998, *Safety of radiation sources and security of radioactive materials*, Interpol, World Customs Organization, Duon, France, 14–18 September 1998, IAEA, www-pub.iaea.org.

New Mexico State University 2005, *Potassium-40*, Argonne National Laboratory, Viewed 15 September 2016, http://phi.nmsu.edu/%7Epvs/teaching/phys593/potassium.pdf

Post, W 2011, *Facts and information about radiation exposure*, The Energy Collective, Viewed 15 September 2016, http://theenergycollective.com/willem-post/53939/radiation-exposure

Strom, DJ, Lynch, TP & Weie, DR 2009, *Radiation doses to Hanford workers from natural Potassium-40*, U.S. Department of Energy, Viewed 15 September 2016, http://www.pnl.gov/main/publications/external/technical_reports/PNNL-18240.pdf

United States Nuclear Regulatory Commission 2016, *Lethal dose (LD)*, United States Nuclear Regulatory Commission, Viewed 15 September 2016, http://www.nrc.gov/reading-rm/basic-ref/glossary/lethal-dose-ld.html

9

Chemistry Is Explosive!

For some reason, most chemistry teachers do not talk about explosives with their students. Similar to illicit drugs, explosives are a topic in chemistry that seems to be avoided. All of the explosives you know – such as TNT (trinitro-toluene) or C-4 – have been made by a chemist. I once applied for a job at a company that, among other things, made high-power explosives for the government. During orientation, we drove past huge concrete bunkers where, my host explained, they made the "energetic materials." Creating explosives is easier than you might imagine. In fact, chemists need to keep a watchful eye to avoid accidently making something explosive. A prime example is ethers, which have the tendency to react with light to form explosive peroxide compounds. To prevent this, ethers need to be stored in metal containers to keep them from being exposed to light. In addition, ethers that have been stored for long periods of time are routinely tested for the presence of peroxides. At one of the schools where I taught, an ancient bottle of an ether was found. The bottle was so old that the liquid ether had evaporated away to leave a large amount of solid, highly explosive peroxide compound. The situation was considered so dangerous that the police were called, and a bomb disposal robot utilized to remove the bottle! By the way, an explosion itself is a chemical process that can be discussed and understood. Questions arise, such as what aspects of a molecule make some explosive and others not? And what creates the destructive force in an explosion? The answers to these questions lie in the study of chemistry.

How Do Bullets Work?

It all starts with a chemical reaction. Gunpowder, also commonly called "black powder," is a mixture of sulfur, charcoal, and potassium nitrate. Although the exact amounts vary, the most common modern proportions in gunpowder by weight are 75% potassium nitrate (saltpeter), 15% charcoal, and 10% sulfur.

Strange Chemistry: The Stories Your Chemistry Teacher Wouldn't Tell You, First Edition. Steven Farmer.
© 2017 John Wiley & Sons, Inc. Published 2017 by John Wiley & Sons, Inc.

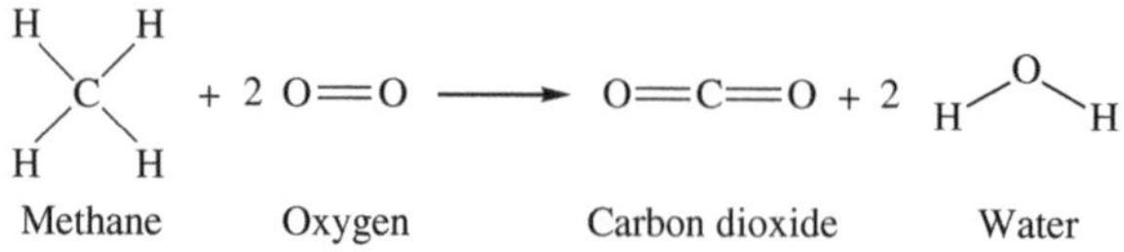

$$H_2C(H)(H) + 2\ O{=}O \longrightarrow O{=}C{=}O + 2\ H{-}O{-}H$$

Methane   Oxygen   Carbon dioxide   Water

Scheme 9.1 The combustion of methane.

The chemical process inside a fired bullet is easier to understand if we first look at the *combustion* reaction of methane. Methane, the fuel, reacts with oxygen, the oxidizer, to produce combustion products (Scheme 9.1).

In this case, the oxygen molecule usually comes from the air and is an essential part of the combustion. Remove the oxygen, and the combustion comes to a halt. Combustions are not usually associated with explosions because their rate of reaction is controlled by the amount of oxygen that is obtainable from the air. Only so much oxygen can be utilized at a time, so the reaction proceeds at a relatively slow rate. You may have noticed how a fire burns more intensely if you blow on it. This is because you are increasing the amount of oxygen in the combustion. Think of it this way: if all the energy contained in a burning log were to be given off in a second rather than hour, the results would be quite spectacular.

So, why is so much energy produced during a combustion reaction? The energy given off during a reaction is related to the strength of the chemical bonds broken and formed. To put it simply, if the bonds formed during the reaction are stronger than the bonds broken, energy is released. When looking at the combustion of methane, we can see that four C—H and two O=O bonds are broken, while two C=O and four O—H bonds are formed. Given that the O—H bond is stronger than the C—H and the C=O bond is stronger than the O=O, we can see that the bonds formed are stronger than those broken, and thus, energy is released.

So what is the difference between a combustion and an explosion reaction? An explosive typically contains nitrogen atoms. During an explosion, the nitrogen atoms recombine to form a triple bond in a $N_{2(g)}$ molecule (the subscript (g) indicates nitrogen *gas*), and this is one of the most stable bonds known. These highly stable bonds cause a tremendous amount of energy to be emitted. On a practical level, racecar mechanics inject nitrous oxide into the cars' engine to put nitrogen into the combustion reaction and increase the energy output. In fact, the triple bond in N_2 is so stable that it is actually difficult to get it to undergo *any* reaction (see the section titled "What is the Most Important Chemical Reaction?" in Chapter 2) (Figure 9.1).

Modern high-powered explosives tend to be organic molecules that contain nitrogen, carbon, and oxygen. The fuel and oxidizer required for the reaction to

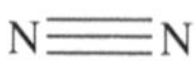

N≡≡≡N   **Figure 9.1** The structure of a N_2 molecule.

occur are already present in the molecules, so oxygen from the air is not necessary. This means that the reaction is only limited by how fast the energy can transfer through the material which can take as little as a millisecond. The virtually instantaneous expulsion of energy is part of what makes these reactions "Explosive." In addition, not requiring oxygen allows explosives to work without the presence of air, even underwater (Figure 9.2). In the case of gunpowder, the exact chemical reaction is fairly complex, but a simplified version is shown in Scheme 9.2.

In this reaction, the sulfur and charcoal act as fuels, and the potassium nitrate is an oxidizer. The speed of the reaction and the energy given off are both important parts of why a bullet works, but the most important factor is pressure. When looking at the gunpowder reaction, solid reactants are transformed into gaseous products (N_2 & CO_2). Because this happens so quickly, and at high temperatures, a tremendous amount of pressure is produced. This is the source

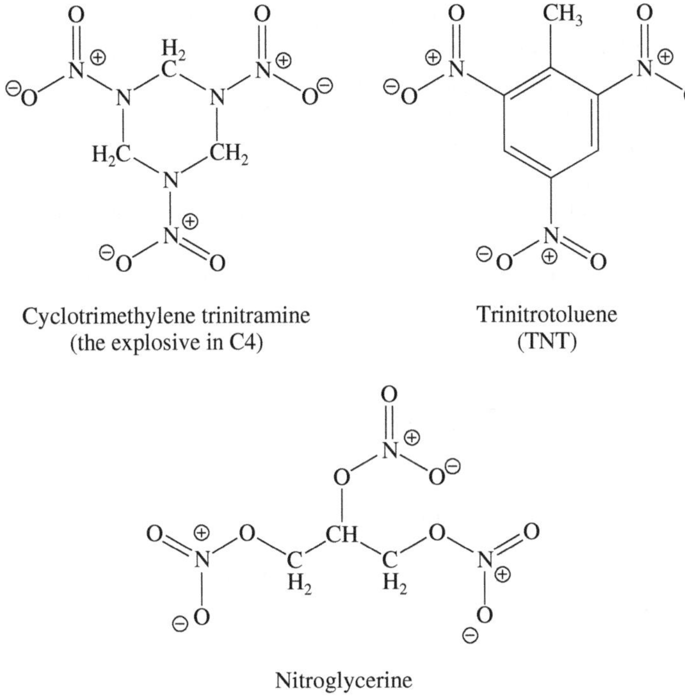

Cyclotrimethylene trinitramine
(the explosive in C4)

Trinitrotoluene
(TNT)

Nitroglycerine

Figure 9.2 The structures of C4, TNT, and nitroglycerin.

$$2\,KNO_{3(s)} + S_{(s)} + 3\,C_{(s)} \rightarrow K_2S_{(s)} + N_{2(g)} + 3\,CO_{2(g)}$$

Scheme 9.2 The chemical reaction of gun powder.

Figure 9.3 A bullet being pushed to over 1000 mph by the mighty N_2 molecule.

of the "blast wave," the primary destructive force of an explosive and the force driving the velocity of a fired bullet. The conversion of a solid or a liquid into a gas results in a tremendous increase in volume. In turn, the increased volume creates pressure that can be harnessed to do work. A steam engine runs because of the pressure created by the conversion of liquid water into steam. Even the combustion engine in an automobile is driven, in part, by the conversion of gasoline and oxygen into gaseous products such as CO_2 that cause the pistons and crankshaft to move.

A 0.45-caliber bullet has about 100 grains of gunpowder. The *grain* is a unit of weight equaling 64.799 mg, meaning that the bullet contains about 6.5 g of gunpowder. Upon combustion, this amount of gunpowder produces about 18.9 l of gas – nearly 3000 times more volume than the solid. Now, imagine the amount of pressure this large volume would create if it were compressed into the small shell of a bullet. This can create pressures over 1000 times that of earth's atmosphere or 14,700 psi (pounds per square inch). For comparison, compressed gas cylinders have pressures of about 1500 psi and a car tires has pressures of about 30 psi. This pressure causes the bullet to be ejected with speeds over 1500 feet per second or over 1000 miles per hour! (Figure 9.3).

What Is the Most Commonly Used Explosive in North America?

The concept of using ammonium nitrate as a commercial explosive was developed in 1956 by Dr Melvin A. Cook, and *ANFO* (*ammonium nitrate/fuel oil*, a widely used bulk industrial explosive mixture), was developed soon after. Liquid forms of explosives, called a *slurry*, proved an invaluable tool for companies extracting ores from open-pit mines because they could be poured into drilled blasting holes. ANFO, unlike dynamite and other nitroglycerin-based materials, is highly stable, can be poured where needed, and is unaffected by the presence of water.

It was discovered in 1955 that mixtures of ammonium nitrate and fine coal dust would give very satisfactory blasting results, and the reaction was improved on by replacing coal dust with fuel oil, for example, diesel and heating oil. This ANFO mixture accounts for an estimated 80% of the roughly

5,500,000,000 pounds of explosive material used annually in North America. ANFO is mainly utilized in the mining and demolition industries.

Ammonium nitrate can actually be found as a natural mineral in the Atacama Desert in Chile; however, almost 100% of the chemical used today is made synthetically. Roughly one-sixth of the ammonium nitrate produced globally goes to producing ANFO explosives.

Ammonium nitrate is actually a fertilizer, so it may seem odd that it is a big player in the discussion of explosive material. The main reason is the nitrogen present in this compound. When we look at the basic chemistry of an ANFO detonation, we see it is a reaction of ammonium nitrate (NH_4NO_3) with a long-chain hydrocarbon (C_nH_{2n+2}) to form nitrogen, carbon dioxide, and water. The normal proportion of the starting materials is two US quarts of fuel oil per 50 pounds of ammonium nitrate. The explosive efficiency of ANFO is approximately 80% that of TNT (Scheme 9.3).

In 2010, ammonium nitrate accounted for about 15% of all nitrogen fertilizers used worldwide, and farmers might each use up to 50 tons per year to fertilize their fields. A 25-kg bag of ammonium nitrate sells for around $20 to a typical consumer, but farmers get it much cheaper due to the bulk amounts they buy. Unfortunately, the relative availability and low cost of both ammonium nitrate and fuel oil has made ANFO popular with terrorists. ANFO was used in the 1993 bombing of the World Trade Center. In addition, roughly 4800 pounds of ANFO was used by Timothy McVeigh to detonate a truck bomb in front of the Alfred P. Murrah Federal Building in Oklahoma City on 19 April 1995. This attack, considered by the US government to be the deadliest act of terrorism in the country prior to the 9/11 attacks, killed 168 people and injured over 600.

Ammonium nitrate fertilizer is the basic constituent of over 90% of the homemade bombs used in Afghanistan and accounts for 80% of US casualties. NATO has estimated that as little as 5% of the nitrate fertilizer entering Afghanistan goes to legitimate use. To combat this problem, ammonium nitrate fertilizer has been made illegal in Afghanistan. Other countries, including Colombia, Denmark, the Philippines, China, and Algeria, have also banned ammonium nitrate fertilizer because of it bomb-making potential. In the United States, the Department of Homeland Security regulates the sale and transfer of ammonium nitrate to prevent its misappropriation or use in an act of terrorism.

Ammonium nitrate was involved in what is generally considered the worst industrial accident in US history, which occurred in 1947 when a fire on the

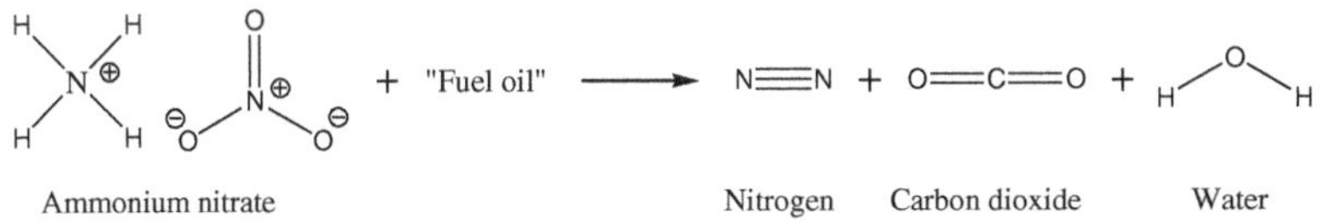

Scheme 9.3 The unbalanced chemical reaction of ANFO.

cargo ship *SS Grandcamp* detonated 2300 tons of ammonium nitrate. The explosion was powerful enough to blow two planes out of the sky and trigger a chain reaction that detonated nearby refineries and neighboring cargo ships. The explosion killed approximately 600 people and injured about 3500 more. Even as recently as 17 April 2013, an explosion at a West, Texas, ammonium nitrate fertilizer plant – later declared a criminal act – injured more than 100 and killed an estimated 5–15 people.

The illicit use of ammonium nitrate explosive is such a problem that manufacturers have started looking for preventative measures. One possible solution is to add iron sulfate to the fertilizer. The iron sulfate would have no effect on the fertilizer, and if someone tried to add fuel oil, the iron would grab the nitrate from ammonium nitrate to form iron nitrate. Similarly, the sulfate would grab the ammonium to form ammonium sulfate. Both ammonium sulfate and iron nitrate are not explosive, even when mixed with a fuel.

What Non-nuclear Substance Is the Most Explosive?

Octanitrocubane (*ONC*) was first synthesized in 1999 by Philip Eaton and Mao-Xi Zhang at the University of Chicago. It was soon discovered that ONC ranks number one as the most explosive non-nuclear substance. Part of the explosive power of ONC comes from its cube-shaped orientation of carbons, called *cubane*. The cube shape forces the carbon atoms to be at 90° angles to each other. The carbon atoms' preferred configuration is to be at 109.5° angles to each other, and this compression to a 90° orientation causes increased energy in the molecule due to "*angle strain.*" This idea is similar to a spring that is bent from its preferred shape. Releasing the spring causes a discharge of energy as it returns to its favored shape. Similarly, molecules that contain a cubane base configuration release energy when a reaction occurs that allows the carbon atoms to assume their preferred angles. In fact, this orientation of carbons was so energetically unfavorable that it was thought cubane would be impossible to make prior to 1964, when it was first synthesized. Similar to cubane, ONC was originally considered impossible to make.

Another reason for the highly energetic character of ONC is the great number of nitro groups per carbon atom. *Nitro* groups (NO_2) are often present in explosives, such as TNT, because they increase the nitrogen and oxygen content of the molecule. ONC contains the cubane framework but with nitro groups instead of the eight hydrogen atoms. As seen by the ONC detonation reaction $C_8(NO_2)_8 \rightarrow 8\,CO_{2(g)} + 4\,N_{2(g)}$, no oxygen is needed as part of the reaction. As discussed previously in relation to explosives, the formation of the stable CO_2 and N_2 molecules releases a tremendous amount of energy. In addition, 12 molecules of gas are released for every 1 ONC molecule, resulting in a powerful shock wave (Figure 9.4).

Figure 9.4 A comparison of the structures of cubane and octanitrocubane.

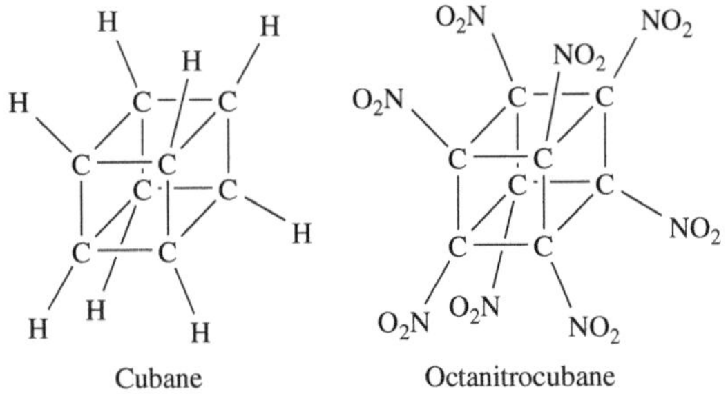

Cubane   Octanitrocubane

Table 9.1 A comparison of the R.E. factors of various explosives.

Explosive	R.E. factor
TNT	1.00
Ammonium nitrate	0.42
Black powder	0.55
Nitroglycerin	1.50
C4	1.34
Octanitrocubane	2.38

Explosives are measured by a *relative effectiveness factor* (*R.E. factor*), which is a measure of an explosive's power relative to TNT by mass. ONC, with an R.E. factor of 2.38, is more than twice as powerful as the compound used in the explosive C-4, which has an R.E. factor of 1.33. ONC is a high-velocity explosive and has a detonation velocity of 10,100 m/s. This makes ONC the fastest known explosive (Table 9.1).

What Poison Is Used as an Explosive in Airbags?

The first airbag patent for use in cars was issued in 1953, but it was not until 1997 that an American Congressional law was established requiring all new passenger cars be equipped with them. The airbags in most vehicles sold in the United States contain about 300 g of a toxic compound called *sodium azide*. In total, the United States uses over 5 million kg of sodium azide per year (Figure 9.5).

Sodium azide is a colorless salt with the chemical formula NaN_3, and it is used as an explosive in airbags. On vehicular impact, the sodium azide is detonated and produces nitrogen gas, which fills the airbags. The nitrogen gas released

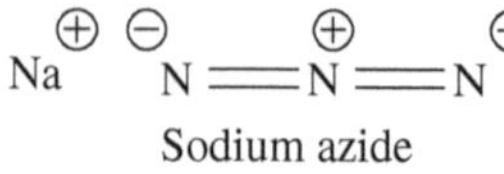

Figure 9.5 The structure of sodium azide.

Na   N≡N≡N

Sodium azide

$$2NaN_3 \rightarrow 2Na + 3N_2$$

Scheme 9.4 The reaction of sodium azide.

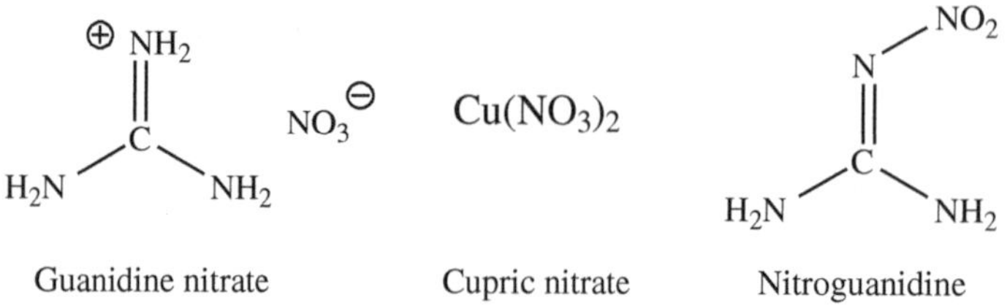

Guanidine nitrate   Cupric nitrate   Nitroguanidine

Figure 9.6 The structures of potential sodium azide substitutes.

inflates the airbag in less than 50 milliseconds and at an average speed of about 150 mph. The basic chemical reaction is shown in Scheme 9.4.

Along with the nitrogen gas, numerous other chemical by-products are released, including carbon dioxide, sodium oxide, and potassium oxide. These gaseous compounds create a potentially dangerous aerosol. Sodium and potassium oxides react with water to form the corresponding hydroxides, which are very powerful bases. There have been documented cases of people receiving chemical burns after an airbag deployment.

In addition to being an explosive, sodium azide is a poison comparable to cyanide in toxicity. Only 0.7 g of it ingested or absorbed through the skin can kill an adult human. Amounts even smaller can cause a variety of symptoms including dizziness, vomiting, convulsions, and death.

Sodium azide kills by preventing the cells of the body from using oxygen, in a process similar to suffocation by carbon monoxide. Simply stated, when cells cannot use oxygen, they die. All cases of fatality due to sodium azide toxicity were the result of respiratory and/or heart failure. In addition, sodium azide changes into a toxic gas, called *hydrazoic acid*, when it comes into contact with water. Furthermore, it can form explosive compounds when it come into contact with solid metals, for example, if it is poured into a drainpipe containing lead or copper. In particular, lead azide $Pb(N_3)_2$ is used as an initiating explosive in blasting caps. The accumulation of azide in laboratory apparatus and drains, where it can react with lead fixtures, has caused explosions when maintenance work has been attempted.

Because of the toxicity of sodium azide, other compounds are currently being considered to inflate airbags. These compounds include guanidine nitrate, cupric nitrate, and nitroguanidine (Figure 9.6).

The toxicity of sodium azide presents an environmental concern when automobiles containing nondeployed airbags are disposed of in such a way that the compound can leach into rivers and drinking water. In fact, sodium azide has been found in groundwater near three manufacturing sites in the United States, resulting in several multimillion-dollar civil lawsuits. Therefore, the Environmental Protection Agency designates sodium azide as a commercial product that is also a hazardous waste when discarded. Typically, airbags are now removed from autos prior to their disposal.

The deadly nature of sodium azide has shown up in the news a number of times. In 1995, a worker at an airbag manufacturer was killed by a sodium azide explosion. In 1996, an overturned truck carrying sodium azide released a toxic gas plume near Salt Lake City, Nevada. In 2000, an Arizona woman was convicted and received the death penalty for murdering her husband by using sodium azide to poison him. Lastly, in April 2010, five people became ill after consuming iced tea from a self-serve urn tainted with sodium azide. How or why the tea became contaminated with sodium azide was never determined.

Explosive Heart Medicine

Nitroglycerin in its most common form is a dense, colorless, oily liquid that detonates if heated to 218 °C or subjected to mechanical shock. It was first made in 1847 by the Italian chemist Ascanio Sobrero, who reacted glycerin with a mixture of sulfuric and nitric acids, noting at the time that the process gave him a powerful headache. *Glycerin* is a substance readily isolated from fats (see the discussion of fats elsewhere in this book). Nitroglycerin was found to be a forceful explosive and difficult to handle because of the ease with which it detonated. Nitroglycerin quickly became popular because it was the first practical explosive produced that was stronger than black powder (gunpowder). In 1867, Alfred Nobel received the patent for a combination that made nitroglycerin safer to handle. Later named *dynamite*, the explosive nitroglycerin base was stabilized by the addition of absorbents such as clay or diatomaceous earth and a small amount of sodium carbonate antacid.

As discussed previously, explosives require an oxidant and a fuel. For nitroglycerin, the nitro groups act as an oxidant while the glycerin backbone acts as a fuel. Similar to most explosives, nitroglycerin contains oxygen, nitrogen, and carbon, and a tremendous amount of energy is released during combustion, which yields product molecules with strong, stable bonds. What makes nitroglycerin particularly explosive is the speed of the detonation. Black powder can only explode as fast as the flame is able to move through the material. Nitroglycerin molecules react virtually simultaneously by a supersonic shock wave started by the first part of the explosion. The speed of reaction coupled

$$4\,C_3H_5(ONO_2)_{3(l)} \rightarrow 12\,CO_{2(g)} + 10\,H_2O_{(g)} + 6\,N_{2(g)} + O_{2(g)}$$

Scheme 9.5 The chemical reaction of nitroglycerin.

with the large number of gaseous product molecules gives extreme force to this shock wave (Scheme 9.5).

Following the discovery in 1887 that amyl nitrite helped alleviate chest pain, Dr. William Murrell experimented with the use of nitroglycerin to reduce blood pressure and alleviate a condition called *angina pectoris*, the medical term for chest pain or discomfort due to coronary heart disease. Angina pectoris is caused by an inadequate flow of blood and oxygen to the heart. Nitroglycerin alleviates this by causing *vasodilation*, or widening of the blood vessels, to increase blood flow to the heart. Today, nitroglycerin is one of the oldest drugs still used to prevent angina pectoris attacks (Figure 9.7).

Amyl nitrite was first synthesized in 1844, at which time it was noticed that it caused facial flushing due to the dilation of blood vessels. In light of this effect, it was theorized to possibly dilate the coronary blood vessels that feed the heart and thus relieve angina pectoris. Once this was proven, amyl nitrite was quickly introduced as a treatment for angina. However, there were severe side effects. The similarity between the side effects of exposure to nitroglycerin and amyl nitrite led Murrell to believe that nitroglycerin could also be used as a treatment for angina. He used pills made from nitroglycerin dissolved in cocoa butter to successfully treat angina but with fewer side effects than amyl nitrite. Aware that nitroglycerin was an explosive, he tested his pills by hammering and burning samples to prove that they were safe. Despite nitroglycerin being used for over 100 years, it was not until the 1990s that it was determined that these effects actually arise when nitroglycerin is converted to nitric oxide (NO), a potent vasodilator.

Medical nitroglycerin (or glyceryl trinitrate, GTN) comes in forms of tablets, sprays, or transdermal patches, with a typical dose being less than 1 mg. Infrequent exposure to nitroglycerin can cause severe headaches. For workers in nitroglycerin manufacturing facilities whose exposure is regular, a phenomenon called "Monday morning headache" can result. During the week, they develop a tolerance for the vasodilating effects of nitroglycerin

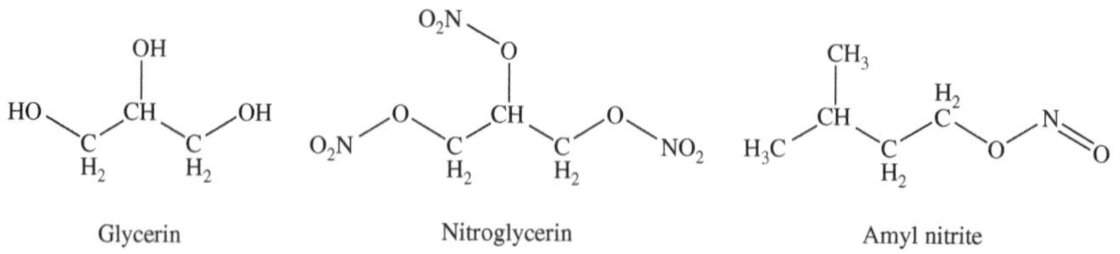

Figure 9.7 A comparison of the structures of glycerin, nitroglycerin, and amyl nitrite.

Figure 9.8 An unusual source of nitroglycerin.

but lose their tolerance over the weekend. Re-exposure on Monday results in the return of the headaches. It is likely that the headaches are caused by the dilation of blood vessels in the brain, which is believed to be the cause of migraine headaches.

Typical side effects of nitroglycerin contact include headache, dizziness, lightheadedness, nausea, and flushing. It has been estimated that the lethal oral dose of this medicine in humans is 200 mg, although some people have survived doses of 1200 mg with no apparent ill effects. There is a documented case where an arborist began to show signs of nitroglycerin overdose after a week of using dynamite to clear stumps. The constant handling of dynamite caused a pair of cotton gloves to become impregnated with nitroglycerin that was subsequently absorbed through the skin of his hands (Figure 9.8).

Further Reading

How do Bullets Work?

Akhavan, J 1998, *The Chemistry of Explosives*, The Royal Society of Chemistry, London.

Ballistics101 2013, *45 ACP ballistics chart*, Ballistics101, Viewed 7 October 2016, http://www.ballistics101.com/45_acp.php

G&A Staff 2011, *Inside the .45 GAP*, Shooting Times, Viewed 7 October 2016, http://www.shootingtimes.com/ammo/ammunition_new_45/

Pauling, L 1962, 'The carbon-carbon triple bond and the nitrogen-nitrogen triple bond', *Tetrahedron*, vol. 17, pp. 229–233.

UC Santa Barbara 2016, *Properties of atoms, radicals, and bonds*, UC Santa Barbara, Viewed 7 October 2016, https://labs.chem.ucsb.edu/zakarian/armen/11—bonddissociationenergy.pdf

The is the Most Commonly Used Explosive in North America

Crowell Moring 2006, *Explosives regulation in the USA, Crowell Moring*, Viewed 24 September 2016, https://www.crowell.com/documents/DOCASSOCFKTYPE_ARTICLES_408.pdf

Cullison, A & Trofimov, Y 2010, *Karzai bans ingredient of Taliban's roadside bombs*, The Wall Street Journal, Viewed 24 September 2016, http://www.wsj.com/articles/SB10001424052748703822404575019042216778962

Homeland Security Newswire 2011, *Materials for fertilizer bombs not regulated*, Homeland Security Newswire, Viewed 24 September 2016, http://www.homelandsecuritynewswire.com/materials-fertilizer-bombs-not-regulated

Pappas, S 2013, *Texas explosion echoes worst industrial accident ever*, NBC News, Viewed 24 September 2016, http://www.nbcnews.com/id/51584341/ns/technology_and_science-science/t/texas-explosion-echoes-worst-industrial-accident-ever/#.VEmL0GfYuSo

What Non-nuclear Substance is the Most Explosive?

Astakhov, AM, Stepanov, RS & Babushkin, AY 1998, 'On the detonation parameters of octanitrocubane', *Combustion Explosion and Shock Waves*, vol. 34, pp. 85–87.

Department of the Army 2007, *Explosive and demolitions*, Field Manual No. 3-34.214 (FM 5-250), Department of the Army.

Dhumal, NR, Patil, UN & Gejji, SP 2004, 'Molecular electrostatic potentials and electron densities in nitroazacubanes,' *Journal of Chemical Physics*, vol. 120, pp. 749–755.

Eaton, PE & Zha, M 2001, 'Octanitrocubane: A new nitrocarbon', *Propellants, Explosives, Pyrotechnics*, vol. 27, pp. 1–6.

Hrovat, DA, Weston, TB, Eaton, PE & Kahr, B 2001, 'A computational study of the interactions among the nitro groups in octanitrocubane', *Journal of the American Chemical Society*, vol. 123, pp. 1289–1293.

Krause, HH 2004, *New Energetic Materials, in Energetic Materials: Particle Processing and Characterization*, Wiley-VCH Verlag GmbH & Co, KGaA, Weinheim, FRG.

Zhang, M, Eaton, PE & Gilardi, R 2000, 'Hepta- and octanitrocubanes', *Angewandte Chemie International Edition*, vol. 39, pp. 401–404.

What Poison is used as an Explosive in Airbags?

Betterton, EA 2003, 'Environmental fate of sodium azide derived from automobile airbags', *Critical Reviews in Environmental Science and Technology*, vol. 33, pp. 423–458.

Betterton, EA, Lowry, J, Ingamells, R & Venner, B 2010, 'Kinetic and mechanism of the reaction of sodium azide with hypochlorite in aqueous solution', *Journal of Hazardous Material*, vol. 182, pp. 716–722.

Centers for Disease Control and Prevention 2012, *Sodium azide poisoning at a restaurant — Dallas County*, Texas, 2010, Centers for Disease Control and Prevention, Viewed 24 September 2016, http://www.cdc.gov/mmwr/preview/mmwrhtml/mm6125a1.htm

Chang, S & Lamm, SH 2003, 'Human health effects of sodium azide exposure: A literature review and analysis', *International Journal of Toxicology*, vol. 22, pp. 175–186.

Corazza, M, Trincone, S & Virgili, A 2004, 'Effects of airbag deployment', *American Journal of Clinical Dermatology*, vol. 5, pp. 295–300.

Lin, SH & Huang, CP 2001, 'Adsorption of hydrazoic acid from aqueous solution by macroreticular resin,' *Journal of Hazardous Materials*, vol. B84, pp. 217–228.

Miljours, S & Braun, CMJ 2003, 'A neuropsychotoxicological assessment of workers in a sodium azide production plant', *International Archives of Occupational and Environmental Health*, vol. 76, pp. 225–232.

Orlando Sentinel 1996, *Hazardous materials spills disrupt Central Utah Towns*, Orlando Sentinel, Viewed 24 September 2016, http://articles.orlandosentinel.com/1996-12-12/news/9612111111_1_central-utah-sulfur-spills-of-hazardous

Explosive Heart Medicine

Beyond Discovery 2003, *From explosives to the gas that heals*, National Academy of Sciences.

Marsh, N & Marsh, A 2000, 'A short history of nitroglycerine and nitric oxide in pharmacology and physiology', *Clinical and Experimental Pharmacology and Physiology*, vol. 27, pp. 313–319.

School of Chemistry 2016, *Nitroglycerin an explosive*, A Vasodilator, University of Bristol, Viewed 6 October 2016, http://www.chm.bris.ac.uk/webprojects2006/ Macgee/Web%20Project/nitroglycerin.htm

The National Institute for Occupational Safety and Health 1994, *Nitroglycerine*, Viewed 6 October 2016, http://www.cdc.gov/niosh/idlh/55630.html

The New York Times 2016, *Nitroglycerin overdose*, The New York Times, Viewed 6 October 2016, http://www.nytimes.com/health/guides/poison/nitroglycerin-overdose/overview.html

10

The Chemistry in *Breaking Bad* and Other Popular Culture

The airing of the television show *Breaking Bad* has given chemistry a whole new reputation. For the first time, a science known as abstract and abstruse – or at the very least, a tough course to pass in school – is explored in terms of its darker, more intriguing applications, such as in the production of illicit drugs, poisons, and explosives. As an organic chemist, I have been inundated with people asking me if I can reproduce some of the chemistry shown in the program. So…could I whip up a batch of methamphetamine, ecstasy, or LSD? Sure! Could I make explosives? Absolutely! Planning and executing chemistry reactions such as these is exactly what you are trained to do as an organic chemist. I like to say that a properly trained organic chemistry could be a menace to society if they wanted to, which is clearly shown in *Breaking Bad*. In fact, a while ago, I was actually offered $10,000 to make a batch of ecstasy for an unsavory friend of a friend. I did not accept the offer course! Organic chemistry has a reputation of being stuffy when, in reality, it is exciting and fundamentally linked to some of the darker things in life.

At one time or another, I have been questioned about virtually all of the chemistry from *Breaking Bad* and several other television shows, movies, and news stories. In this chapter, I hope to satisfy everyone's curiosity to some extent.

How Does Methamphetamine Act as a Stimulant?

In 1919, a Japanese chemist named Nagai Nagayoshi was the first to produce methamphetamine. During World War II, the US military administered it to troops to keep them alert on long missions. Today, methamphetamine is still marketed legally as a prescription drug (Desoxyn®) and used to treat obesity, attention deficit hyperactivity disorder, and narcolepsy. However, since 1971, methamphetamine has been classified as a Schedule II controlled substance, which means that it has currently accepted medical uses in the United States, and also a high potential for abuse. The United Nations World Drug Report calls methamphetamine the most-abused hard drug on earth; the world's 26

Strange Chemistry: The Stories Your Chemistry Teacher Wouldn't Tell You, First Edition. Steven Farmer.
© 2017 John Wiley & Sons, Inc. Published 2017 by John Wiley & Sons, Inc.

million methamphetamine addicts equals the combined number of cocaine and heroin users.

There are 1.4 million methamphetamine users in the United States alone. During the first decade of the 2000s, illicit methamphetamine use reached epidemic levels in this country. From 2002 to 2004, approximately 877,000 Americans tried methamphetamine illegally for the first time. In 2004 alone, there were more than 130,000 reported emergency department visits related to methamphetamine use and production, and authorities responded to over 18,000 reports of illegal methamphetamine laboratories.

To understand how methamphetamine works, we must consider that our brains are made up of nerve cells, or neurons, with spaces called synapses in between them. One neuron communicates with another by sending chemical neurotransmitters across these synapses. Neurotransmitters released at the end of one neuron travel across the synapse and bind to specific receptors on the next neuron, triggering a nerve impulse. All functions of the brain and nervous system are based on the movement of specific neurotransmitters across the synapses.

In the brain, different regions controlling different bodily functions – such as speech, reasoning, or motor coordination – are affected by specific neurotransmitters. The neurotransmitter dopamine plays an important part in the areas of the brain which are responsible for cognition, motivation, voluntary movement, and sleep. Methamphetamine increases the amount of dopamine in the brain, in part, by mimicking the structure of dopamine. Excessive amounts of dopamine overly stimulate these parts of the brain and lead to the effects associated with methamphetamine use, such as insomnia, hyperactivity, and repetitive behaviors. Other molecules, including amphetamine, MDMA (Ecstasy), and methylphenidate (Ritalin®), also mimic the structure of dopamine and have similar effects (Figures 10.1 and 10.2).

More importantly, dopamine stimulates the pleasure system of the brain. Dopamine-based exhilaration is at least partially responsible for sensations of pleasure. Dopamine is released in the brain during enjoyable experiences such as eating, having sex, spending time with loved ones, and receiving

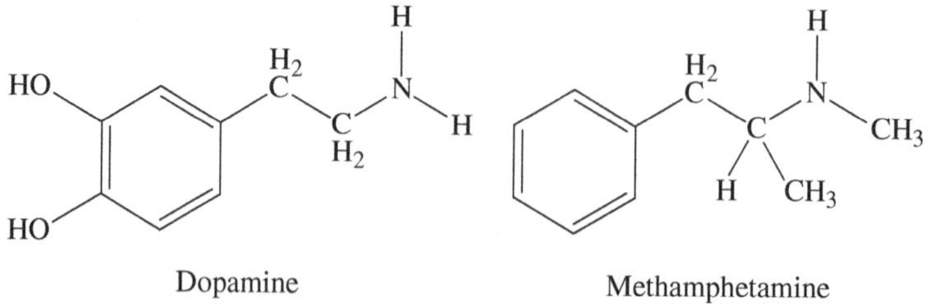

Figure 10.1 A comparison of the structures of dopamine and methamphetamine.

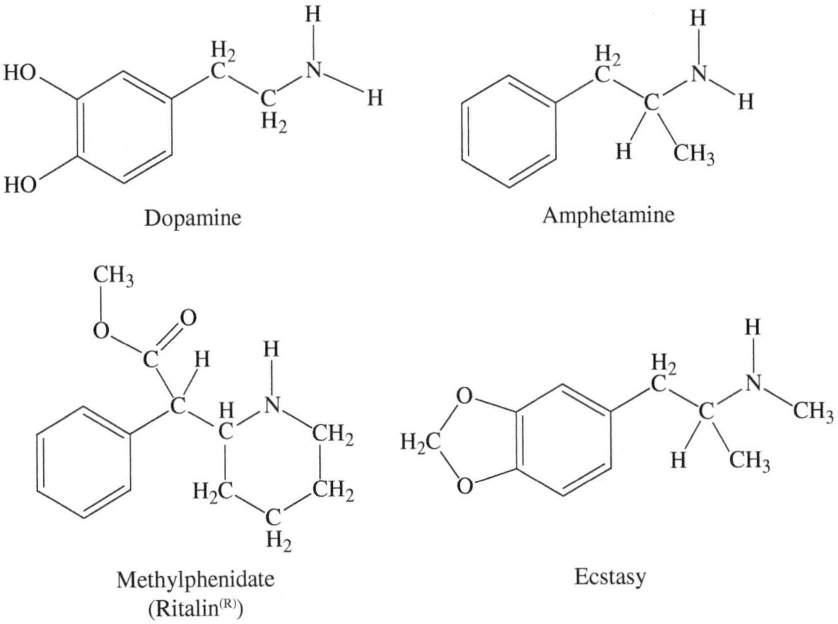

Figure 10.2 A comparison of molecules that are structurally similar to methamphetamine.

compliments. By increasing dopamine levels in the pleasure system of the brain, by more than ten times as natural rewards, methamphetamine creates a euphoric high. This overstimulation of the pleasure system is part of what makes methamphetamine so addicting.

Many mental illnesses are often caused by a "chemical imbalance" of one or more neurotransmitters. In the case of dopamine, excessive amounts are linked to conditions such as hyperactivity, obsessive–compulsive disorder, and schizophrenia. Too little dopamine, in contrast, can result in depression. Treatment of these types of disorders usually involves pharmaceuticals that regulate dopamine levels in the brain.

Schizophrenia is attributed to an overstimulation of the dopamine receptors in the areas of the brain responsible for cognitive control and emotional response. Evidence for the *excess dopamine* theory of schizophrenia comes largely from the fact that some drugs that reduce dopamine's ability to stimulate the brain also alleviate many symptoms of schizophrenia. It should be noted that because many stimulants, such as methamphetamines, stimulate these same portions of the brain they are capable of inducing symptoms that are virtually indistinguishable from schizophrenia. At high doses, methamphetamine can bring about a drug-induced psychosis that causes the user to experience paranoia, confusion, and hallucinations. Fortunately, these psychotic episodes pass when the user stops ingesting methamphetamine and gets some sleep.[1]

Why would methamphetamine be used to treat attention deficit hyperactivity disorder (ADHD)? Dopamine appears to be an important neurotransmitter in the areas of the brain responsible for regulating attention. Amphetamines and methamphetamines stimulate these areas by increasing dopamine levels, thereby allowing the patient to concentrate on a specific task.

What Is "Pseudo," and How Is It Related to Methamphetamine?

Pseudoephedrine is a chemical originally isolated from a plant in 1889. The herb *Ephedra sinica*, known in Chinese as Ma Huang, has been used in traditional Chinese medicine for 5000 years and is a natural source of pseudoephedrine.

Pseudoephedrine is primarily used as a decongestant and is an active ingredient in common over-the-counter medicines such as Sudafed®, Claritin®, Actifed®, Contac®, Zyrtec®, Theraflu®, and Aleve®. As you can see by comparing their chemical structures, pseudoephedrine is very similar to adrenaline. Therefore, it is not surprising that pseudoephedrine-containing formulations also cause side effects of nervousness, restlessness, and insomnia (Figure 10.3).

Pseudoephedrine's principal mechanism of action is its effect on the adrenergic receptor system in nerves. *Adrenaline* is a neurotransmitter and is part of the human body's acute stress response system, also called the "fight or flight" response. Pseudoephedrine and adrenaline affect nerves and therefore the entire body in a similar manner. Adrenaline increases blood flow to the

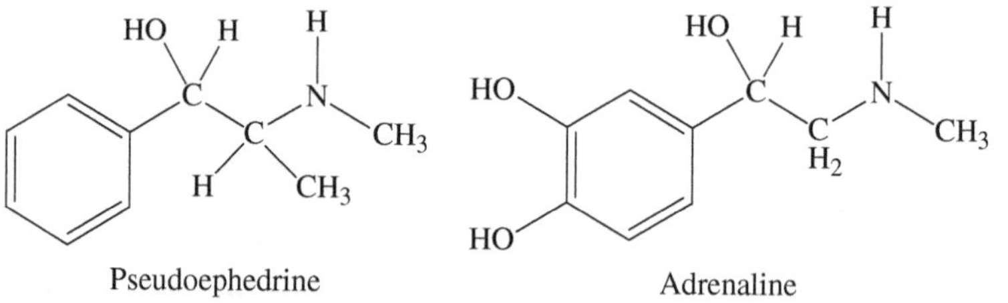

Pseudoephedrine   Adrenaline

Figure 10.3 A comparison of the structures of pseudoephedrine and adrenaline.

---

1 I am often asked which drug of abuse I feel is the most insidious. I admit almost every illicit drug has had some adverse effect on someone I know, but one truly stands out, methamphetamine. It is evil in a crystalline form and I have seen it ruin multiple people's lives. I have felt the pain of realizing a good friend stole money from me to buy it. I have had to try to talk down a friend who was in the grips of methamphetamine psychosis. I saw him crying, saying that the police were chasing him everywhere and hiding his bed. Just think of this if anyone ever offers you some methamphetamine.

muscles and oxygen to the lungs by raising the heart rate, contracting blood vessels (vasoconstriction), and dilating air passages. Pseudoephedrine acts as a decongestant by also causing vasoconstriction. Once constricted, the blood vessels allow less fluid to leave and enter the nose, throat, and sinus linings, which results in decreased inflammation of nasal membranes, decreased mucus production, and a decrease in the symptoms of nasal congestion. It may be hard to imagine but adrenaline would technically act as a decongestant. It is unclear how adrenaline would affect a runny nose but you may have heard of someone's mouth going dry during times of fear or stress.

Because of the similarities in their chemical structures, it is rather simple to convert pseudoephedrine into methamphetamine. Methamphetamine "cooks" have developed many variations on the basic recipe for doing this. The manufacturing route, known as the "Moscow method" (featured in *Breaking Bad*), involves reacting pseudoephedrine with hydroiodic acid (HI), which is produced by reacting iodine with red phosphorus in water (Scheme 10.1).

Using pseudoephedrine to synthesize methamphetamine became such a serious problem in the United States that Congress passed the Combat Methamphetamine Epidemic Act of 2005 (CMEA). This act controls pseudoephedrine sales by imposing purchase limits, placing products behind the sales counter rather than on the shelves, documenting all sales, and verifying customer identification. In fact, Oregon (2006) and Mississippi (2010) now classify pseudoephedrine-containing products as Schedule III substances and are available only by prescription. The CMEA and other legislation had an immediate effect on curtailing illicit methamphetamine manufacturing. The DEA reported 18,091 methamphetamine manufacturing investigations in 2004 but only 8181 in 2006. Furthermore, the number of methamphetamine-related emergency department visits nationwide dropped from 109,655 in 2005 to 79,924 in 2006.

A gaping hole still remaining in the CMEA is the lack of a mechanism for alerting retailers when a customer has purchased pseudoephedrine-containing products from other stores. This loophole has been exploited by methamphetamine cooks in a new tactic called *meth smurfing*. In meth

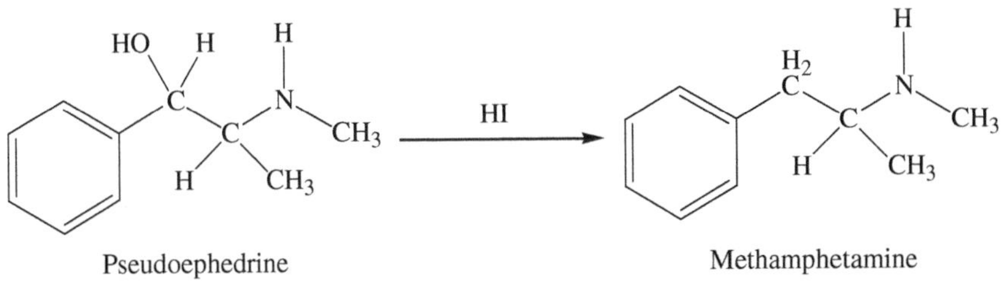

Scheme 10.1 The synthesis of methamphetamine from pseudoephedrine.

smurfing, a group of people buy small amounts of pseudoephedrine from multiple locations and then sell the products to methamphetamine cooks.

The use of pseudoephedrine to synthesize methamphetamine is discussed in the early episodes of *Breaking Bad*. Jesse – who calls pseudoephedrine "pseudo" – circumvents the CMEA purchase limitations by utilizing a large group of "meth smurfing" buyers whom he calls "Smurfs." Fortunately, the CMEA's tactics kept "meth smurfing" from supplying Jesse and Walter with enough pseudoephedrine to make large batches of methamphetamine. Eventually, they resort to a different method for making methamphetamine, which is called the reductive amination method and uses phenyl acetone and methylamine as the starting materials.

The federal government closely monitors virtually all chemicals that can be used to produce methamphetamine. Any business or organization in possession of these chemicals is required by law to protect them from theft or loss. In another episode of *Breaking Bad*, Walter and Jesse must steal a drum of methylamine to supply their manufacturing. In addition, finding a large source of methylamine is a recurring quest in the show's final season (Figure 10.4; Scheme 10.2).[2]

Figure 10.4 Sadly, a true story.

---

2 As part of my work as an organic chemist, I was allowed to obtain a small sample of methylamine. The DEA required that it was kept in a locked cabinet when not in use and a logbook was kept which recorded its use. A representative of the DEA showed up one day to review this logbook and make sure that none of it had "disappeared."

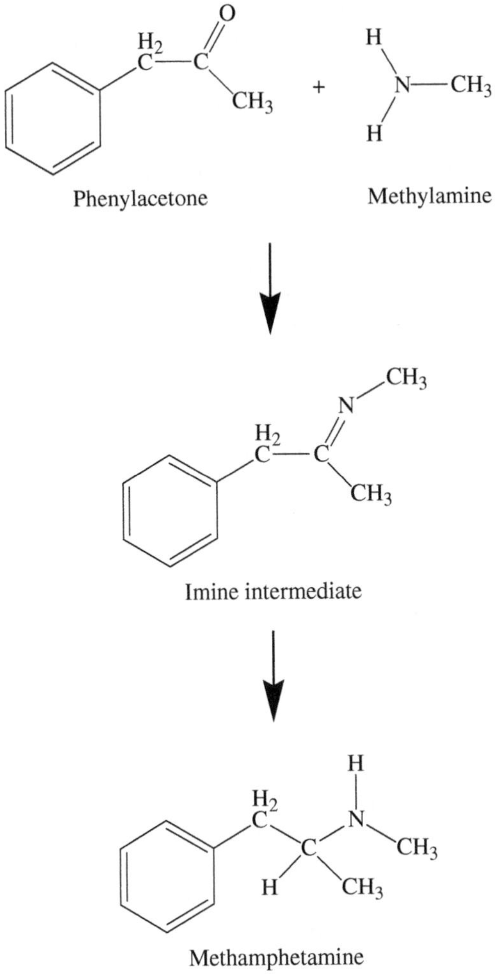

Scheme 10.2 The synthesis of methamphetamine from phenylacetone.

What Is Ricin?

Ricin is a poison that was frequently mentioned in *Breaking Bad*. It appeared repeatedly and was used as a poison in methamphetamine, food, and even tea. Ricin occurs naturally in the beans of the castor plant (*Ricinus communis*), a shrub native to tropical Africa, and is also found growing wild along highways in the United States. The castor plant has green or reddish leaves and produces clusters of seed pods with three seeds per pod. The seeds have a glossy coat and spots that are black, brown, gray, or white. The seeds, or beans, of the castor plant are the source of castor oil, which is used in lubricants, paints, and varnishes, and also as a purgative medicine. Castor oil, however, does not contain ricin.

Although the castor plant was once widely grown, the presence of toxic ricin has limited its cultivation. All parts of the plant are poisonous; however, the highest concentration of ricin is present in the seeds and pods. Ricin levels in the seeds range from 1.16% to 6.25% by weight. The poison, a protein with a molecular weight of about 63,000 g/mol, is one of the most toxic natural substances in the world. When ricin is isolated from the castor bean and turned into a powder, it becomes particularly dangerous; only 500 µg – the weight of a grain of sand – can kill a person. Pure ricin is almost 300 times more toxic than cyanide.

Ricin works by entering the cells of the body and preventing them from making needed proteins. Without these essential proteins, cells die, and eventually the liver, spleen, and kidneys stop working, ultimately resulting in death. Dying from ricin poisoning can occur within 36–72 hours of exposure, depending on the dose taken and whether the poison was inhaled, ingested, or injected. Most of the known fatalities from ricin poisoning are individuals who ate the castor beans. Adults need to eat about 20 beans before the digestive system becomes incapable of destroying the toxin. In some cases, however, death has followed ingestion of only two castor beans.

The symptoms of ricin poisoning include vomiting, diarrhea, dehydration, low blood pressure, blood in the urine, seizures, and eventually death. There is no antidote for ricin poisoning. Spoiler Alert! This is why Walter White was able to tell Lydia Rodarte-Quayle of her ricin poisoning in the chilling *Breaking Bad* series finale. There was nothing she could do about it.

Ricin is more than just a television show prop and has appeared on many other occasions in the media. Perhaps the most notorious case of ricin poisoning occurred in 1978, when Bulgarian writer and journalist Georgi Markov died after being attacked by a man with an umbrella that had been rigged to inject a 2-mm titanium/ricin-containing pellet into Markov's skin.

Because ricin can be powdered, it has been placed in envelopes and mailed. In 2013, the former actress Shannon Richardson sent ricin-laced letters to President Barack Obama and New York City mayor Michael Bloomberg. Richardson, best known for small roles on *The Vampire Diaries* and *The Walking Dead*, confessed to sending the letters in the hope of framing her estranged husband. In July of 2014, she was convicted of producing and possessing a biological toxin and sentenced to 18 years' imprisonment.

The Thalidomide Disaster

In the *Breaking Bad* episode "Cat's in the Bag," Walter White was giving a lecture about chirality in chemistry. Chirality describes how certain molecules can

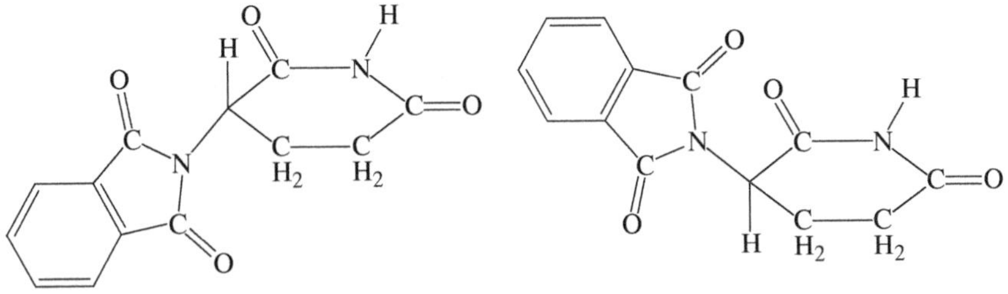

Left-handed thalidomide   Right-handed thalidomide

Figure 10.5 A comparison of the structures of right- and left-handed thalidomides.

come in two forms that are nonsuperimposable mirror images of each other. Thalidomide is an example, as it exists in a right- and left-handed form. White was using thalidomide to show that although these two forms of thalidomide look similar; their biological effects are quite different. The right-handed version of thalidomide is a sedative and antidote to morning sickness; however, the left-handed version is associated with causing birth defects. Unfortunately, when thalidomide is used clinically, it is present in both its versions, so early administration of the drug to patients brought about tragic results (Figure 10.5).

Thalidomide was developed by a German company, Grünenthal, in the 1950s. Grünenthal researchers called thalidomide nontoxic because very high doses did not kill rats, mice, rabbits, cats, or dogs and did not seem to produce any immediate side effects. Although it was originally released as a sedative, it was eventually also prescribed for insomnia, coughs, colds, and headaches. At one point, thalidomide was considered safe enough to be sold over-the-counter. The drug became available for purchase in 45 countries; by 1961, it was the best-selling sedative in Germany. The problems started when thalidomide was marketed as "the drug of choice" for treatment of morning sickness and nausea in pregnant women. Unfortunately, Grünenthal never conducted animal studies investigating the effect of thalidomide on fetuses. It turns out that thalidomide is a *teratogen*, a drug or other substance that interferes with fetal development. In 1961, thalidomide was taken off the market, but not before approximately 10,000 children were born with major birth defects. Affected babies often had missing or deformed limbs that looked like flippers – a phenomenon called *phocomelia*, which means "seal limb." Although many victims of these birth defects died in their first year of life, roughly 50% of them survived. As of 2010, roughly €500 million have been paid to the "children of thalidomide."

During normal embryonic development, cells along the sides of the embryo turn into buds that eventually stretch out to form arms and legs. Thalidomide interferes by killing the developing blood vessels, thereby preventing the limbs from taking their final shape.

In the United States, pharmacologist Frances Oldham Kelsey, M.D. was assigned by the FDA to review an application for thalidomide to be sold in this country. She refused approval, despite intense pressure from the drug's manufacturer, stating that further studies were needed. For her role in keeping thalidomide off the market in 1962, President John F. Kennedy gave her the President's Award for Distinguished Federal Civilian Service. In response to the thalidomide tragedy, in October of 1962, Congress unanimously passed the Kefauver Harris Amendment, which requires companies to demonstrate the effectiveness of new drugs and report adverse reactions to the FDA.

Despite its risks to embryos and fetuses, thalidomide is still being prescribed to treat leprosy, Crohn's disease, and some types of cancer. Of course, female patients are made aware of the dangers of becoming pregnant while taking the drug.

What Is Phosphine Gas, and Why Is It a Potential Murder Weapon?

In the pilot episode of *Breaking Bad*, phosphine gas was introduced at the end, when Walter White heated up a pan of water and threw in a bottle of red phosphorus, creating the gas that instantly poisoned criminals Krazy 8 and Emilio Koyama.

Phosphine (PH_3) is a highly toxic, colorless, and odorless gas (although sometimes it can smell like garlic). Its main application is as an insecticide for the fumigation of grains, animal feed, and leaf-stored tobacco. In addition, it is used as a rodenticide, as well as in the semiconductor industry to introduce phosphorus into silicon crystals (Figure 10.6).

Phosphine gas is extremely poisonous. Human exposure to air that is a mere 0.0035% phosphine gas causes instantaneous nausea and respiratory distress; exposure to air that is 0.02% phosphine gas is immediately life threatening. Phosphine inhibits the critical cellular enzymes and proteins involved in metabolic processes, especially the ones responsible for moving oxygen through the body. It also causes an increase in highly reactive cellular peroxides, causing injury to cells and severe organ damage.

Symptoms of a mild exposure include severe lung irritation, cough, shortness of breath, abnormally low blood pressure, headache, dizziness, drowsiness, fatigue, nausea, and bluish discoloration of the skin (cyanosis). Severe exposure can result in the accumulation of fluid in the lungs (pulmonary edema), collapse of blood vessels, cardiac arrest, coma, and death.

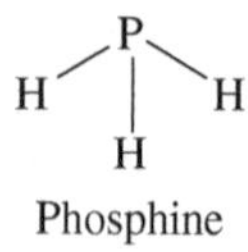

Figure 10.6 The structure of phosphine.

Phosphine

Heating red phosphorus in the presence of certain acids produces phosphine gas. Because the most common ingredients of methamphetamine are red phosphorus, hydroiodic acid, and pseudoephedrine, phosphine gas is an unintended and potentially lethal by-product of meth production. The Centers for Disease Control and Prevention (CDC) reported that between 1996 and 1999, more than 150 first responders and hospital personnel were exposed to phosphine gas from illegal methamphetamine laboratories. In August of 1996, three people working in a methamphetamine laboratory in Los Angeles County died of phosphine poisoning.

Phosphine gas can be produced in a chemical reaction with a phosphide salt such as aluminum phosphide (AlP), magnesium phosphide (Mg_3P_2), or zinc phosphide (Zn_3P_2). The most commonly used reactant is AlP, which produces phosphine when it reacts with water: ($AlP + 3 H_2O \rightarrow Al(OH)_3 + PH_3$). AlP can be used to fumigate grains, seeds, nuts, flours, pasta, cereals, dried fruit, candy, tobacco, clothing, wood, and hay against pests. AlP is also used to kill burrowing rodents such as gophers, ground squirrels, prairie dogs, and moles.

After ingestion, AlP produces phosphine gas by reacting with stomach acid ($AlP + 3 HCl \rightarrow AlCl_3 + PH_3$). The phosphine gas is then absorbed into the body, with lethal consequences. In fact, pesticide ingestion is one of the leading methods of suicide in the world. Every year, about 300,000 people worldwide – more than one-third of all suicides – die from intentional pesticide poisoning. Phosphides, AlP in particular, are among the most common suicide agents used, because they can be easily purchased, and there is no antidote. According to one study, AlP was found to be the most popular means of suicide in Northern India.

Unfortunately, because phosphine gas-generating pesticides are widely used, accidental exposures also occur. The American Association of Poison Control Centers reported 133 such exposures to phosphine gas-generating pesticides in 2012. In February of 2010, two young children in Layton, Utah, were killed by exposure to phosphine gas generated from AlP pellets placed in burrows around their house to kill rodents. The AlP pellets were mixed with water to release phosphine gas, which apparently migrated from the soil into their home.

In another odd circumstance, household pets can be a source of the toxic gas. Animals that ingest Zn_3P_2 insecticide often regurgitate, releasing phosphine gas into the air, and veterinary hospital staff members treating such animals can be poisoned from this exposure. From 2006 to 2011, the CDC reported phosphine gas poisonings at four veterinary hospitals treating dogs that had ingested Zn_3P_2.

Acetylcholine, Pesticides, and Nerve Gas

Acetylcholine (ACh), discovered by Henry Hallett Dale in 1914, was the first of the body's neurotransmitters to be identified. In 1936, Dale went on to be

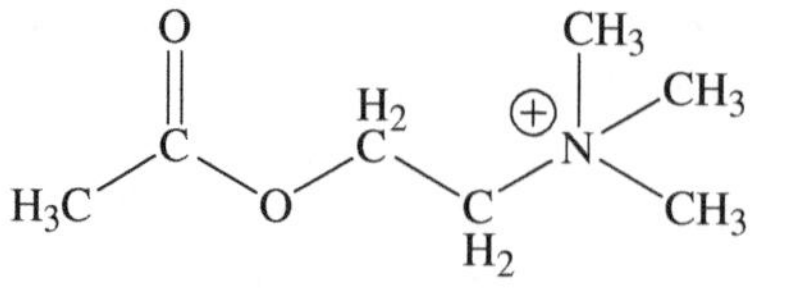

Figure 10.7 The structure of acetylcholine.

Acetylcholine

awarded the Nobel Prize in Physiology/Medicine for this discovery. ACh is the most common neurotransmitter in the body and has a major role in the nervous system. In the peripheral nervous system, ACh is involved in muscle movement, including the cardiac muscle. In the brain, ACh is believed to be involved in learning, memory, dreaming, and mood (Figure 10.7).

Neurons that are affected by ACh are called *cholinergic*. As with other neurotransmitters, arrival of the nerve impulse at the end of a cholinergic neuron causes the release of ACh into the synaptic gap. Subsequently, the ACh diffuses to the next neuron, where it binds to receptor proteins and thereby allows the neuron to generate an electrical potential. For cholinergic neurons to receive another nerve impulse, ACh must be released from the receptor proteins to which it is bound. This only happens if the concentration of ACh in the synaptic gap, or space between the two neurons, is very low. If the concentration of ACh in the synapse remains high enough, the cholinergic neurons cannot fire again, and they are rendered overstimulated and useless. Because of the prevalence of ACh in the body, overstimulation of cholinergic neurons produces a wide variety of severe symptoms, including loss of muscle control, muscle paralysis, excessive salivation, vomiting, diarrhea, excessive sweating, uncontrolled urination, heart failure, respiratory arrest, tremors, convulsions, and death.

The effects of heightened ACh levels were highlighted in an episode of the television show *Eureka* entitled "Noche de Sueno." In the episode, after a truck carrying toxic waste crashes in Eureka, the townsfolk start having shared dreams. Eventually, it is discovered that the shared dreams were causing some of the townsfolk to not sleep, which in turn leads to increased levels of ACh. Eventually, the ACh levels of some of the dreamers become so high that paralysis and even death occurs. The dreamers are saved when it is determined that the shared dreams are caused by an experiment. ACh is often mentioned in the various *Star Trek* television series in relation to its ability to affect muscles or cognitive abilities. In the *Star Trek: the Next Generation* series finale, "All Good Things…" they use Captain Picard's increase in hippocampus ACh levels to show that he had quickly gained 2 days' worth of memories, indicating that he was most likely traveling through time.

In fact, the lethal effect of many poisons is due to their mimicking the structure of ACh, causing overstimulation of cholinergic neurons and thus paralysis and possibly death. *Succinylcholine* (SUX) is structurally similar enough to ACh to also cause muscle paralysis but, when carefully administered, provides

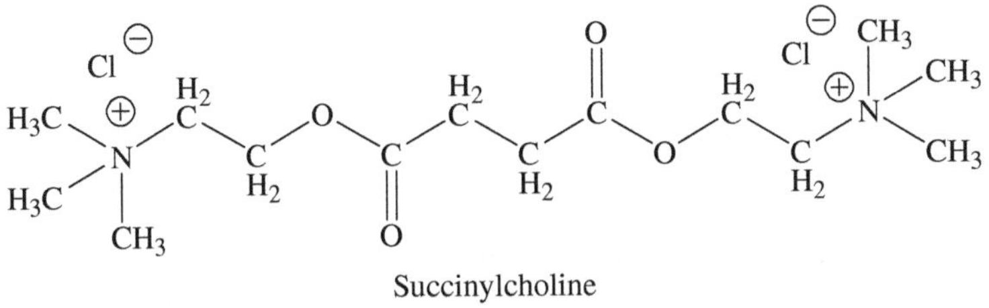

Succinylcholine

Figure 10.8 The structure of succinylcholine.

profound muscle relaxation. SUX is an important medication in the health field because muscle relaxation is required during many medical procedures including intubation of the trachea, treatment of a dislocated shoulder, and many surgeries. The administration of SUX without proper medical supervision, however, can cause death. In fact, SUX is commonly used in the lethal injections for death sentences (Figure 10.8).

For a time, SUX was considered an almost perfect poison for murder because it is quickly broken down by enzymes in the body and therefore very difficult to trace in autopsies. There have been multiple famous examples of the use of SUX to commit murder in the news and various television series. The TV series *Forensic Files* actually discussed two famous real-life cases that involved murders using SUX. The first episode was entitled "Nursery Crimes" and focused on Genene Jones, a former pediatric nurse who killed somewhere between 1 and 46 children in her care using injected drugs, including SUX. Her suspected motive was to induce medical crises in her patients with the intention of reviving them afterward, in order to receive praise and attention. After the suspicious death of a child under Jones' care, an investigation discovered puncture marks in a bottle of SUX at the hospital. Contents of the apparently full bottle were later found to be diluted, implying that some of the SUX had been removed. An autopsy using a newly developed testing method showed that the child's tissues contained SUX. This case is unique because it is the first time the presence of SUX was used to provide a murder conviction. In 1985, Jones was sentenced to 99 years in prison for the murders. The second episode of *Forensic Files* was entitled "Political Thriller." In 2006, Nevada politician Kathy Augustine died mysteriously during a re-election campaign. Augustine had served in the Nevada Senate and was Nevada's first female State Controller from 1999 until she was murdered in 2006. Suspicions were raised when the medical examiner could not explain the cause of her death, or the tiny puncture wounds discovered during the autopsy. Upon investigation, an FBI toxicology test found SUX in Augustine's system. The investigation then focused on Augustine's husband Chaz Higgs, a critical care nurse who had ready access to SUX. In 2007, Higgs was convicted of murder and sentenced to life in prison. SUX also makes an appearance in the seventh season finale of

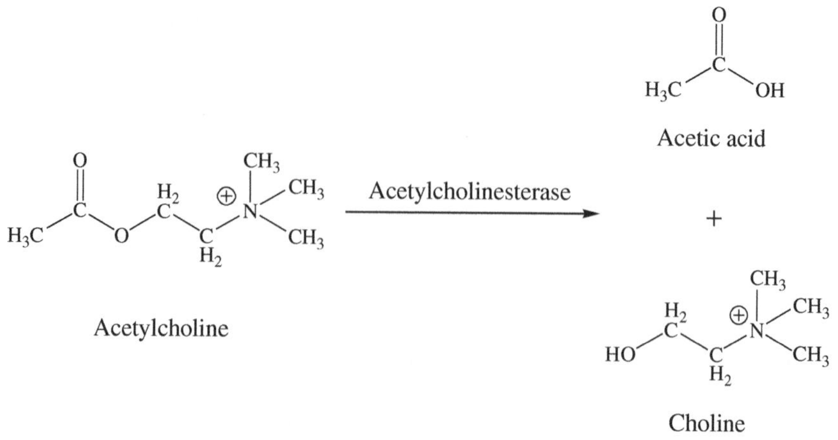

Scheme 10.3 The decomposition of acetylcholine.

the television show *Law & Order: Criminal Intent* entitled "Frame," in which character Nicole Wallace kills Frank Goren with an injection of SUX after having sex with him.

The concentrations of ACh in the body are maintained, in part, by an enzyme called *acetylcholinesterase*. This enzyme catalyzes a reaction that removes ACh from the synaptic gap by converting it into acetic acid and choline. If acetylcholinesterase activity is inhibited, this reaction stops and the synaptic concentration of ACh will become higher than normal. Many poisons induce their toxic effect by inhibiting acetylcholinesterase, leading to an accumulation of ACh (Scheme 10.3).

Organophosphates are molecules that contain the elements carbon and phosphorus. For over 50 years, many organophosphates have been commonly used as insecticides in both households and agriculture. Although organophosphates were first discovered in 1854, their toxicity was not established until the 1930s. The first organophosphate insecticide ever developed was tetraethyl pyrophosphate, which was created in Germany during World War II as part of a nerve gas development program. Organophosphate insecticides became widely used after World War II, with some of the most commonly known compounds being malathion and parathion. As of 2007, organophosphate insecticides held nearly 30% of the world's $16.7 billion insecticides market. Many large chemical companies such as DOW and BASF produce organophosphate insecticides on a large scale. In the United States alone, about 33 million pounds (35% of all insecticides) of organophosphate insecticides were applied in 2007, with chlorpyrifos (DOW trade names: Lorsban® and Dursban®) being the most commonly used. The widespread use and deadly nature of organophosphate insecticides is not without its consequences.

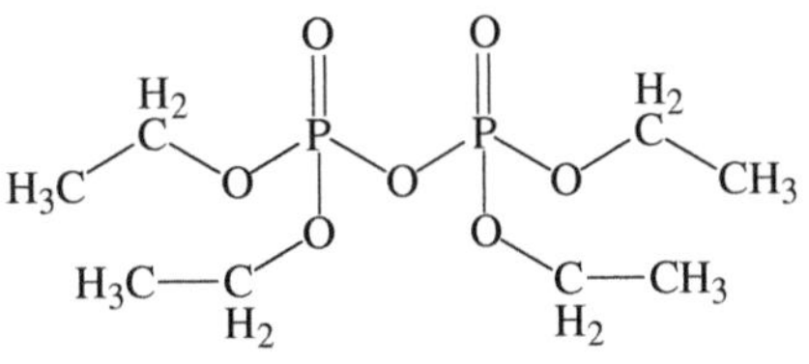

Tetraethyl pyrophosphate

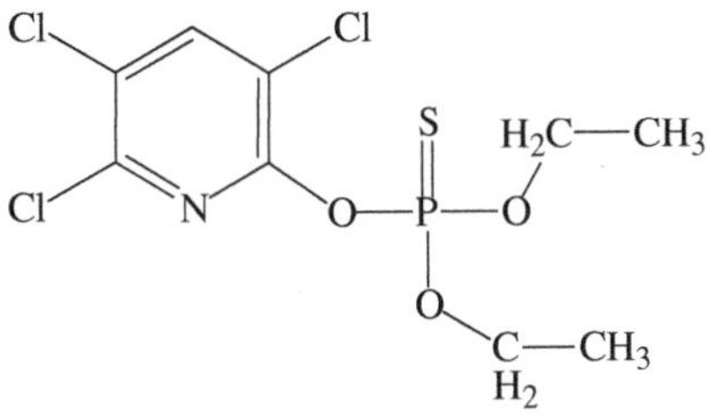

Chlorpyrifos

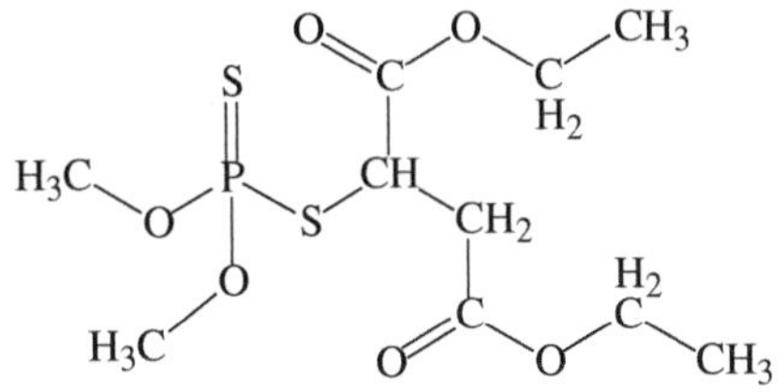

Malathion

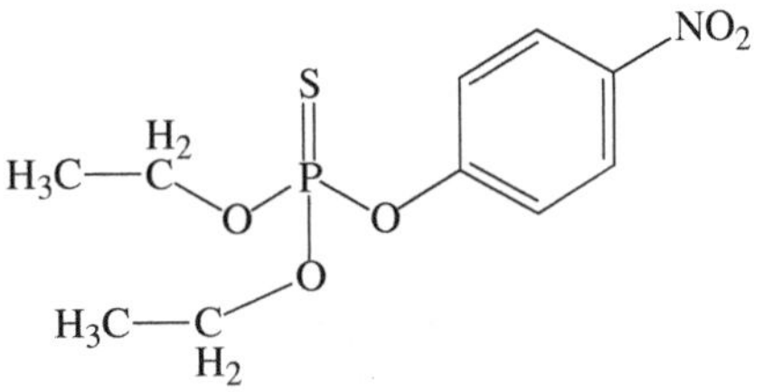

Parathion

Figure 10.9 The structures of some organophosphate insecticides.

The Environmental Protection Agency estimated that the general US population consumes 12 μg of chlorpyrifos per day directly from food residue. In addition, it has been estimated that organophosphate insecticides exposures, either intentional or unintentional, are responsible for approximately 200,000 deaths worldwide (Figure 10.9).

Another example of organophosphates insecticides poisoning was shown in the television show *House MD*, in the episode "Poison." Two high school students become ill with symptoms of low heart rate, abdominal pain, dizziness, and seizures. House thinks it is organophosphate poisoning; however, the typical treatments of atropine and pralidoxime have no effect. Upon investigation, it is determined that the students had bought and worn stolen jeans contaminated with the organophosphate insecticide, phosmet. The two students are cured using an enzyme called a *hydrolase*, which is specific for making acetylcholinesterase active again (Figure 10.10).

Poisoning with organophosphate insecticides has also been portrayed in popular media many times. In an episode entitled "Loophole" of the television show *Law & Order: Special Victims Unit*, a woman is taken to the hospital after being exposed to organophosphate poison. At the hospital, she is told that her cholinesterase levels are half what she should have. She is promptly treated with atropine and is told she should be better in 24 hours.

The development and production of extremely toxic *organophosphate nerve agents* (also called nerve gas) started in pre-World War II Germany in an effort to find novel insecticides. Although Germany had stockpiles of nerve agent

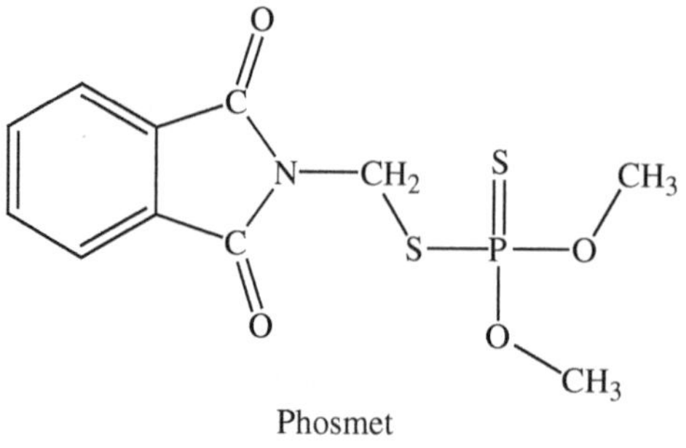

Figure 10.10 The structure of phosmet.

Phosmet

munitions during World War II, it is unclear why they were not used. In the closing days of the war, the United States and its allies discovered these stockpiles and started manufactured nerve agent munitions of their own. The only known use of nerve agents in battle was during the Iraq-Iran conflict. It should be noted that nerve agents have been outlawed by most of the nations in the world with the signing of the Chemical Weapons Convention (CWC) arms control treaty. This treaty outlaws the production, stockpiling, and use of chemical weapons. Almost every country in the world has agreed to this treaty, although there are a few remaining holdouts including Angola, Egypt, North Korea, and South Sudan.

One of the first and probably most well-known nerve gases, *Sarin*, was originally developed as a pesticide in Germany in 1938. By the mid-1950s, intensive research into organophosphate insecticides by Imperial Chemical Industries, a British chemical company, produced a new group of deadly compounds known as the V-agents. The deadliest V-agent was *VX*, the most potent of all known nerve gases. Considered among the most toxic substances ever made by man, VX is approximately 10-fold more poisonous than Sarin and 700 times more toxic than cyanide. In fact, 0.000612 g of VX could be expected to kill an average person. This is less than the weight of a half a grain of rice! VX was actually marketed as an insecticide for a short time before being determined too toxic for safe use. Subsequently, the British government became interested and investigated VX as a possible nerve gas (Figure 10.11).

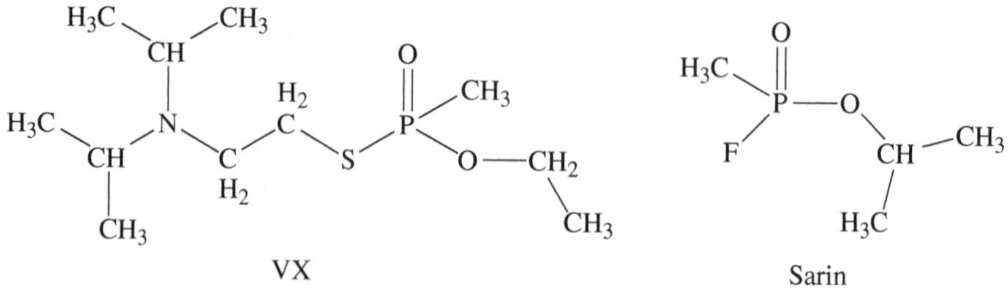

VX   Sarin

Figure 10.11 A comparison of the structures of VX and sarin.

Organophosphates exert their main toxicological effect through chemically reacting with acetylcholinesterase in the central nervous system. This reaction actually changes the structure of acetylcholinesterase, causing it to become inactive. Once the acetylcholinesterase becomes inactive, ACh starts to accumulate in the body and this provides all of the symptoms associated with over-stimulated cholinergic neurons (Scheme 10.4).

The nonpharmacologic treatment for organophosphates poisoning usually involves ventilation. However, pharmacologic treatments are usually also given. The drug *atropine* is considered the cornerstone in the treatment of organophosphate poisoning. Atropine works because it can displace ACh from the protein receptors of cholinergic neurons and block ACh from being bound. This reduces the effects of the accumulated ACh on the cholinergic neurons. In addition, anticonvulsants such as Valium® can be given to diminish muscular seizures. Although these treatments may work, they are both only temporary solutions because the inactivated acetylcholinesterase will inevitably cause more ACh to accumulate. The drug *pralidoxime* is a useful antidote for the toxic effects of acetylcholinesterase inhibitors such as the organophosphates. It has been shown to convert inactivated acetylcholinesterase back to its active form through a chemical reaction. Pralidoxime interacts with acetylcholinesterase especially well because it contains a positively charged nitrogen, just like ACh. Once the acetylcholinesterase has been reactivated, the ACh levels can be maintained at a normal level (Scheme 10.5).

Nerve gas has been a prominent plot device in movies and television due to its deadly nature and bad reputation. An excellent example of the use of nerve gas can be seen in the James Bond film *Goldfinger*. In the movie, Auric Goldfinger, the villain, kills a room full of gangsters with a nerve gas called Delta 9. Later, as part of operation "Grand Slam", Goldfinger plans to break into the US gold repository at Fort Knox by using Delta 9 to incapacitate all the guards.

In the movie *Executive Decision*, a hijacked plane is carrying a bomb loaded with nerve gas, along with a threat to detonate it over US airspace. The movie *Saw II* used the threat of a house being filled with nerve gas as the main plot element. In order to survive, the eight victims each had to retrieve an antidote to the nerve gas from one of several traps around the house.

VX nerve gas is also center stage in the plot of the movie *The Rock*. In the movie, a renegade brigadier general and his military group steal 15 rockets of deadly VX gas, and take 81 tourists hostage on the prison island of Alcatraz. They threaten to release the gas over San Francisco unless the Pentagon pays them 100 million dollars. The FBI director sends a team consisting of FBI chemical weapons expert Stanley Goodspeed (Nicholas Cage) and former M16 agent John Mason (Sean Connery). Together they break into Alcatraz and neutralize the chemical bombs. In the movie, Nicolas Cage's character actually describes the effects VX gas has as a cholinesterase inhibitor.

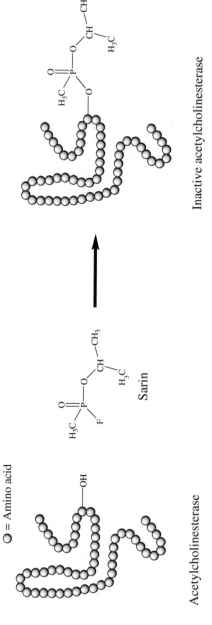

Scheme 10.4 The inactivation of acetylcholinesterase with Sarin gas.

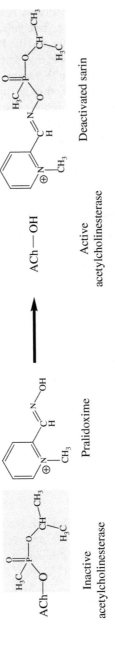

Scheme 10.5 The reactivation of acetylcholinesterase with pralidoxime.

The entire fifth season of the television show *24* centered on VX nerve gas. In the show, Russian terrorists steal 20 canisters of VX nerve gas. The stolen nerve gas was to be released on Russian soil as an excuse to invoke military clauses of an antiterrorism treaty allowing American petroleum interests to be secured in Central Asia. However, this plan backfires, and the Russian terrorists deploy the nerve gas on multiple high-value targets in the United States, killing thousands within a shopping mall and the Counter Terrorism Unit (CTU) building.

The episode appropriately entitled "Chemistry," of the television crime drama *Legends*, deals with the kidnapping of a high school chemistry teacher and his family by the Chechen mafia. Because the teacher was once a Russian chemical weapons designer, it is feared that the kidnappers will use him to manufacture VX nerve gas.

Although banned by convention, nerve gas has, unfortunately, been used in real life. Members of the religious cult Aum Shinrikyo actually made and released their own sarin gas. On 27 June 1994, they released the gas in a residential area, affecting about 500 people and causing 7 deaths. On 20 March 1995, they released sarin gas in the Tokyo subway, affecting approximately 5000 people and killing 12. The founder of the Aum Shinrikyo, Shoko Asahara, was found to be the mastermind of the attacks and sentenced to death by hanging.

A more recent example of the use of sarin nerve gas is the infamous Ghouta chemical attack, which occurred on 21 August 2013, in the Ghouta suburb of Damascus, Syria. As part of a civil war, the standing Syrian government fired nerve-gas-containing rockets into the Ghouta suburb, which was heavily populated with rebels. It is estimated that there were over 1000 fatalities due to this attack. After massive political pressure following the attack, Syria has agreed to abide by the CWC and has begun to destroy its stockpile of chemical weapons.

Further Reading

How Does Methamphetamine Act as a Stimulant? What is "Pseudo," and How is it Related to Methamphetamine?

Britton, MG, Empey, DW, John, GC, McDonnell, KA & Hughe, DTD 1978, 'Histamine challenge and anterior nasal rhinometry: Their use in the assessment of pseudoephedrine and triprolidine as nasal decongestant in subjects with hayfever', *British Journal of Clinical Pharmacology*, vol. 6, pp. 51–55.

Cunningham, JK, Callaghan, RC, Tong, D, Liu, M, Li, H & Lattyak, WJ 2012, 'Changing over-the-counter ephedrine and pseudoephedrine products to prescription only: Impacts on methamphetamine clandestine laboratory seizures', *Drug and Alcohol Dependence*, vol. 126, pp. 55–64.

Lloyd, A, Russell, M, Blanes, L, Doble, P & Roux, C 2013, 'Lab-on-a-chip screening of methamphetamine and pseudoephedrine in samples from clandestine laboratories', *Forensic Science International*, vol. 228, pp. 8–14.

Pellati, F & Benvenuti, S 2008, 'Determination of ephedrine alkaloids in Ephedra natural products using HPLC on a pentafluorophenylpropyl stationary phase', *Journal of Pharmaceutical and Biomedical Analysis*, vol. 48, pp. 254–263.

Rigdon, BA 2012, 'Pharmacists on the front lines in the fight against meth: A state comparison of the laws regulating the retail sale of pseudoephedrine', *The Journal of Legal Medicine*, vol. 33, pp. 253–283.

What is Ricin?

Audi, J, Belson, M, Patel, M, Schier, J & Osterloh, J 2009, 'Ricin poisoning: A comprehensive review', *JAMA*, vol. 294, pp. 2342–2351.

McKeon, TA, Auld, D, Brandon, DL, Leviatov, S & He, X 2014, 'Toxin content of commercial castor cultivars', *Journal of the American Oil Chemists' Society*, vol. 91, pp. 1515–1519.

Musshoff, F & Madea, B 2009, 'Ricin poisoning and forensic toxicology', *Drug Testing and Analysis*, vol. 1, pp. 184–191.

The Thalidomide Disaster

Ando, Y, Fuse, E & Figg, WD 2002, 'Thalidomide metabolism by the CYP2C subfamily', *Clinical Cancer Research*, vol. 8, pp. 1964–1973.

Grunenthal 2016, *Thalidomide chronology*, Grunenthal.

Silverman, WA 2002, *The schizophrenic career of a "monster drug"*, Viewed 24 September 2016, http://pediatrics.aappublications.org/content/110/2/404.full

Vargesson, N 2009, 'Thalidomide-induced limb defects: resolving a 50-year-old puzzle', *BioEssays*, vol. 31, pp. 1327–1336.

Wilkinson, E 2009, *Thalidomide survivors to get £20m*, BBC News, Viewed 24 September 2016, http://news.bbc.co.uk/2/hi/8428838.stm

Zimmer, C 2010, *Answers begin to emerge on how thalidomide caused defects*, New York Times, Viewed 24 September 2016, http://www.nytimes.com/2010/03/16/science/16limb.html?ref=science&pagewanted=all

What Is Phosphine Gas, and Why Is It a Potential Murder Weapon?

DGunnell, D & Eddleston, M 2003, 'Suicide by intentional ingestion of pesticides: a continuing tragedy in developing countries', *International Journal of Epidemiology*, vol. 32, pp. 902–909.

Efrati, I & Hasson, Nir 2014, *Two toddlers die after Jerusalem home sprayed for pests*, The Jerusalem Post, Viewed 23 January 2014, http://www.jpost.com/National-News/Family-of-6-believed-poisoned-by-pesticides-in-Jerusalem-home-toddler-dies-338999

Gurjar, M, Baronia, AK, Azim, A & Sharma, K 1999, 'Three fatalities involving phosphine gas, produced as a result of methamphetamine manufacturing', *Journal of Forensic Sciences*, vol. 44, pp. 647–652.

KSL.com 2010, *Family loses 2nd child in suspected pesticide poisoning*, KSL.com, Viewed 24 September 2016, http://www.ksl.com/?nid=148&sid=9629232

Lemoine, TJ, *et al.* 2011, 'Unintentional fatal phosphine gas poisoning of a family', *Pediatric Emergency Care*, vol. 27, pp. 869–871.

Office of Environmental Health Hazard Assessment 2003, *Phosphine*, Office of Environmental Health Hazard.

O'Malley, M, Fong, H, Sánchez, ME, Roisman, R, Nonato, Y & Mehler, L 2011, 'Inhalation of phosphine gas following a fire associated with fumigation of processed pistachio nuts', *Journal of Emergencies, Trauma, and Shock*, vol. 4, pp. 378–384.

Schwartz, A, Walker, R, Sievert, J, Calvert, GM & Tsai, RJ 2012, 'Occupational phosphine gas poisoning at veterinary hospitals from dogs that ingested zinc phosphide', *Morbidity and Mortality Weekly Report*, vol. 61, pp. 286–288.

The National Institute for Occupational Safety and Health 2015, *PHOSPHINE: Lung damaging agent*, Viewed 9 October 2015, Centers for Disease Control and Protection, Viewed 24 September 2016, http://www.cdc.gov/niosh/ershdb/EmergencyResponseCard_29750035.html

US Environmental Protection Agency 2000, *Phosphine*, US Environmental Protection Agency, Viewed 24 September 2016, http://www.epa.gov/ttnatw01/hlthef/phosphin.html

World Library of Toxicology 2010, *Suicides from pesticide ingestion*, World Library of Toxicology, Viewed 24 September 2016, http://toxipedia.org/display/wlt/Suicides+from+Pesticide+Ingestion

Acetylcholine, Pesticides, and Nerve Gas

Colovic, MB, Krstic, DZ, Lazarevic-Pasti, TD, Bondzic, AM & Vasic, VM 2013, 'Acetylcholinesterase inhibitors: Pharmacology and toxicology', *Current Neuropharmacology*, vol. 11, pp. 315–335.

Dorkins, HR 1982, 'Suxamethonium: The development of a modern drug from 1906 to the present day', *Medical History*, vol. 26, pp. 145–168.

Environmental Health and Medicine Education 2010, *Cholinesterase inhibitors: Including insecticides and chemical warfare nerve agents Part 2: What are cholinesterase inhibitors?*, Agency for Toxic Substances and Disease Registry, Viewed 12 December 2016, https://www.atsdr.cdc.gov/csem/csem.asp?csem=11&po=5

Environmental Health and Medicine Education 2010, *Cholinesterase inhibitors: Including insecticides and chemical warfare nerve agents Part 4 – Section 11 management strategy 3: Medications 2-PAM (2-pyridine aldoxime methylchloride) (Pralidoxime)*, Agency for Toxic Substances and Disease Registry, Viewed 12 December 2016, https://www.atsdr.cdc.gov/csem/csem.asp?csem=11&po=23

Fawcett, WP, Aracava, Y, Adler, M, Pereira, EFR & Albuquerque, EX 2009, 'Acute toxicity of organophosphorus compounds in guinea pigs is sex- and age-dependent and cannot be solely accounted for by acetylcholinesterase inhibition', *The Journal of Pharmacology and Experimental Therapeutics*, vol. 328, pp. 516–524.

Grube, A, Donaldson, D, Kiely, T & Wu, L 2011, *Pesticides industry sales and usage 2006 and 2007 market estimates*, US Environmental Agency, Viewed 1 October 2016, https://www.epa.gov/sites/production/files/2015-10/documents/market_estimates2007.pdf

Thiermann, H, Worek, F & Kehe, K 2013, 'Limitations and challenges in treatment of acute chemical warfare agent poisoning', *Chemico-Biological Interactions*, vol. 206, pp. 435–443.

US Environmental Agency 2011, *Chlorpyrifos: Preliminary human health risk assessment for registration review*, US Environmental Agency, Viewed 1 October 2016, http://www.regulations.gov/document?D=EPA-HQ-OPP-2008-0850-0025

World Health Organization 1990, *Public health impact of pesticides used in agriculture*, World Health Organization, Viewed 1 October 2016, http://apps.who.int/iris/bitstream/10665/39772/1/9241561394.pdf

11

Why You Should Not Use Illegally Made Drugs: The Organic Chemistry Reason

This discussion gets its own chapter and has a special place in my heart because it was the first science-based, real-life story I was ever told. One day in my junior high school biology class, the teacher suddenly decided to play a film for us instead of the usual lecture. The incident of the "frozen addicts" had just occurred in Northern California, where I grew up, and the story was making headlines in the area. Our teacher was a gruff, older man who gave us a stern warning that what had happened to the frozen addicts could easily happen to us if we did not "watch out!" Viewing the film, I was horrified to see what had occurred. In fact, I have never forgotten it. Roughly 20 years later, I began teaching chemistry courses at UC Davis. Trying to improve myself as a teacher, I thought of all the excellent teachers I had had, realizing that almost all of them would take time out of their lectures to tell interesting stories pertaining to the course material. This was when I decided to also start telling stories in classes, and instantly I thought of the story of the "Frozen Addicts." Although it had been over two decades later, the story was still fresh in my mind. The student response I received from this account was so profound, I started looking for other stories to tell, eventually leading me to the selections in this book. In asking students who have heard dozens of my stories, they almost universally told me that the "Frozen Addicts" was the most remarkable.

Why You Shouldn't Use Illegally Made Drugs

The short answer to the question of why you should never use illegally made drugs is simply that they are not monitored by the US Food and Drug Administration (FDA). Although almost everyone has heard of the FDA, it is likely that most do not fully understand or appreciate how important its function is to our society. The primary function of the FDA is to protect public health by insuring the safety, effectiveness, and sanitation of human and veterinary drugs, medical devices, food products, and cosmetics. There are some exceptions, such as meat from livestock, poultry, and some egg products, these being regulated by the US

Strange Chemistry: The Stories Your Chemistry Teacher Wouldn't Tell You, First Edition. Steven Farmer.
© 2017 John Wiley & Sons, Inc. Published 2017 by John Wiley & Sons, Inc.

Department of Agriculture, as well as alcohol-containing products, which are regulated by the Bureau of Alcohol, Tobacco, Firearms, and Explosives.

Already mentioned in this book are numerous examples of what can happen if FDA approval is circumvented. In order for a drug to receive FDA approval, extensive testing must be performed to show that taking the drug will not be harmful. In the discussion of thalidomide, the disaster was prevented from affecting the United States because a drug inspector did not feel that sufficient testing had been performed to show the safety of the drug. In the story about bath salts, FDA approval was circumvented by labeling the product "Not for human consumption," so the drugs themselves were never tested to prove their safety. In addition, the fact that illicit drug manufacturers are constantly changing the structure of the bath salts is frightening. None of these new drug variations are being tested for safety, and though the chemical changes are subtle sometimes, even small changes can have a significant impact on a drug's effect.

For the purpose of this discussion, we will focus on how the FDA monitors the sanitation of drugs, particularly the importance of the agency's role in preventing the *adulteration of drugs* and presence of impurities. Prior to the creation of the FDA, the adulteration of food products and drugs was a significant problem in the United States. For instance, ground-up rice was used instead of brown sugar, flour was frequently cut with chalk, and honey was replaced with glucose and brown coloring. This problem still exists today, with artificial vanilla being used to replace real vanilla and high fructose corn syrup being used to sweeten honey. In fact, it was an extreme example of this type of adulteration that prompted the passing of many of the regulations empowering the FDA. In 1937, *sulfanilamide* was a drug used to treat streptococcal infections, available in tablet and powder form. Due to a consumer demand for a liquid formulation of sulfanilamide, the S.E. Massengill Company in Bristol, Tennessee, began to look for an appropriate solvent. The company's chief chemist found that sulfanilamide would dissolve in the solvent diethylene glycol. After testing the mixture for flavor, appearance, and fragrance, the company began shipping their product for sale across the country. Unfortunately, the Massengill Company missed one very important fact: diethylene glycol is actually metabolized in the body to ethylene glycol (antifreeze) and is therefore a deadly poison. At the time, the food and drugs laws did not require that safety studies be done on new drugs, so the oversight was not discovered until numerous deaths had been reported. Before the product could be removed, more than 100 people had died in 15 states.

Street drug adulteration, in particular, has been the suspected cause of numerous deaths. As mentioned earlier, bath salts are commonly used to replace the more expensive MDMA in ecstasy tablets. In addition, the oxycodone in painkillers bought on the street is commonly replaced with the much more potent fentanyl, a large part of the reason opioid overdose deaths is becoming an epidemic in the United States.

The other part of what the FDA monitors is the purity of drugs. Adulteration implies that a human knowingly added or substituted a component of a drug. *Impurities* are unwanted compounds that are produced or left over during the synthesis of a molecule. In fact, it is very rare that a molecule is synthesized perfectly pure. Generally, other variations of the molecule are also generated, and they need to be removed by additional purification steps. Although these impurities are sometimes benign, they can also be highly dangerous. In illegally made drugs, the manufacturers rarely perform adequate purification because the process will reduce the amount of product and therefore the amount of profit. Street manufactures only worry that the drug produces the desired effect and care very little for possible harmful effects of impurities. A prime example was explained in the Krokodil story. After its illicit synthesis, Krokodil users typically injected the compound as is, without any further purification. The acids and other molecules used for the synthesis were not removed, and their direct injection caused erosion of the users' flesh. Illegally made methamphetamine is usually loaded with various side products. The DEA (Drug Enforcement Administration) can actually tell what starting materials and synthesis pathway was used just through testing for the presence of certain impurities.[1]

You may not realize it, but the FDA monitors every batch of drugs that has been designated for human consumption. Manufacturers are expected to follow specific guidelines set out by the FDA called *Current Good Manufacturing Practices* (cGMPs). If the FDA is not convinced that these guidelines are met, the batch of the drug is not approved and not allowed to be given to humans. In short, cGMPs ensure the identity, strength, quality, and purity of drug products. They make certain that strong quality management systems are in place, which helps to prevent instances of contamination, mix-ups, deviations, failures, and errors. Specific cGMPs vary depending on the actual product and the company making it. Although a passionate teacher, the author did have an opportunity to work at pharmaceutical company for about a year as a synthetic organic chemist. As one of the highlights of his chemistry career, he was the lead scientist in creating a 3 kg batch of a drug designated for human consumption. Of course, specific cGMPs were followed and FDA approval was obtained. The level of precision and sanitation expected by the FDA was quite amazing. But knowing that all drugs designated for human consumption must go through a similar process should bring everyone significant piece of mind whenever taking pharmaceuticals. To give you an idea of the level of precision expected in the cGMPS followed during this synthesis, some examples will be discussed. Because some of you may be unfamiliar with organic synthesis, baking a batch

---

1 I remember looking at a sample of methamphetamine that was on display by the local police department at a job fair. I was horrified to see that the sample was a dirty gray color, whereas pure methamphetamine is clean and white. I learned that the gray color had been caused by the presence of mercury, used as part of the synthesis process.

of cookies will be used as an analogy. As far as the analogy is concerned, cooking is actually quite similar to organic synthesis, as both involve mixing certain ingredients and heating them for a defined temperature and period of time.

1) A designated room was used for creating the drug batch. The room was scrupulously cleaned using a defined set of procedures. Upon completion, test wipes were performed and sent out to an independent testing facility to show that no trace contaminants are present in the room.

   In our cookie-baking scheme, this is like having several kitchens in your house, but being limited to just one particular kitchen for baking cookies. No other cooking would be allowed in this kitchen. Then, prior to baking the cookies, you thoroughly clean the kitchen and afterward invite your anal-retentive neighbor over to determine if you did a good enough job.

2) The recipe for creating the drug batch was already optimized, had been run multiple times, and no deviation was allowed. Absolutely no substitutions or changes beyond those defined in the recipe were permitted.

   In terms of the cookies, this is like receiving a family recipe from a relative, one that has been developed over time and is quite good. However, you consider, perhaps, substituting margarine for butter or adding an extra one-fourth cup of sugar to make it sweeter. Nope. None of this would be allowed. The given recipe must be followed exactly.

3) All new glassware was purchased to make this one drug batch. Once the job was finished, the glassware can never be used again for a cGMP project. In your kitchen, this would require that you purchase a new bowl, stirring spoon, and baking pan every time you want to make a batch of cookies.

4) All of the starting material used for making the drug batch was tested to prove its identity and purity. If one of the starting materials was shown to have even the slightest amount of a contaminant in it, the material had to be purified prior to starting the process, and analytic testing employed to prove that the impurity had been removed.[2]

   So now you are at the store buying some flour to make your cookies. You might think that having the word "flour" written on the front of the bag would be good enough. Not for the FDA. You would need to actually test the contents of this bag and scientifically prove that it actually contains flour, free from any type or amount of contaminant, before you could use it to make your cookies. If you checked and found a grain of rice, you would have to sift the whole bag to make sure that any additional grains were removed.

5) All measures had to be witnessed by two people. If 5.0 g of something was needed, one person had to weigh out the 5.0 g and initial a documentation

---

2 It is not an exaggeration when I say that I had to scientifically prove that a bottle of water actually contained water for this drug batch. When I complained to a coworker, he quickly pointed out that there are numerous clear liquids and there is no way of knowing if the manufacturer made a mistake filling the bottle.

notebook. Then a second person came along, reweighed the sample, confirmed that it weighed 5.0 g, and also initialed the notebook.

For you, this means you will need at least one assistant to make your cookies. All measurements need to be confirmed. After you measure out your four cups of sugar, you need that second person to confirm that you did not make a mistake.

6) External sanitation inspections occurred randomly during the creation of the drug batch. An independent testing company randomly showed up to test the cleanliness of the room being used. At one point, a person came by the laboratory, stuck a large piece of cellophane tape on the wall, removed it, and later counted the number of dust particles captured on it using a microscope. If there were too many dust particles, the room's ventilation filtration would have been deemed inadequate, and the drug batch likely canceled. Imagine, in the middle of making your batch of cookies, your anal-retentive neighbor pops in to see if you have been cleaning up any mess you make while mixing up the cookie batter.

7) Once the drug was made, it underwent further extensive testing to confirm its identity and purity. If, for example, there were any leftover solvents – even in the minutest amount – detected in the sample, the entire batch would be discarded and production redone from the start.[3]

Now that you have finished baking your batch of cookies, you must send out sample cookies to be tested for impurities. If a tiny piece of eggshell is found under a microscope, you will be instructed to dump the entire batch in the garbage.

8) Finally, upon completion, the laboratory notebook and all analyses were sent to the FDA for review and approval. If the FDA found any issue whatsoever, the batch would not have been approved for human consumption. Companies that perform cGMP projects typically have employees whose sole job is to keep up-to-date with FDA cGMP expectations and assure that they are being followed. Multiple Ph.D. chemists in the company read through the laboratory notebook, looking for mistakes. Keep in mind, if the FDA approval is received, it is just for this one batch. If more of the drug is needed, the whole process is repeated. Back to your cookies… After reviewing your delicious baked cookies, your anal-retentive neighbor looks over your recipe and sees that the oven was set for 325° instead of the 350° specified by the family recipe. He orders you to hand over the entire batch for disposal.

The above steps to following cGMP guidelines are presented as an overview and not nearly all inclusive of the actual specifications required by the FDA.

---

3 For the drug batch I was working on, it came to light that a minute amount of solvent was trapped in crystals and could not be removed. Despite the fact that the amount was infinitesimal, the FDA refused to give approval and the entire drug batch had to be thrown out. After repeating the procedure, we were able to completely remove the solvent and the FDA was satisfied.

However, they give you a sense of how scrupulous this agency is with regard to manufacture of items for human consumption. Now think about illicit drug manufacturers. How closely do you think these individuals follow similar protocol? The answer is not at all. Typically, the person making the drug has had perhaps one college-level class in organic chemistry, they get many of their materials from the hardware store, they are usually high on the drug itself while making it, and purifications are usually not performed. An excellent example was displayed in the television series *Breaking Bad*. Although the character Walter White was meticulous with his procedure, he used glassware stolen from a high school chemistry laboratory, which could contain residues of other chemicals. The materials he used were often purchased in a hardware store and definitely not intended for use in chemical synthesis. The 99% methamphetamine purity boasted on the program would probably not have received FDA approval, which expects 100% purity. Even Walter admitted in the show that the blue color of the methamphetamine was due to an impurity. The next obvious question is: What is the impurity? Is it highly toxic? Is it a carcinogen? Notice that these questions are never answered. The presence of an impurity may not seem that bad right now but the next story will show why so much effort is put forth to ensure the lack of impurities in drugs.

The Tragic Case of the Frozen Addicts

The cautionary story of the Frozen Addicts starts in 1982 in northern California, when a man was brought to an emergency room completely immobile – frozen. Seeing that the male was motionless and expressionless, the physicians initially did not know if he had any brain function. After noticing there was some slight movement in his hands, they wrapped the finger of his right hand around a pencil and slipped a yellow notepad under it. The man was able to write his name, but it took him 5 minutes. Very soon after, five more patients were discovered with the same presenting symptoms, so doctors were naturally curious to find out the cause. For all appearances, these patients had advanced cases of Parkinson's disease, but there were problems with that diagnosis. Parkinson's disease comes on gradually, does not strike people overnight, and it usually only occurs after the age of 50. All of patients involved were in their 30s and 40s. When it was determined that all five had been injecting heroin-like substances obtained on the street, the focus turned to these drugs.

In the 1980s, the manufacturers of illicit drugs hit on a brilliant concept. Instead of making existing drugs such as PCP (phencyclidine) and heroin, they began to synthesize entirely new derivatives of those drugs. Because the drugs were being manufactured, it would be quite easy to make subtle changes to the structure. These "designer drugs" would provide the desired effects because of the similarity of structure, but because the structure was new, it would be completely legal. With the derivative compounds not specified as controlled substances, they could be made, sold, possessed, and used with impunity. Designer drugs were hard to detect, completely untested, and created to effectively avoid being deemed illegal.

The drug at the heart of this story is called *1-methyl-4-phenyl-propionoxy-piperidine* or *MPPP* for short. It was originally synthesized in 1947, and illicit drug manufacturers realized that since it was a close analog of *meperidine* (better known as Demerol®), it would probably give a similar heroin-like high when injected. What the illegal manufactures did not realize is that the synthesis of MPPP required multiple steps, all of which needed to be done perfectly. If too much acid or too high a temperature was used, for example, other side products would be produced as well. The investigation into the cause of the Frozen Addicts quickly turned to these possible side products. Scientists analyzed samples of the MPPP powder being sold on the street and found that there was an unidentified contaminant mixed in. In fact, tests showed that MPPP being sold on the streets of California contained significant amounts of this impurity, with one sample even consisting of 97% of the impurity! (Figure 11.1).

How can such a subtle change allow for the formation of a side product? With heat or acid, MPPP breaks apart to form a positively charged carbon atom called a *carbocation* and a negatively charged oxygen atom called a *carboxylate*. In the case of MPPP, this decomposition is relatively easy, only requiring a slight amount of heat, because the negative charge of the carboxylate is adjacent to a double bond and stabilized by a process called *conjugation*. The negative charge is also stabilized by being on an oxygen, the second most electronegative element on the Periodic Table. *Electronegativity* is a measure of an atom's ability to

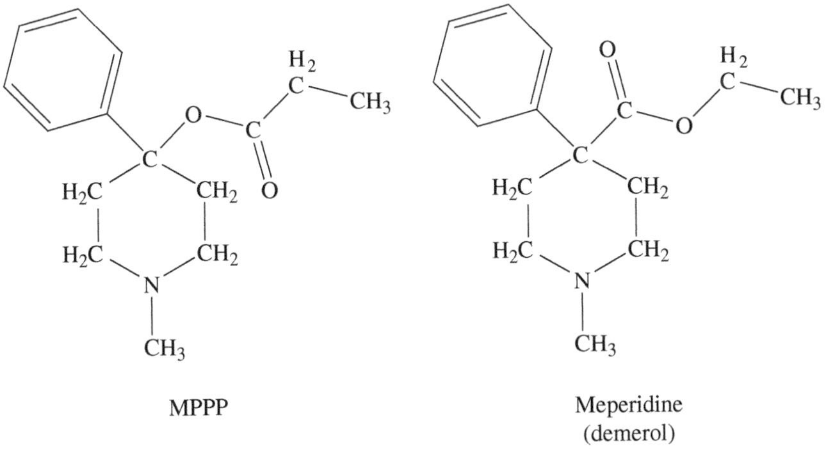

MPPP

Meperidine
(demerol)

Figure 11.1 A comparison of the structures of MPPP and meperidine.

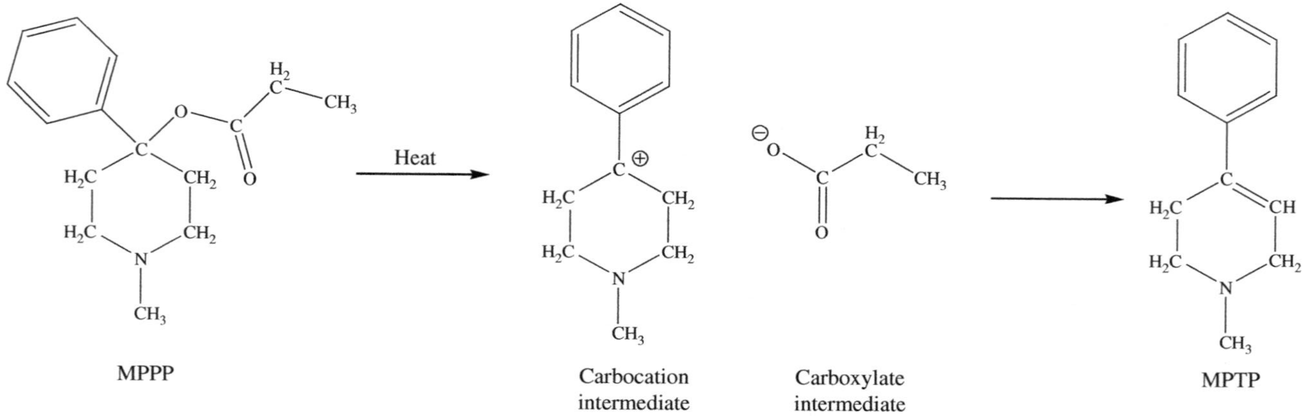

Scheme 11.1 The formation of MPTP from MPPP.

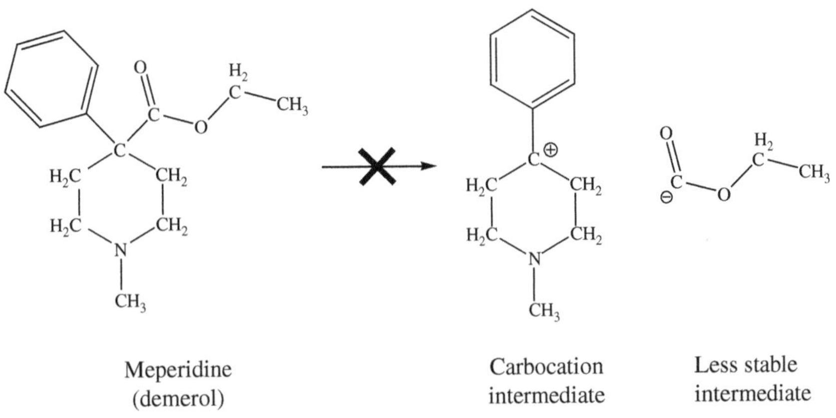

Meperidine (demerol)   Carbocation intermediate   Less stable intermediate

Scheme 11.2 The decomposition of meperidine.

attract electrons in a bond and its ability to stabilize a negative charge. Once the carbocation intermediate of this decomposition is formed, a hydrogen from an adjacent carbon can be lost to form a double bond, thereby creating the impurity, *1-methyl-4-phenyl-1,2,3,6-tetrahydropyridine (MPTP)*. It is this impurity, MPTP, which was causing the addicts to become frozen (Scheme 11.1).

When looking at Demerol®, you can see that a negatively charged species created during a similar decomposition does not have the same stabilizing factors, which makes it much more difficult for this reaction to occur. The negative charge is not adjacent to a double bond and is not stabilized by conjugation. In addition, the negative charge is on a carbon, which is less electronegative than oxygen. Overall, Demerol® tends to not decompose, which prevents it from forming the impurity MPTP during its synthesis (Scheme 11.2).

Once MPTP enters the human brain, it is metabolized by the enzyme *monoamine oxidase* (MAO) to become *1-methyl-4-phenylpyridinium* or *MPP+*. Because of its structure and electrical charge, MPP+ fits into the same neuron systems that handles dopamine. The brain starts delivering MPP+ into critical neurons, where it becomes trapped and starts killing cells. Scientists referred to this as the Trojan Horse Effect. In particular, MPP+ causes nerve cell death in a small area at the base of the brain called the *substantia nigra*. These cells are responsible for the production of the dopamine neurotransmitter essential for the normal control of movement. With repeated exposure to MPTP and thus MPP+, the number of neurons in this area is decreased, causing a corresponding decrease in dopamine production and, in turn, a decline in motor function. In the case of the Frozen Addicts, it caused complete paralysis. Because brain cells cannot be regenerated, the condition of the six Frozen Addicts was permanent (Figure 11.2).

Doctors began to work on trying to treat the addicts' condition. In the past, patients with a rare form of Parkinson's caused by a fever called

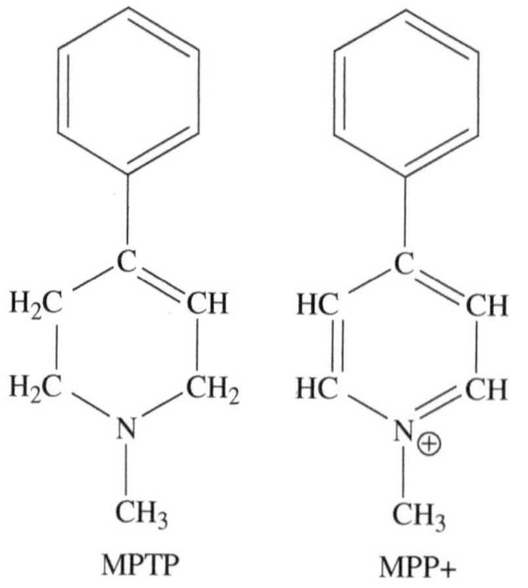

Figure 11.2 A comparison of the structures of MPTP and MPP+.

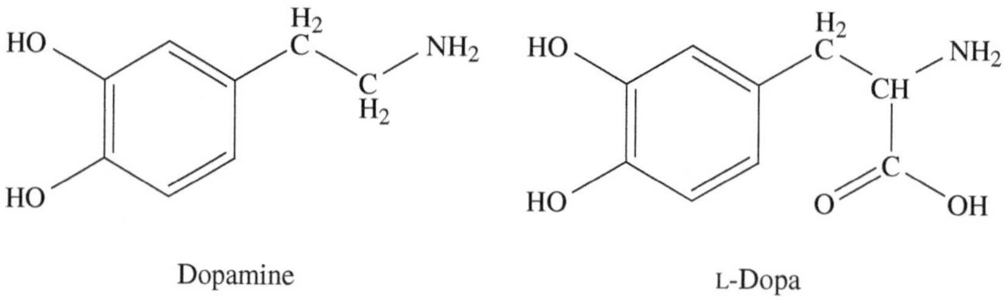

Figure 11.3 A comparison of the structures of dopamine and L-Dopa.

post-encephalitic parkinsonism (PEP) were also completely frozen. Their plight was the subject of Oliver Sack's book "Awakenings" and the corresponding movie with Robin Williams and Robert De Niro. For patients with PEP, the drug L-*Dopa* allowed them to become "unfrozen." L-Dopa is used by cells in the brain to create dopamine, so its administration would compensate for the loss of brain cells that create dopamine. The administration of L-Dopa did help the Frozen Addicts but not without consequences. The patients quickly developed a tolerance for L-Dopa and typically needed higher doses, but with higher doses came extreme side effects, some of which were worse than being frozen (Figure 11.3).

Dopamine is used in *limbic system* of the brain, which is involved with emotion. In fact, an excess of dopamine in the brain is a widely held theory for the cause of schizophrenia. Unfortunately, all of the Frozen Addicts began to suffer from vivid and terrifying schizophrenic-like hallucinations, probably from overstimulation of the limbic system by excessive doses of L-Dopa. In the case of one patient, she would pick up a kitchen knife and run screaming after an imaginary intruder. Another side effect was extreme dyskinesia, causing the addicts to writhe uncontrollably and drool. In addition, the L-Dopa would occasionally, suddenly switch off, leaving the addicts again unable to move. In one instance, two of the unfrozen addicts were robbing a house and became frozen while escaping, thus making them easily apprehended by police.

Due to the developed high tolerance and extreme side effects, the patients would repeatedly have to stop taking L-Dopa and become frozen again until their tolerance lessened and they could start taking the drug again at a lower dose. These addicts were stuck between the two extremes. With medication, they could move but with extreme hallucinations and uncontrollable movements. Without medication, they were frozen and unable to move.

The sad plight of the Frozen Addicts ended up providing numerous benefits to society, however. MPTP was quickly determined to be a powerful neurotoxin and began to be treated with the respect it deserved. At the time, MPTP was actually commercially available and used in organic chemistry research. After the story of the Frozen Addicts became widely known, numerous chemists who had used MPTP as part of their research came forward to report Parkinson's-like symptoms.

It was discovered that MPTP destroyed the substantia nigra of the patients' brains, which is exactly the part of the brain that slowly dies in Parkinson's disease. Scientists had been searching for a toxin that could destroy certain brain cells, thereby producing a complex of signs and symptoms indistinguishable from Parkinson's disease. The discovery that MPTP exposure mimics Parkinson's disease allowed it to be used as an animal model. Prior to the discovery of MPTP, research into the causes of Parkinson's disease was severely limited. Researchers had to rely on what they observed in humans. Being able to artificially create Parkinson's in certain animals allows for the testing of new drugs and treatments for the disease. Since its discovery, MPTP has been used to develop animal models for testing new therapies in the human disease, and investigations of the mechanisms of MPTP toxicity have also provided insights into the possible causes of Parkinson's disease. The discovery of MPTP has been heralded as one of the biggest breakthroughs in modern Parkinson's research.

Due to the lack of success of L-Dopa administration in the frozen addicts, doctors undertook the first clinical *stem cell trials* to help alleviate the symptoms of three of these individuals. These trials, which took place in Sweden, involved transplanting substantia nigra cells from aborted fetuses into the area of the brain damaged by MPTP. This would represent the first time doctors had attempted to repair brain damage by implanting brain tissue from aborted fetuses. The operations were a success, providing the first unequivocal evidence that implanted fetal cells can do the work of brain cells that have died. After the operations, three of the frozen addicts were able to live independently and lead almost normal lives. Currently, implanted fetal tissue is being considered as a possible treatment for Parkinson's disease.

Of the six frozen addicts, three have died, one was murdered, one has not been heard from, while one is still being afflicted by exposure to MPTP over

30 years later. Although an extreme example, the plight of the Frozen Addicts clearly shows the risks to anyone who takes illicitly made drugs. Without FDA clearance, the drugs could contain almost anything, from baby laxative to hydrochloric acid. In addition, illicit drug manufacturers' idea of changing a drug slightly to circumvent laws is problematic. This slight change in a drug could have huge consequences in terms of its effects. An extra carbon or two could possibly make the drug a thousand times more potent; it is virtually impossible to tell. This is why the FDA requires extensive safety testing prior to a drug being provided to the public. Finally, organic synthesis is a complex process that requires decades to truly master. The fact that most illicit drug manufacturers have little or no training in organic chemistry techniques only gives rise to problems. Part of organic chemistry training is to understand the nature of the molecules and be able to anticipate any problems with a synthesis pathway. Honestly, the labile nature of MPPP would have been obvious to a Ph.D. chemist trained in organic chemistry. Moreover, they would have the experience necessary to perform the delicate synthesis without making the MPTP impurity.

Further Reading

Borchers, AT, Hagie, F, Keen, CL & Gershwin ME 2007, 'The history and contemporary challenges of the US Food and Drug Administration', *Clinical Therapeutics*, vol. 29, pp. 1–16.

Kolata, G 1992, *Success reported using fetal tissue to repair a brain*, New York Times, Viewed 23 September 2016, http://www.nytimes.com/1992/11/26/us/success-reported-using-fetal-tissue-to-repair-a-brain.html

Langston, J & Palfreman, J 2014. *The Case of the Frozen Addicts: How the Solution of a Medical Mystery Revolutionized the Understanding of Parkinson's Disease*, IOS Press, Amsterdam.

Locklear, M 2016, *How tainted drugs "froze" young people—but kickstarted Parkinson's research*, Viewed 23 September 2016, http://arstechnica.com/science/2016/05/medical-mystery-how-tainted-drugs-froze-young-people-but-kickstarted-parkinsons-research/

US Food and Drug Administration 2015, *Facts about the Current Good Manufacturing Practices (CGMPs)*, US Food and Drug Administration, Viewed 23 September 2016, http://www.fda.gov/Drugs/DevelopmentApprovalProcess/Manufacturing/ucm169105.htm

Index

Strange Chemistry: The Stories Your Chemistry Teacher Wouldn't Tell You, First Edition. Steven Farmer.
© 2017 John Wiley & Sons, Inc. Published 2017 by John Wiley & Sons, Inc.

9 781119 265269